Heisenberg Calculus and Spectral Theory of Hypoelliptic Operators on Heisenberg Manifolds

of the
American Mathematical Society

Number 906

Heisenberg Calculus and Spectral Theory of Hypoelliptic Operators on Heisenberg Manifolds

Raphaël S. Ponge

July 2008 • Volume 194 • Number 906 (first of 4 numbers) • ISSN 0065-9266

American Mathematical Society
Providence, Rhode Island

2000 *Mathematics Subject Classification.* Primary 58J40, 58J50; Secondary 58J35, 32V10, 35H10, 53D10.

Library of Congress Cataloging-in-Publication Data

Ponge, Raphaël S., 1972–
 Heisenberg calculus and spectral theory of hypoelliptic operators on Heisenberg manifolds / Raphaël S. Ponge
  p. cm. — (Memoirs of the American Mathematical Society, ISSN 0065-9266 ; v. 194, no. 906)
 Includes bibliographical references.
 ISBN 978-0-8218-4148-8 (alk. paper)
  1. Hypoelliptic operators. 2. Spectral theory (Mathematics) 3. Calculus. 4. Differentiable manifolds. I. Title.
QA329.42.P65 2008
515′.7242—dc22   2008008508

Memoirs of the American Mathematical Society

This journal is devoted entirely to research in pure and applied mathematics.

Subscription information. The 2008 subscription begins with volume 191 and consists of six mailings, each containing one or more numbers. Subscription prices for 2008 are US$675 list, US$540 institutional member. A late charge of 10% of the subscription price will be imposed on orders received from nonmembers after January 1 of the subscription year. Subscribers outside the United States and India must pay a postage surcharge of US$38; subscribers in India must pay a postage surcharge of US$43. Expedited delivery to destinations in North America US$53; elsewhere US$130. Each number may be ordered separately; *please specify number* when ordering an individual number. For prices and titles of recently released numbers, see the New Publications sections of the *Notices of the American Mathematical Society*.

Back number information. For back issues see the *AMS Catalog of Publications*.

Subscriptions and orders should be addressed to the American Mathematical Society, P. O. Box 845904, Boston, MA 02284-5904, USA. *All orders must be accompanied by payment.* Other correspondence should be addressed to 201 Charles Street, Providence, RI 02904-2294, USA.

Copying and reprinting. Individual readers of this publication, and nonprofit libraries acting for them, are permitted to make fair use of the material, such as to copy a chapter for use in teaching or research. Permission is granted to quote brief passages from this publication in reviews, provided the customary acknowledgment of the source is given.

Republication, systematic copying, or multiple reproduction of any material in this publication is permitted only under license from the American Mathematical Society. Requests for such permission should be addressed to the Acquisitions Department, American Mathematical Society, 201 Charles Street, Providence, Rhode Island 02904-2294, USA. Requests can also be made by e-mail to reprint-permission@ams.org.

Memoirs of the American Mathematical Society (ISSN 0065-9266) is published bimonthly (each volume consisting usually of more than one number) by the American Mathematical Society at 201 Charles Street, Providence, RI 02904-2294, USA. Periodicals postage paid at Providence, RI. Postmaster: Send address changes to Memoirs, American Mathematical Society, 201 Charles Street, Providence, RI 02904-2294, USA.

© 2008 by the American Mathematical Society. All rights reserved.
Copyright of this publication reverts to the public domain 28 years
after publication. Contact the AMS for copyright status.
This publication is indexed in *Science Citation Index*®, *SciSearch*®, *Research Alert*®,
CompuMath Citation Index®, *Current Contents*®/*Physical, Chemical & Earth Sciences*.
Printed in the United States of America.

∞ The paper used in this book is acid-free and falls within the guidelines
established to ensure permanence and durability.
Visit the AMS home page at http://www.ams.org/

10 9 8 7 6 5 4 3 2 1   13 12 11 10 09 08

Contents

Chapter 1. Introduction — 1
1.1. Heisenberg manifolds and their main differential operators — 2
1.2. Intrinsic approach to the Heisenberg calculus — 3
1.3. Holomorphic families of $\mathbf{\Psi}_H$DOs — 8
1.4. Heat equation and complex powers of hypoelliptic operators — 9
1.5. Spectral asymptotics for hypoelliptic operators — 13
1.6. Weyl asymptotics and CR geometry — 14
1.7. Weyl asymptotics and contact geometry — 15
1.8. Organization of the memoir — 15

Chapter 2. Heisenberg manifolds and their main differential operators — 17
2.1. Heisenberg manifolds — 17
2.2. Main differential operators on Heisenberg manifolds — 22

Chapter 3. Intrinsic Approach to the Heisenberg Calculus — 29
3.1. Heisenberg calculus — 29
3.2. Principal symbol and model operators. — 37
3.3. Hypoellipticity and Rockland condition — 42
3.4. Invertibility criteria for sublaplacians — 52
3.5. Invertibility criteria for the main differential operators — 56

Chapter 4. Holomorphic families of $\mathbf{\Psi}_H$DOs — 65
4.1. Almost homogeneous approach to the Heisenberg calculus — 65
4.2. Holomorphic families of $\mathbf{\Psi}_H$DOs — 67
4.3. Composition of holomorphic families of Ψ_HDOs — 69
4.4. Kernel characterization of holomorphic families of $\mathbf{\Psi}$DOs — 73
4.5. Holomorphic families of $\mathbf{\Psi}$DOs on a general Heisenberg manifold — 76
4.6. Transposes and adjoints of holomorphic families of Ψ_HDOs — 78

Chapter 5. Heat Equation and Complex Powers of Hypoelliptic Operators — 81
5.1. Pseudodifferential representation of the heat kernel — 81
5.2. Heat equation and sublaplacians — 87
5.3. Complex powers of hypoelliptic differential operators — 93
5.4. Rockland condition and the heat equation — 97
5.5. Weighted Sobolev Spaces — 102

Chapter 6. Spectral Asymptotics for Hypoelliptic Operators — 107
6.1. Spectral asymptotics for hypoelliptic operators — 107
6.2. Weyl asymptotics and CR geometry — 111
6.3. Weyl asymptotics and contact geometry — 119

Appendix A. Proof of Proposition 3.1.18   123

Appendix B. Proof of Proposition 3.1.21   127

Appendix. Bibliography   131
   References   131

Abstract

This memoir deals with the hypoelliptic calculus on Heisenberg manifolds, including CR and contact manifolds. In this context the main differential operators at stake include the Hörmander's sum of squares, the Kohn Laplacian, the horizontal sublaplacian, the CR conformal operators of Gover-Graham and the contact Laplacian. These operators cannot be elliptic and the relevant pseudodifferential calculus to study them is provided by the Heisenberg calculus of Beals-Greiner and Taylor.

The Heisenberg manifolds generalize CR and contact manifolds and their name stems from the fact that the relevant notion of tangent space in this setting is rather that of a bundle of graded two-step nilpotent Lie groups. Therefore, the idea behind the Heisenberg calculus, which goes back to Stein, is to construct a pseudodifferential calculus modelled on homogeneous left-invariant convolution operators on nilpotent groups.

The aim of this monograph is threefold. First, we give an intrinsic approach to the Heisenberg calculus by finding an intrinsic notion of principal symbol in this setting, in connection with the construction of the tangent groupoid in [**Po6**]. This framework allows us to prove that the pointwise invertibility of a principal symbol, which can be restated in terms of the so-called Rockland condition, actually implies its global invertibility.

Second, we study complex powers of hypoelliptic operators on Heisenberg manifolds in terms of the Heisenberg calculus. In particular, we show that complex powers of such operators give rise to holomorphic families in the Heisenberg calculus. To this end, due to the lack of microlocality of the Heisenberg calculus, we cannot make use of the standard approach of Seeley, so we rely on an alternative approach based on the pseudodifferential representation of the heat kernel in [**BGS**]. This has some interesting consequences related to hypoellipticity and allows us to construct a scale of weighted Sobolev spaces providing us with sharp estimates for the operators in the Heisenberg calculus.

Third, we make use of the Heisenberg calculus and of the results of this monograph to derive spectral asymptotics for hypoelliptic operators on Heisenberg manifolds. The advantage of using the Heisenberg calculus is illustrated by reformulating in a geometric fashion these asymptotics for the main geometric operators on CR and contact manifolds, namely, the Kohn Laplacian and the horizontal sublaplacian in the CR setting and the horizontal sublaplacian and the contact Laplacian in the contact setting.

Received by the Editor September 13, 2005

2000 *Mathematics Subject Classification.* Primary 58J40, 58J50; Secondary 58J35, 32V10, 35H10, 53D10.

Key words and phrases. Heisenberg calculus, complex powers, heat equation, spectral asymptotics, analysis on CR and contact manifolds, hypoelliptic operators.

Research partially supported by NSF grant DMS 0409005.

CHAPTER 1

Introduction

This memoir deals with the hypoelliptic calculus on Heisenberg manifolds, including CR and contact manifolds. In this context the main differential operators at stake include the Hörmander's sum of squares, the Kohn Laplacian, the horizontal sublaplacian, the CR conformal operators of Gover-Graham and the contact Laplacian. These operators cannot be elliptic and the relevant pseudodifferential calculus to study them is provided by the Heisenberg calculus of of Beals-Greiner [**BG**] and Taylor [**Tay**].

The Heisenberg manifolds generalize CR and contact manifolds and their name stems from the fact that the relevant notion of tangent space in this setting is rather that of a bundle of graded two-step nilpotent Lie groups. Therefore, the idea behind the Heisenberg calculus, which goes back to Stein, is to construct a pseudodifferential calculus modelled on homogeneous left-invariant convolution operators on nilpotent groups.

Our aim in this monograph is threefold. First, we give an intrinsic approach to the Heisenberg calculus by defining an intrinsic notion of principal symbol in this setting, in connection with the construction of the tangent groupoid in [**Po6**]. This framework allows us to prove that the pointwise invertibility of a principal symbol, which can be restated in terms of the so-called Rockland condition, actually implies its global invertibility.

These results have been already used in [**Po9**] to produce new invariants for CR and contact manifolds, extending previous results of Hirachi [**Hi**] and Boutet de Monvel [**Bo2**]. Moreover, since our approach to the principal symbol connects nicely with the construction of the tangent groupoid of a Heisenberg manifold in [**Po6**], this presumably allows us to make use of global K-theoretic arguments in the Heisenberg setting, as those involved in the proof of the (full) Atiyah-Singer index theorem ([**AS1**], [**AS2**]). Therefore, this part of the memoir can also be seen as a step towards a reformulation of the Index Theorem for hypoelliptic operators on Heisenberg manifolds.

Second, we study complex powers of hypoelliptic operators on Heisenberg manifolds in terms of the Heisenberg calculus. In particular, we show that complex powers of such operators give rise to holomorphic families in the Heisenberg calculus. To this end, due to the lack of microlocality of the Heisenberg calculus, we cannot make use of the standard approach of Seeley, so we rely on an alternative approach based on the pseudodifferential representation of the heat kernel in [**BGS**]. This has some interesting consequences related to hypoellipticity and allows us to construct a scale of weighted Sobolev spaces providing us with sharp estimates for the operators in the Heisenberg calculus.

These results are important ingredients in [**Po11**] to construct an analogue for the Heisenberg calculus of the noncommutative residue trace of Wodzicki ([**Wo1**],

[**Wo2**]) and Guillemin [**Gu1**] and to study the zeta and eta functions of hypoelliptic operators. In turn this has several geometric consequences. In particular, this allows us to make use of the framework of Connes' noncommutative geometry, including the local index formula of [**CM**].

Third, we make use of the Heisenberg calculus and of the results of this monograph to derive spectral asymptotics for hypoelliptic operators on Heisenberg manifolds. The advantage of using the Heisenberg calculus is illustrated by reformulating in a geometric fashion these asymptotics for the main geometric operators on CR and contact manifolds, namely, the Kohn Laplacian, the horizontal sublaplacian and the Gover-Graham operators in the CR setting and the horizontal sublaplacian and the contact Laplacian in the contact setting.

On the other hand, although the setting of this monograph is the hypoelliptic calculus on Heisenberg calculus, it is believed that the results herein can be extended to more general settings such as the hypoelliptic calculus on Carnot-Carathéodory manifolds which are equiregular in the sense of [**Gro**].

Following is a more detailed description of the contents of this memoir.

1.1. Heisenberg manifolds and their main differential operators

A Heisenberg manifold (M, H) consists of a manifold M together with a distinguished hyperplane bundle $H \subset TM$. This definition covers many examples: Heisenberg group and its quotients by cocompact lattices, (codimension 1) foliations, CR and contact manifolds and the confolations of Elyahsberg and Thurston.

In this setting the relevant tangent structure for a Heisenberg manifold (M, H) is rather that of a bundle GM of two-step nilpotent Lie groups (see [**BG**], [**Be**], [**EM**], [**EMM**], [**FS1**], [**Gro**], [**Po6**], [**Ro2**]).

The main examples of differential operators on Heisenberg manifolds are the following.

(a) Hörmander's sum of squares on a Heisenberg manifold (M, H) of the form,

$$(1.1.1) \qquad \Delta = \nabla^*_{X_1}\nabla_{X_1} + \ldots + \nabla^*_{X_m}\nabla_{X_m},$$

where the (real) vector fields $X_1, \ldots, X_m$ span H and ∇ is a connection on a vector bundle $\mathcal{E}$ over M and the adjoint is taken with respect to a smooth positive measure on M and a Hermitian metric on $\mathcal{E}$.

(b) Kohn Laplacian $\Box_{b;p,q}$ acting on (p,q)-forms on a CR manifold M^{2n+1} endowed with a CR compatible Hermitian metric (not necessarily a Levi metric).

(c) Horizontal sublaplacian $\Delta_{b;k}$ acting on horizontal differential forms of degree k on a Heisenberg manifold (M, H). When M^{2n+1} is a CR manifold the horizontal sublaplacian preserves the bidegree and so we can consider its restriction $\Delta_{b;p,q}$ to foms of bidegree (p,q).

(d) Gover-Graham operators $\Box_\theta^{(k)}$, $k = 1, \ldots, n+1, n+2, n+4, \ldots$ on a strictly pseudoconvex CR manifold M^{2n+1} endowed with a CR compatible contact form θ (so that θ defines a pseudohermitian structure on M). These operators have been constructed by Gover-Graham [**GG**] as the CR analogues of the conformal GJMS operators of [**GJMS**]. In particular, they are differential operators which tranforms conformally under a conformal change of contact form and for $k = 1$ we recover the conformal sublaplacian of Jerison-Lee [**JL1**].

(e) Contact Laplacian on a contat manifold M^{2n+1} associated to the contact complex of Rumin [**Ru**]. This complex acts between sections of a graded subbunbdle

$(\oplus_{k\neq n}\Lambda_R^k)\oplus\Lambda_{R,1}^n\oplus\Lambda_{R,2}^n$ of horizontal forms. The contact Laplacian is a differential operator of order 2 in degree $k\neq n$ and of order 4 in degree n.

In the examples (a)–(c) the operators are instances of sublaplacians. More precisely, a sublaplacian is a second order differential $\Delta: C^\infty(M,\mathcal{E})\to C^\infty(M,\mathcal{E})$ which near any point $a\in M$ is of the form,

$$(1.1.2) \qquad \Delta = -\sum_{j=1}^d X_j^2 - i\mu(x)X_0 + \sum_{j=1}^d a_j(x)X_j + b(x),$$

where $X_0, X_1, \ldots, X_d$ is a local frame of TM such that $X_1, \ldots, X_d$ span H and the coefficients $\mu(x)$ and $a_1(x),\ldots,a_d(x), b(x)$ are local sections of $\operatorname{End}\mathcal{E}$.

1.2. Intrinsic approach to the Heisenberg calculus

Although the differential operators above may be hypoelliptic under some conditions, they are definitely not elliptic. Therefore, we cannot rely on the standard pseudodifferential calculus to study these operators.

The substitute to the standard pseudodifferential calculus is provided by the Heisenberg calculus, independently introduced by Beals-Greiner [**BG**] and Taylor [**Tay**] (see also [**Bo1**], [**CGGP**], [**Dy1**], [**Dy2**], [**EM**], [**FS1**], [**RS**]). The idea in the Heisenberg calculus, which goes back to Elias Stein, is the following. Since the relevant notion of tangent structure for a Heisenberg manifold (M,H) is that of a bundle GM of 2-step nilpotent graded Lie groups, it stands for reason to construct a pseudodifferential calculus which at every point $x\in M$ is well modelled by the calculus of convolution operators on the nilpotent tangent group G_xM.

The result is a class of pseudodifferential operators, the Ψ_HDOs, which are locally ΨDOs of type $(\frac{1}{2},\frac{1}{2})$, but unlike the latter possess a full symbolic calculus and makes sense on a general Heisenberg manifold. In particular, a Ψ_HDO admits a parametrix in the Heisenberg calculus if, and only if, its principal symbol is invertible, and then the Ψ_HDO is hypoelliptic with a gain of derivatives controlled by its order (see Section 3.1 for a detailed review of the Heisenberg calculus).

1.2.1. Intrinsic notion of principal symbol.
In [**BG**] and [**Tay**] the principal symbol of a Ψ_HDO is defined in local coordinates only, so the definition *a priori* depends on the choice of these coordinates. In the special case of a contact manifold, an intrinsic definition have been given in [**EM**] and [**EMM**] as a section over a bundle of jets of vector fields representing the tangent group bundle of the contact manifold. This approach is similar to that of Melrose [**Me2**] in the setting of the b-calculus for manifolds with boundary.

In this paper we give an intrinsic definition of the principal symbol, valid for an arbitrary Heisenberg manifold, using the results of [**Po6**].

Let (M^{d+1},H) be a Heisenberg manifold. As shown in [**Po6**] the tangent Lie group bundle of (M,H) can be described as the bundle $(TM/H)\oplus H$ together with the grading and group law such that, for sections X_0, Y_0 of TM/H and sections X', Y' of H, we have

$$(1.2.1) \qquad t.(X_0+X') = t^2 X_0 + tX', \quad t\in\mathbb{R},$$

$$(1.2.2) \qquad (X_0+X').(Y_0+Y') = X_0+Y_0+\frac{1}{2}\mathcal{L}(X',Y')+X'+Y',$$

where $\mathcal{L}:H\times H\longrightarrow TM/H$ is the intrinsic Levi form such that
$$\mathcal{L}(X',Y')=[X',Y'] \quad \text{in } TM/H. \tag{1.2.3}$$

In suitable coordinates, called Heisenberg coordinates, this description of GM is equivalent to previous descriptions of GM in terms of the Lie group of a nilpotent Lie algebra of jets of vector fields. A consequence of this equivalence is a tangent approximation result for Heisenberg diffeormorphisms stating that in Heisenberg coordinates such a diffeomorphism is well approximated by the induced isomorphisms between the tangent groups (see [**Po6**, Prop. 2.21]).

Let $\mathcal{E}$ be a vector bundle over M. For $m \in \mathbb{C}$ let $\Psi_H^m(M,\mathcal{E})$ denote the class of Ψ_HDOs of order m acting on the sections of $\mathcal{E}$. Furthermore, let $\mathfrak{g}^*M$ be the linear dual of the Lie algebra bundle $\mathfrak{g}M$ of GM, with canonical projection $\pi:\mathfrak{g}^*M\to M$, and let us define $S_m(\mathfrak{g}^*M,\mathcal{E})$ as the space of sections $p_m(x,\xi)$ in $C^\infty(\mathfrak{g}^*M\setminus 0, \operatorname{End}\pi_*\mathcal{E})$ such that $p_m(x,\lambda.\xi)=\lambda^m p_m(x,\xi)$ for any $\lambda>0$.

The key to our definition of the principal symbol is the aforementioned approximation result for Heisenberg diffeomorphisms of [**Po6**]. More precisely, it allows us to carry out in Heisenberg coordinates a proof of the invariance by Heisenberg diffeomorphisms of the Heisenberg calculus allong similar lines as that in [**BG**] (see Appendix A). The upshot is that it yields a change of variable formula for the principal symbol in Heisenberg coordinates showing that the latter can be intrinsically defined as a section over $\mathfrak{g}^*M\setminus 0$. Therefore, we obtain:

PROPOSITION 1.2.1. *For any $P \in \Psi_H^m(M,\mathcal{E})$, $m \in \mathbb{C}$ there exists a unique symbol $\sigma_m(P) \in S_m(\mathfrak{g}^*M,\mathcal{E})$ such that, for any $a \in M$, the symbol $\sigma_m(P)(a,.)$ agrees in trivializing Heisenberg coordinates centered at a with the principal symbol of P at $x=0$.*

The symbol $\sigma_m(P)(x,\xi)$ is called the principal symbol of P. In local coordinates it can be explicitly related to the principal symbol in the sense of [**BG**] (see Eqs. (3.2.8)–(3.2.9)). In general the two definitions don't agree, but they do when P is a differential operator or the bundle H is integrable. In any case we have a linear isomorphism $\sigma_m: \Psi_H^m(M,\mathcal{E})/\Psi_H^{m-1}(M,\mathcal{E}) \xrightarrow{\sim} S_m(\mathfrak{g}^*M,\mathcal{E})$.

As a consequence of this intrinsic definition of principal symbol we can define the model operator of a Ψ_HDO $P \in \Psi_H^m(M,\mathcal{E})$ at a point $a \in M$ as the left-invariant Ψ_HDO P^a on G_aM with symbol $\sigma_m(a,.)$, that is, the left-convolution operator with the inverse Fourier transform of $\sigma_m(a,.)$ (see Definition 3.2.7).

These notions principal symbol and of model operators show that the Heisenberg calculus is well approximated by the calculus of left-invariant pseudodifferential operators on the tangent groups G_aM, $a \in M$.

First, as it follows from the results of [**BG**], for any $a \in M$ the convolution product on the group G_aM defines a bilinear product,
$$*^a: S_{m_1}(\mathfrak{g}_a^*M) \times S_{m_2}(\mathfrak{g}_a^*M) \longrightarrow S_{m_1+m_2}(\mathfrak{g}_a^*M). \tag{1.2.4}$$

This product depends smoothly on a in such way to give rise to a bilinear product,
$$*: S_{m_1}(\mathfrak{g}^*M,\mathcal{E}) \times S_{m_2}(\mathfrak{g}^*M,\mathcal{E}) \longrightarrow S_{m_1+m_2}(\mathfrak{g}^*M,\mathcal{E}), \tag{1.2.5}$$
$$p_{m_1}*p_{m_2}(a,\xi)=[p_{m_1}(a,.)*^a p_{m_2}(a,.)](\xi) \quad \forall p_{m_j}\in S_{m_j}(\mathfrak{g}^*M,\mathcal{E}). \tag{1.2.6}$$

It then can be shown that the above product correspond to the product of Ψ_HDOs at the level of principal symbols and that the model operator of a product of two Ψ_HDOs is the product of the corresponding model operators (see Proposition 3.2.9).

On the other hand, we also can carry out in Heisenberg coordinates versions of the proofs that the transpose and adjoints of Ψ_HDOs are again Ψ_HDOs (see Appendix B). As a consequence we can identify their principal symbols and see that the model operators at a point of the transpose and the adjoint of a Ψ_HDO are respectively the transpose and the adjoint of its model operator (see Propositions 3.2.11 and 3.2.12).

1.2.2. Rockland condition, parametrices and hypoellipticity. It follows from the results of [**BG**] that for a Ψ_HDO $P \in \Psi_H^m(M,\mathcal{E})$ the existence of a parametrix in $\Psi_H^{-m}(M,\mathcal{E})$ is equivalent to the invertibility of its principal symbol $\sigma_m(P)$. Moreover, when $\Re m \geq 0$ this implies that P is hypoelliptic with gain of $\frac{1}{2}\Re m$ derivatives.

In general it may be difficult to determine the invertibility of the principal symbol of a Ψ_HDO, because the product (1.2.5) for symbols is not anymore the pointwise product of symbols. Nevertheless, this problem can be understood in terms of a representation theoretic criterion, the so-called Rockland condition.

If P is a homogeneous left invariant ΨDO on a nilpotent graded group G then to any unitary representation π we can associate an (unbounded) operator π_P on the representation space $\mathcal{H}_\pi$ such that the domain of its closure contains the space $C^\infty(\pi)$ of smooth vectors of π. The Rockland condition then requires that for any non-trivial irreducible unitary representation π of G the closure $\overline{\pi_P}$ is injective on $C^\infty(\pi)$.

It is a remarkable result that the Rockland condition for P is equivalent to its hypoellipticity (see [**Ro1**], [**HN1**], [**HN2**], [**CGGP**]). Moreover, it can be shown that P is hypoelliptic iff it admits a left ΨDO inverse (see [**Fo**], [**Ge1**], [**CGGP**]). It then follows that P admits a two-sided ΨDO inverse if, and only if, P and P^t satisfies the Rockland condition.

In the setting of the Heisenberg calculus we say that a Ψ_HDO $P \in \Psi_H^m(M,\mathcal{E})$ satisfies the Rockland condition at a point a when the model operator P^a satisfies the Rockland condition on G_aM. It then follows that the principal symbol $\sigma_m(P)$ is invertible at $x = a$, i.e., $\sigma_m(P)(a,.)$ admits an inverse in $S_{-m}(\mathfrak{g}_a^*M,\mathcal{E}_a)$ with respect to the product $*^a$, if, and only if, P and P^t satisfies the Rockland condition at a.

If $P \in \Psi_H^m(M,\mathcal{E})$ is such that P and P^t satisfy the Rockland condition at every point then, as mentioned above, for each point $a \in M$ we get an inverse $q^a \in S_{-m}(\mathfrak{g}_a^*M,\mathcal{E}_a)$ for $\sigma_m(P)(a,.)$. However, in order to obtain an inverse for $\sigma_m(P)$ in $S_m(\mathfrak{g}^*M,\mathcal{E})$ we still have to check that the family $(q^a)_{a \in M}$ varies smoothly with a.

By using an idea of Christ [**Ch2**] it has been shown in [**CGGP**] that given a smooth family of homogeneous left invariant ΨDOs $(P_u)_{u \in U}$ on a fixed nilpotent homogeneous group G such that P^u and $(P^u)^t$ satisfy the Rockland condition for every u then the family of inverses depend smoothly on u.

In this memoir we show that in the Heisenberg setting this result is also true when the group varies from point to point. Namely, we prove:

THEOREM 1.2.2. *Let $P : C^\infty(M,\mathcal{E}) \to C^\infty(M,\mathcal{E})$ be a $\Psi_H DO$ of order m. Then the following are equivalent:*

(i) P and P^t satisfy the Rockland condition at every point of M;

(ii) The principal symbol of P is invertible.

Moreover, when $m = 0$ both (i) and (ii) are equivalent to:

(iii) For any $a \in M$ the model operator P^a is invertible on $L^2(G_aM, \mathcal{E}_a)$.

In substance this theorem states that in the Heisenberg the pointwise invertibility of a principal symbol is equivalent to its invertibility. The proof elaborates on the ideas of [**Ch2**] and [**CGGP**] and is divided into two steps.

In the first step we prove the theorem in the case $m = 0$. In this case, the equivalence of (i) and (iii) follows from a result of Głowacki [**Gł2**] and it is immediate that (ii) implies (iii). Therefore, we only have to prove that (iii) implies (ii). The arguments are based on the ideas of [**Ch2**] as in [**CGGP**], but instead of relying on the result of Christ [**Ch1**] on the L^p boundedness of zero'th order convolutions operators on nilpotent graded groups, we rely on its earlier version due to Knapp-Stein [**KS**], which can be more conveniently generalized to the setting of families of groups.

The second step is the reduction to the case $m = 0$. This is similar to what is done in [**CGGP**], but instead of making use of the commutative approximation of the identity on a fixed nilpotent group of [**Gł1**], we make use of integer powers of a sublaplacian with an invertible principal symbol (such an operator always exists thanks to the results of [**BG**]).

On the other hand, Theorem 1.2.2 has several interesting consequences. First, if P satisfies the Rockland condition at every point and we have $\Re m \geq 0$ then P is hypoelliptic with gain of $\frac{1}{2}\Re m$ derivatives (Proposition 3.3.20).

Second, even though the representation theory of G_aM may vary as a ranges over points of M the Rockland condition is an open condition. More precisely, we prove:

PROPOSITION 1.2.3. *Let $P : C_c^\infty(M, \mathcal{E}) \to C^\infty(M, \mathcal{E})$ be a $\Psi_H DO$ of integer order m with principal symbol $p_m(x, \xi)$ and let $a \in M$.*

1) If P satisfies the Rockland condition at a then there exists an open neighborhood V of a such that P satisfies the Rockland condition at every point of V.

2) If $p_m(a, \xi)$ is invertible in $S_m(\mathfrak{g}_a^*M, \mathcal{E}_a)$ then there exists an open neighborhood V of a such that $p_{m|V}$ is invertible on $S_m(\mathfrak{g}^*V, \mathcal{E})$.

Finally, if $(p_{\nu \in B})_{\nu \in B}$ is a smooth family with values in $S_m(\mathfrak{g}^*M, \mathcal{E})$ parametrized by a manifold B such that p_ν admits an inverse $p_\nu^{(-1)}$ in $S_{-m}(\mathfrak{g}^*M, \mathcal{E})$ for any $\nu \in B$, then the family $(p_\nu^{(-1)})_{\nu \in B}$ too depends smoothly on B (see Proposition 3.3.22).

1.2.3. Invertibility criteria for sublaplacians. As alluded to above the sublaplacians cover the important examples that are the Hörmander's sum of squares, the Kohn Laplacian or the horizontal sublaplacian. If $\Delta : C^\infty(M, \mathcal{E}) \to C^\infty(M, \mathcal{E})$ is a sublaplacian then as shown in [**BG**] the Rockland condition can be formulated in terms of the Levi form (1.2.3) as follows.

For $a \in M$ let $2n$ be the rank of the Levi form $\mathcal{L}_a$ and, using the same notation as in (1.1.2), consider the singular set

(1.2.7) $$\Lambda_a = (-\infty, -\frac{1}{2}\sum_{j=1}^d |\lambda_j|] \cup [\frac{1}{2}\sum_{j=1}^d |\lambda_j||L(a)|, \infty) \quad \text{if } 2n < d,$$

(1.2.8) $$\Lambda_a = \{\pm\frac{1}{2}\sum_{j=1}^d (1+2\alpha_j)|\lambda_j|; \alpha_j \in \mathbb{N}^d\} \quad \text{if } 2n = d,$$

where $\lambda_1, \ldots, \lambda_d$ are the eigenvalues of $\mathcal{L}(a)$ with respect to the frame $X_0, \ldots, X_d$ in (1.1.2). Then the Rockland conditions at a for Δ and Δ^t are both equivalent to the single condition,

$$(1.2.9) \qquad \mathrm{Sp}\,\mu(a) \cap \Lambda_a = \emptyset.$$

In fact, when the condition (1.2.9) holds at every point, we evan can derive an explicit formula for the inverse of the principal symbol of Δ. This is carried out in [**BG**] in the scalar case only, but we really need to deal with the system case in order to study sublaplacians acting on forms. For instance, the Kohn Laplacian locally is scalar modulo lower order terms if, and only if, the Levi form diagonalizes in a smooth eigenframe, which needs not exist in general.

In Section 3.4, after having recalled the arguments of [**BG**] in the scalar case, we explain how to extend them for systems of sublaplacians. In particular, this allows us to complete the treatment of the Kohn Laplacian in [**BG**] (see below).

1.2.4. Invertibility criteria for the main differential operators on Heisenberg manifolds. In Section 3.5 we work out the previous invertibility criteria for the principal symbols of the main examples of operators on Heisenberg manifolds. In particular, we recover in a unified fashion several known hypoellipticity results.

(a) *Hörmander's sum of squares*. For a sum of squares as in (1.1.1) the condition (1.1.2) is equivalent to have $\mathrm{rk}\,\mathcal{L}_a \neq 0$, so that the invertibility of the principal symbol of Δ is equivalent to the condition,

$$(1.2.10) \qquad H + [H, H] = TM.$$

This is exactly the bracket condition of Hörmander [**Hö2**] for a codimension 1 distribution $H \subset TM$.

(b) *Kohn Laplacian*. In the case of the Kohn Laplacian acting on (p, q)-forms on a CR manifold M^{2n+1} the condition (1.1.2) reduces to Kohn's $Y(q)$-condition. For instance when M is κ-strictly pseudoconvex this reduces to have $q \neq \kappa$ and $q \neq n - \kappa$.

(c) *Horizontal sublaplacian*. For the horizontal sublaplacian $\Delta_{b;k}$ acting on horizontal forms of degree k on a Heisenberg manifold the relevant condition to look at is a condition that we call condition $X(k)$: given a point $a \in M$ and letting $2n$ be the rank of the Levi form $\mathcal{L}$ at a, we say that the condition $X(k)$ is satisfied at a when we have

$$(1.2.11) \qquad k \notin \{n, n+1, \ldots, d-n\}.$$

More precisely, we show that the condition (1.1.2) reduces to the condition $X(k)$ and so $\Delta_{b;k}$ has an invertible principal symbol if, and only if, the condition $X(k)$ holds at every point (see Proposition 3.5.4).

For $k = 0$ we get $\mathrm{rk}\,\mathcal{L}_a \neq 0$ which is equivalent to the condition (1.2.10) (in fact $\Delta_{b;0}$ is a sum of squares modulo lower order terms). When M^{2n+1} is a contact manifold the condition $X(k)$ exactly means that we must have $k \neq n$, so that we recover the hypoellipticity results of [**Ta**] and [**Ru**], but in the non-contact case our invertibility criterion for the horizontal sublaplacian seems to be new.

When M^{2n+1} is a CR manifold and we consider the horizontal sublaplacian $\Delta_{b;p,q}$ acting on (p,q)-forms we can refine the $X(k)$ condition into the $X(p,q)$ condition (see Proposition 3.5.6). For instance when M^{2n+1} is κ-strictly pseudoconvex it means that we must have $(p, q) \neq (\kappa, n - \kappa)$ and $(p, q) \neq (n - \kappa, \kappa)$.

(d) Gover-Graham operators. On a strictly pseudoconvex CR manifold M^{2n+1} the Gover-Graham operators $\square_\theta^{(k)}$, $k=1,2,\ldots,n+1,n+2,n+4,\ldots$, are products of sublaplacians modulo lower order terms. Except for $k=n+1$ all the sublaplacians that are involved have invertible principal symbols, so except for the value $k=n+1$ the principal symbol of $\square_\theta^{(k)}$ is invertible (see Proposition 3.5.7).

(e) Contact Laplacian. It has been shown by Rumin [**Ru**] that in every degree the contact Laplacian satisfies the Rockland condition at every point, so it follows from Theorem 1.2.2 that in every degree its principal symbol is invertible.

1.3. Holomorphic families of Ψ_HDOs

In order to deal with complex powers of hypoelliptic operators we define holomorphic families of Ψ_HDOs and check their main properties in Chapter 4.

In a local Heisenberg chart $U \subset \mathbb{R}^{d+1}$ the definition of a holomorphic family of Ψ_HDOs parametrized by an open $\Omega \subset \mathbb{C}$ is similar to that of the definition of a holomorphic family of ΨDOs in [**Wo1**, 7.14] and [**Gu2**, p. 189] (see also [**KV**]). In particular, we allow the order of the family of Ψ_HDOs to vary analytically.

Most of the properties of Ψ_HDOs extend *mutatis mutandis* to the setting of holomorphic families of Ψ_HDOs. In particular, the product of two holomorphic families of Ψ_HDOs is again a holomorphic family of Ψ_HDOs (Proposition 4.3.6).

There is, however, a difficulty when trying to extend the definition to general Heisenberg manifolds. More precisely, the proof of the invariance of the Heisenberg calculus by Heisenberg diffeomorphisms relies on a characterization of the distribution kernels of Ψ_HDOs by means of a suitable class of distributions $\mathcal{K}^*(U \times \mathbb{R}^{d+1}) = \sqcup_{m\in\mathbb{C}}\mathcal{K}^m(U \times \mathbb{R}^{d+1}) \subset \mathcal{D}'(U \times \mathbb{R}^{d+1})$. Each distribution $K \in \mathcal{K}^m(U \times \mathbb{R}^{d+1})$ admits an asymptotic expansion, in the sense of distributions,

$$(1.3.1) \qquad K \sim \sum_{j\geq 0} K_{m+j}, \qquad K_l \in \mathcal{K}_l(U \times \mathbb{R}^{d+1}),$$

where $\mathcal{K}_l(U\times\mathbb{R}^{d+1})$ consists of distributions that are smooth for $y \neq 0$ and homogeneous of degree l if $l \notin \mathbb{N}$ and are homogeneous of degree l up to logarithmic terms otherwise (see [**BG**] and Chapter 3). In particular, the definition of $\mathcal{K}_l(U \times \mathbb{R}^{d+1})$ depends upon whether l is an integer or in not, which causes trouble for defining holomorphic families with values in $\mathcal{K}^*(U \times \mathbb{R}^{d+1})$ when the order crosses integers.

This issue is resolved by means of a new description of the class $\mathcal{K}^*(U\times\mathbb{R}^{d+1})$ in terms of what we call *almost homogeneous* distributions. The latter are homogenous modulo smooth terms and under the Fourier transform they correspond to the almost homogeneous symbols considered in [**BG**].

Since the definition of an almost homogeneous dsitribution of degree l does not depend on whether l is an integer or not, there is no trouble anymore to define holomorphic families of almost homogeneous kernels. Therefore, we can make use of the characterization of $\mathcal{K}^*(U \times \mathbb{R}^{d+1})$ in terms of almost homogenous distributions to define holomorphic families with values in $\mathcal{K}^*(U \times \mathbb{R}^{d+1})$ (see Definition 4.4.3).

We show that the distribution of kernel holomorphic families of Ψ_HDOs can be characterized in terms of holomorphic families with values in $\mathcal{K}^*(U \times \mathbb{R}^{d+1})$. This allows us to extend the arguments in the proof of the invariance by Heisenberg diffeomorphisms of the Heisenberg calculus to prove that holomorphic families of Ψ_HDOs too are invariant under Heisenberg diffeomorphisms (Proposition 4.5.2).

As a consequence we can define holomorphic families of Ψ_HDOs on an arbitrary Heisenberg manifold independently of the choice of a covering by Heisenberg charts.

Let us also mention that the almost homogeneous approach to the Heisenberg calculus can also be used to constructing a class of Ψ_HDOs with parameter containing the resolvents of hypoelliptic Ψ_HDOs (see [**Po12**]).

1.4. Heat equation and complex powers of hypoelliptic operators

One of the main goals of this memoir is to obtain complex powers of hypoelliptic operators on Heisenberg manifolds as holomorphic families of Ψ_HDOs along with some applications to hypoellipticity.

It has been shown by Mohammed [**Mo2**] that the complex powers of invertible positive hypoelliptic operators with multicharacteristics are ΨDOs in the class constructed in [**BGH**], but in the Heisenberg setting we would like to obtain them as holomorphic families of Ψ_HDOs. To this end we cannot follow the standard approach of Seeley [**Se**] due to the lack of microlocality of the Heisenberg calculus. Instead we make use of the pseudodifferential representation of the heat kernel of [**BGS**], which is especially suitable for dealing with positive differential operators (we will deal with the general case in [**Po12**] using another approach).

Let us also mention that a similar approach to complex powers has been used independently by Mathai-Melrose-Singer [**MMS**] and Melrose [**Me3**] in the context of projective pseudodifferential operators on Azamaya bundles.

From now on we let (M^{d+1}, H) be a compact Heisenberg manifold equipped with a smooth density > 0 and let $\mathcal{E}$ be a Hermitian vector bundle over M.

1.4.1. Pseudodifferential representation of the heat kernel.
Consider a selfadjoint differential operator $P : C^\infty(M, \mathcal{E}) \to C^\infty(M, \mathcal{E})$ which is bounded from below and has an invertible principal symbol, so that the heat kernel $k_t(x, y)$ of P is smooth for $t > 0$.

Recall that the heat semigroup e^{-tP} allows us to invert the heat operator $P + \partial_t$. Conversely, constructing a suitable pseudodifferential calculus nesting parametrices for $P + \partial_t$ allows us to derive the small time heat kernel asymptotics for P.

In the elliptic setting this approach was carried out by Greiner [**Gre**] and the relevant pseudodifferential calculus is the Volterra calculus (see [**Gre**], [**Pi**]). The latter consists only in a modification of the classical pseudodifferential calculus in order to take into account the parabolicity and the Volterra property with respect to the time variable of the heat equation. In particular, Greiner's approach holds in fairly greater generality and has many applications (see, e.g., [**BGS**], [**BS1**], [**BS2**], [**Gre**], [**Kr1**], [**Kr2**], [**KSc**], [**Me2**], [**Pi**], [**Po4**], [**Po7**]).

The Greiner's approach has been extended to the Heisenberg calculus in [**BGS**], with the purpose of deriving the small time heat kernel asymptotics for the Kohn Laplacian on CR manifolds. In particular, a class of Volterra Ψ_HDOs is obtained which contains parametrices for the heat operator $P + \partial_t$. As a consequence, once the principal symbol of $P + \partial_t$ is invertible in this calculus, the inverse of $P + \partial_t$ is a Volterra Ψ_HDO which, in turn, yields a pseudodifferential representation of the heat kernel of P. More precisely, we have:

THEOREM 1.4.1 ([**BGS**]). *Let $P : C^\infty(M, \mathcal{E}) \to C^\infty(M, \mathcal{E})$ be a selfadjoint differential operator of even Heisenberg order v which is bounded from below and* such that the principal symbol of $P + \partial_t$ is invertible in the Volterra-Heiseneberg calculus. Then:

1) The inverse $(P + \partial_t)^{-1}$ is a Volterra $\Psi_H DO$;

2) The heat kernel $k_t(x, y)$ of P has an asymptotics in $C^\infty(M, (\operatorname{End}\mathcal{E})\otimes|\Lambda|(M))$ of the form

$$(1.4.1) \qquad k_t(x,x) \sim_{t \to 0^+} t^{-\frac{d+2}{v}} \sum t^{\frac{2j}{v}} a_j(P)(x),$$

where the density $a_j(P)(x)$ is locally computable in terms of the symbol $q_{-v-2j}(x,\xi,\tau)$ of degree $-v - 2j$ of any Volterra-$\Psi_H DO$ parametrix for $P + \partial_t$.

This framework is recalled in Section 5.1 and we can extend to this setting the intrinsic approach of Chapter 3.

1.4.2. Heat equation and sublaplacians. Let $\Delta : C^\infty(M, \mathcal{E}) \to C^\infty(M, \mathcal{E})$ be a selfadjoint sublaplacian which is bounded from below. In [**BGS**] the authors construct explicitly an inverse in the Volterra-Heisenberg calculus for the principal symbol of $\Delta + \partial_t$ when $\mathcal{E}$ is the trivial line bundle and when at every point $a \in M$, with the notation of (1.2.7)–(1.2.8), we have

$$(1.4.2) \qquad |\mu(a)| < \frac{1}{2}\sum_{j=1}^{d}|\lambda_j|.$$

Since Δ is selfadjoint, and so $\mu(a)$ is real, the above condition is the same as (1.2.9) when $\operatorname{rk}\mathcal{L}_a < d$, but when $\operatorname{rk}\mathcal{L}_a = d$ this is a stronger condition.

In fact, the explicit formulas of [**BGS**] can be extended to the case where $\mathcal{E}$ is an arbitrary vector bundle and where Δ satisfies the weaker condition (1.1.2) at every point (see Proposition 5.2.9). As a consequence Theorem 1.4.1 holds for the Kohn Laplacian even when the Levi form is not diagonalizable.

Moreover, as we actually can invert the principal symbol of $\Delta + \partial_t$ in a refined class of symbols, for any integer $k = 2, 3, \ldots$ we can invert the principal symbol of $\Delta^k + \partial_t$ in the Volterra-Heisenberg calculus (see Proposition 5.2.12).

1.4.3. Complex powers. Let $P : C^\infty(M, \mathcal{E}) \to C^\infty(M, \mathcal{E})$ be a positive selfadjoint differential operator of even Heisenberg order v and assume that the principal symbol of P is invertible, i.e., P satisfies the Rockland condition at every point. Thanks to the spectral theorem we can define the complex powers P^s, $s \in \mathbb{C}$, of P as unbounded operators on $L^2(M, \mathcal{E})$ which are bounded for $\Re s \leq 0$.

Moreover, for $\Re s < 0$ the Mellin formula holds,

$$(1.4.3) \qquad P^s = \Gamma(s)^{-1} \int_0^\infty t^s (1 - \Pi_0(P)) e^{-tP} \frac{dt}{t},$$

where $\Pi_0(P)$ denotes the orthogonal projection onto the kernel of P. Combining this formula with the pseudodifferential representation of the heat kernel of P in terms of the Volterra-Heisenberg calculus allows us to prove:

THEOREM 1.4.2. *Assume that the principal symbol of $P + \partial_t$ is an invertible Volterra-Heisenberg symbol. Then the complex powers P^s, $s \in \mathbb{C}$, of P form a holomorphic 1-parameter group of $\Psi_H DOs$ such that $\operatorname{ord} P^s = ms \; \forall s \in \mathbb{C}$.*

In particular, this theorem holds for the following sublaplacians:

(a) A sum of squares of the form (1.1.1), provided that the bracket condition (1.2.10) holds;

(b) The Kohn Laplacian on a CR manifold acting on (p,q)-forms under condition $Y(q)$;

(c) The horizontal Laplacian on a Heisenberg manifold acting on horizontal forms of degree k under condition $X(k)$;

(d) The horizontal Laplacian on a CR manifold acting on (p,q)-forms under condition $X(p,q)$.

In fact, partly by making use of Theorem 1.4.2, we will show that when the condition (1.2.10) holds the principal symbol of $P + \partial_t$ is automatically invertible in the Volterra-Heisenberg symbol (see below). Therefore, we obtain:

THEOREM 1.4.3. *If the bracket condition (1.2.10) holds then the complex powers P^s, $s \in \mathbb{C}$, of P form a holomorphic 1-parameter group of $\Psi_H DOs$ such that* $\mathrm{ord} P^s = ms \ \forall s \in \mathbb{C}$.

In particular, Theorem 1.4.3 is valid for the contact Laplacian on a contact manifold. In this context this allows us to fill a technical gap in [**JK**] concerning the proof of the fact that the complex powers of the contact Laplacian give rise to $\Psi_H DOs$ which is an important step in the proof there of the Baum-Connes conjecture for $SU(n,1)$ (see [**Po10**]).

1.4.4. Rockland condition and heat equation. Theorem 1.4.2 has several interesting applications related to hypoellipticity.

First, Theorem 1.2.2 can be extended to $\Psi_H DOs$ with non-integer orders as follows.

THEOREM 1.4.4. *Assume that the bracket condition (1.2.10) holds. Then for any $P \in \Psi_H^m(M, \mathcal{E})$, $m \in \mathbb{C}$, the following are equivalent:*

(i) *The principal symbol of P is invertible;*

(ii) *P and P^t satisfy the Rockland condition at every point $a \in M$.*

(iii) *P and P^* satisfy the Rockland condition at every point $a \in M$.*

As a consequence of this theorem we can prove that when the condition (1.2.10) holds any $P \in \Psi_H^m(M, \mathcal{E})$ with $\Re m \geq 0$ satisfying the Rockland condition at every point is hypoelliptic with gain of $\frac{1}{2}\Re m$ derivative(s) (see Proposition 5.4.2).

Next, let $P: C^\infty(M, \mathcal{E}) \to C^\infty(M, \mathcal{E})$ be a selfadjoint differential operator of even Heisenberg order v. We shall say that the principal symbol of P is positive when it can be put into the form $\overline{q_{\frac{v}{2}}} * q_{\frac{v}{2}}$ for some symbol $q_{\frac{v}{2}}$ homogeneous of degree $\frac{v}{2}$. Then, by making use of Theorem 1.4.2 and by extending to the Volterra-Heisenberg setting the arguments of the proof of Theorem 1.2.2 we prove:

THEOREM 1.4.5. *Assume that the bracket condition (1.2.10) holds and that P satisfies the Rockland condition at every point.*

1) P is bounded from below if, and only if, it has a positive principal symbol.

2) If P has a positive principal symbol, then the principal symbol $P + \partial_t$ is invertible in the Volterra-Heisenberg calculus.

This proves that, when the condition (1.2.10) holds, in Theorem 1.4.1 we can replace the invertibility condition on the principal symbol of $P+\partial_t$ by the validity of the Rockland condition for P at every point. Consequently, we see that the results of [**BGS**] actually hold for a wide class of operators. In particular, Theorem 1.4.1 is valid for the contact Laplacian on a contact manifold.

1.4.5. Weighted Sobolev spaces.

As another application of Theorem 1.4.2, under the bracket condition (1.2.10) we can construct a scale of Weighted Sobolev spaces $W_H^s(M, \mathcal{E})$, $s \in \mathbb{C}$, providing us with sharp regularity estimates for $\Psi_H\text{DOs}$.

Let $\Delta_{\nabla, X} : C^\infty(M, \mathcal{E}) \to C^\infty(M, \mathcal{E})$ be a sum of squares as in (1.1.1). Since the bracket condition (1.2.10) holds, Theorem 1.4.2 tells us that the complex powers $(1+\Delta_X)^s$, $s \in \mathbb{C}$, give rise to an analytic 1-parameter group of invertible $\Psi_H\text{DOs}$.

For $s \in \mathbb{R}$ the weighted Sobolev space $W_H^s(M, \mathcal{E})$ is defined as the space of distributional sections $u \in \mathcal{D}'(M, \mathcal{E})$ such that $(1+\Delta_{\nabla, X})^{\frac{s}{2}} u$ is in $L^2(M, \mathcal{E})$ together with the Hilbertian norm,

$$(1.4.4) \qquad \|u\|_{W_H^s} = \|(1+\Delta_{\nabla, X})^{\frac{s}{2}} u\|_{L^2}, \qquad u \in W_H^s(M, \mathcal{E}).$$

It can be shown that, up to the choice of an equivalent Hilbertian norm, this definition does not depend on the choices of the vector fields $X_1, \ldots, X_m$ and of the connection ∇ and that when s is a positive integer it agrees with the previous definition of the Weighted Sobolev spaces of Folland-Stein [**FS1**] (see Section 5.5).

Moreover, the spaces $W_H^s(M, \mathcal{E})$ can be nicely compared to the standard Sobolev spaces $L_s^2(M, \mathcal{E})$. More precisely, we show that we have the following continuous embeddings,

$$(1.4.5) \qquad \begin{array}{ll} L_s^2(M) \hookrightarrow W_H^s(M) \hookrightarrow L_{s/2}^2(M) & \text{if } s \geq 0, \\ L_{s/2}^2(M) \hookrightarrow W_H^s(M) \hookrightarrow L_s^2(M) & \text{if } s < 0. \end{array}$$

On the other hand, these Sobolev spaces are suitable for studying $\Psi_H\text{DOs}$, for we have:

PROPOSITION 1.4.6. *Let $P : C^\infty(M, \mathcal{E}) \to C^\infty(M, \mathcal{E})$ be a $\Psi_H DO$ of order m and set $k = \Re m$. Then, for any $s \in \mathbb{R}$, the operator P extends to a continuous linear mapping from $W_H^{s+k}(M, \mathcal{E})$ to $W_H^s(M, \mathcal{E})$.*

As a consequence we get sharp regularity results for $\Psi_H\text{DOs}$ satisfying the Rockland condition:

PROPOSITION 1.4.7. *Let $P : C^\infty(M, \mathcal{E}) \to C^\infty(M, \mathcal{E})$ be a $\Psi_H DO$ of order m such that P satisfies the Rockland condition at every point and set $k = \Re m$. Then for any $u \in \mathcal{D}'(M, \mathcal{E})$ we have*

$$(1.4.6) \qquad Pu \in W_H^s(M, \mathcal{E}) \implies u \in W_H^{s+k}(M, \mathcal{E}).$$

In fact, for any $s' \in \mathbb{R}$ we have the estimate,

$$(1.4.7) \qquad \|u\|_{W_H^{s+k}} \leq C_{ss'}(\|Pu\|_{W_H^s} + \|u\|_{W_H^{s'}}), \qquad u \in W_H^{s+k}(M, \mathcal{E}).$$

When P is a differential operator of Heisenberg order v the properties (1.4.6) and (1.4.7) correspond to the maximal hypoellipticity of [**HN3**].

In addition, the weighted Sobolev spaces $W_H^s(M, \mathcal{E})$ can be localized, so that it makes sense to say that a distributional section is W_H^s near a point, and we can prove a localized version of Proposition 1.4.7 (see Proposition 5.5.14).

Finally, we also give a version of Proposition 1.4.6 for holomorphic families of $\Psi_H\text{DOs}$ and in particular for complex powers of positive differential operator satisfying the Rockland condition (Propositions 5.5.15 and 5.5.16 for the detailed statements).

1.5. Spectral asymptotics for hypoelliptic operators

Another main goal of this monograph is to make use of the Heisenberg calculus to derive spectral asymptotics for hypoelliptic operators on Heisenberg manifolds and in particular to get explicit geometric expressions for the leading terms of these asymptotics for the main geometric differential operators on CR and contact manifolds.

1.5.1. Heat equation and spectral asymptotics. Consider a selfadjoint differential operator $P: C^\infty(M, \mathcal{E}) \to C^\infty(M, \mathcal{E})$ of even Heisenberg order v which is bounded from below and such that the principal symbol of $P + \partial_t$ is invertible in the Volterra-Heisenberg calculus. Then the heat kernel asymptotics (1.4.1) holds at the level of densities, so that as $t \to 0^+$ we have

$$(1.5.1) \qquad \operatorname{Tr} e^{-tP} \sim t^{-\frac{d+2}{m}} \sum t^{\frac{2j}{m}} A_j(P), \qquad A_j(P) = \int_M \operatorname{tr}_\mathcal{E} a_j(P)(x).$$

Next, let $\lambda_0(P) \leq \lambda_1(P) \leq \ldots$ denote the eigenvalues of P counted with multiplicity and let $N(P; \lambda)$ denote its counting function, that is,

$$(1.5.2) \qquad N(P; \lambda) = \#\{k \in \mathbb{N}; \ \lambda_k(P) \leq \lambda\}, \qquad \lambda \geq 0.$$

In addition, define

$$(1.5.3) \qquad \nu_0(P) = \Gamma(1 + \frac{d+2}{m})^{-1} A_0(P).$$

Then we obtain:

PROPOSITION 1.5.1. *1) We have $\nu_0(P) > 0$.*

2) As $\lambda \to \infty$ we have $N(P; \lambda) \sim \nu_0(P) \lambda^{\frac{d+2}{m}}$.

3) As $k \to \infty$ we have $\lambda_k(P) \sim \left(\frac{k}{\nu_0(P)}\right)^{\frac{m}{d+2}}$.

Once it is proved that $\nu_0(P)$ is > 0 we can make use of Karamata's Tauberian theorem to deduce from (1.5.1) the asymptotics for $N(P; \lambda)$ and $\lambda_k(P)$. Thus the bulk the proof is to establish the positivity of $\nu_0(P)$, which is carried out via spectral theoretic considerations.

By relying on other pseudodifferential calculi several authors have also obtained Weyl asymptotics in the more general setting of hypoelliptic operators with multicharacteristics (see [**II**], [**Me1**], [**MS**], [**Mo1**], [**Mo2**]). Nevertheless, as far as the Heisenberg setting is concerned, the approach using the Volterra-Heisenberg calculus has two main advantages.

First, the pseudodifferential analysis is significantly simpler. In particular, the Volterra-Heisenberg calculus yields for free the heat kernel asymptotics once the principal symbol of the heat operator is shown to be invertible, for which it is enough to use the Rockland condition when the condition (1.2.10) holds.

Second, since the Volterra-Heisenberg calculus fully takes into account the underlying Heisenberg geometry of the manifold and is invariant by change of Heisenberg coordinates, we can get explicit geometric expressions for the coefficient $\nu_0(P)$ in the case of the main geometric differential operators on CR and contact manifolds (see below for the precise formulas).

1.6. Weyl asymptotics and CR geometry

Let M^{2n+1} be a compact κ-strictly pseudoconvex CR manifold and let θ be a contact form whose associated Levi form has signature $(n-\kappa, \kappa, 0)$, so that θ defines a pseudohermitian structure on M. We endow M with a Levi metric compatible with θ. Then the volume of M with respect to this Levi metric is independent of the choice of the Levi form and is equal to

$$\text{vol}_\theta M = \frac{(-1)^\kappa}{n!} \int_M \theta \wedge d\theta^n. \tag{1.6.1}$$

We call $\text{vol}_\theta M$ the pseudohermitian volume of (M, θ) and we relate it to the Weyl asymptotics (1.5.3) for the Kohn Laplacian and the horizontal sublaplacian as follows.

For $\mu \in (-n, n)$ we let

$$\nu(\mu) = (2\pi)^{-(n+1)} \int_{-\infty}^{\infty} e^{-\mu \xi_0} \left(\frac{\xi_0}{\sinh \xi_0}\right)^n d\xi_0. \tag{1.6.2}$$

Then for the Kohn Laplacian we prove:

THEOREM 1.6.1. *Let $\Box_{b;p,q}$ be the Kohn Laplacian acting on (p, q) forms with $q \neq \kappa$ and $q \neq n - \kappa$. Then as $\lambda \to \infty$ we have*

$$N(\Box_{b;p,q}; \lambda) \sim \alpha_{n\kappa pq}(\text{vol}_\theta M)\lambda^{n+1}, \tag{1.6.3}$$

where $\alpha_{n\kappa pq}$ is equal to

$$\binom{n}{p} \sum_{\max(0, q-\kappa) \leq k \leq \min(q, n-\kappa)} \frac{1}{2} \binom{n-\kappa}{k} \binom{\kappa}{q-k} \nu(n - 2(\kappa - q + 2k)). \tag{1.6.4}$$

In particular $\alpha_{n\kappa pq}$ is a universal constant depending only on n, κ, p and q.

In the strictly pseudoconvex case, i.e., when $\kappa = 0$, this theorem follows from the computation of $A_0(\Box_{b;p,q})$ in [**BGS**], but for the case $\kappa \geq 1$ this seems to be a new result.

Next, in the CR setting the horizontal sublaplacian preserves the bidegree and, in the same way as with the Kohn Laplacian, we prove:

THEOREM 1.6.2. *Let $\Delta_{b;p,q} : C^\infty(M, \Lambda^{p,q}) \to C^\infty(M, \Lambda^{p,q})$ be the horizontal sublaplacian acting on (p, q)-forms with $(p, q) \neq (\kappa, n - \kappa)$ and $(p, q) \neq (n - \kappa, \kappa)$. Then as $\lambda \to \infty$ we have*

$$N(\Delta_{b;p,q}; \lambda) \sim \beta_{n\kappa pq}(\text{vol}_\theta M)\lambda^{n+1}, \tag{1.6.5}$$

where $\beta_{n\kappa pq}$ is equal to

$$\sum_{\substack{\max(0,q-\kappa) \leq k \leq \min(q,n-\kappa) \\ \max(0,p-\kappa) \leq l \leq \min(p,n-\kappa)}} 2^n \binom{n-\kappa}{l}\binom{\kappa}{p-l}\binom{n-\kappa}{k}\binom{\kappa}{q-k} \nu(2(q-p) + 4(l-k)). \tag{1.6.6}$$

In particular $\beta_{n\kappa pq}$ is a universal constant depending only on n, κ, p and q.

Finally, suppose that M is strictly pseudoconvex, i.e., $\kappa = 0$, and for $k = 1, \ldots, n+1, n+2, n+4, \ldots$ let $\Box_\theta^{(k)}$ be the Gover-Graham operator of order k. Then we have:

THEOREM 1.6.3. *Assume $k \neq n+1$. Then there exists a universal constant $\nu_n^{(k)} > 0$ depending only on n and k such that as $\lambda \to \infty$ we have*

(1.6.7) $$N(\Box_\theta^{(k)}; \lambda) \sim \nu_n^{(k)} (\mathrm{vol}_\theta M) \lambda^{\frac{n+1}{k}}.$$

1.7. Weyl asymptotics and contact geometry

Let (M^{2n+1}, H) be a compact orientable contact manifold. Let θ be a contact form and let J be a calibrated almost complex structure on H so that $d\theta(X, JX) = -d\theta(JX, X) > 0$ for any section X of $H \setminus 0$. We then endow M with the Riemannian metric $g_{\theta,J} = d\theta(.,J.) + \theta^2$. The volume of M with respect to $g_{\theta,J}$ depends only on θ and is equal to:

(1.7.1) $$\mathrm{vol}_\theta M = \frac{1}{n!} \int_M d\theta^n \wedge \theta.$$

We call $\mathrm{vol}_\theta M$ the contact volume of M.

We can relate the Weyl asymptotics for the horizontal sublaplacian to the contact volume to get:

THEOREM 1.7.1. *Let $\Delta_{b;k} : C^\infty(M, \Lambda_\mathbb{C}^k H^*) \to C^\infty(M, \Lambda_\mathbb{C}^{k+1} H^*)$ be the horizontal sublaplacian on M in degree k with $k \neq n$. Then as $\lambda \to \infty$ we have*

(1.7.2) $$N(\Delta_{b;k}; \lambda) \sim \gamma_{nk} (\mathrm{vol}_\theta M) \lambda^{n+1}, \qquad \gamma_{nk} = \sum_{p+q=k} 2^n \binom{n}{p}\binom{n}{q} \nu(p-q).$$

In particular γ_{nk} is universal constant depending on n and k only.

Note that when M is a strictly pseudoconvex CR manifold the asymptotics (1.7.2) is compatible with (1.6.5) because the contact volume differs from the pseudohermitian volume by a factor of 2^{-n}.

Finally, we can also deal with the contact Laplacian as follows.

THEOREM 1.7.2. *1) Let $\Delta_{R;k} : C^\infty(M, \Lambda^k) \to C^\infty(M, \Lambda^k)$ be the contact Laplacian in degree k with $k \neq n$. Then there exists a universal constant $\nu_{nk} > 0$ depending only on n and k such that as $\lambda \to \infty$ we have*

(1.7.3) $$N(\Delta_{R;k}; \lambda) \sim \nu_{nk} (\mathrm{vol}_\theta M) \lambda^{n+1}.$$

2) For $j = 1, 2$ consider the contact Laplacian $\Delta_{R;n} : C^\infty(M, \Lambda_j^n) \to C^\infty(M, \Lambda_j^n)$. Then there exists a universal constant $\nu_n^{(j)} > 0$ depending only on n and j such that as $\lambda \to \infty$ we have

(1.7.4) $$N(\Delta_{R;nj}; \lambda) \sim \nu_n^{(j)} (\mathrm{vol}_\theta M) \lambda^{\frac{n+1}{2}}.$$

1.8. Organization of the memoir

The rest of the memoir is organized as follows. In Chapter 2 we start by recalling the main definitions and examples concerning Heisenberg manifolds and their tangent Lie group bundles. Then we review the constructions of the main differential operators on Heisenberg manifolds: sum of squares, Kohn Laplacian, horizontal sublaplacian, Gover-Graham operators and the contact Laplacian.

In Chapter 3 after a detailed review of the main known facts about the Heisenberg calculus, we give an intrinsic definition of the principal symbol and model operators of a $\Psi_H\mathrm{DO}$ and check their main properties. Then we prove Theorem 1.2.2 and its consequences. We conclude the chapter by a closer look at the main differential operators on Heisenberg manifolds.

In Chapter 4 we define holomorphic families of $\Psi_H\mathrm{DO}$ and study their main properties. In particular, we make use of an almost homogeneous approach to the Heisenberg calculus.

Chapter 5 is devoted to complex powers of positive hypoelliptic differential operators in connection with the heat equation. After having recalled the pseudodifferential representation of the heat kernel of such an operator in terms of the Volterra-Heisenberg calculus of [**BGS**], we use it to establish Theorem 1.4.2. Then we make use of Theorem 1.4.2 to extend Theorem 1.2.2 to $\Psi_H\mathrm{DOs}$ with non-integer orders and to prove Theorem 1.4.5. Eventually, we construct the weighted Sobolev spaces $W_H^s(M,\mathcal{E})$, $s \in \mathbb{R}$, and check their main properties. In particular, we prove that they yield sharp regularity results for $\Psi_H\mathrm{DOs}$.

In Chapter 6, we deal with spectral asymptotics for hypoelliptic operators on Heisenberg manifolds. First, we derive general spectral asymptotics for such operators on a general Heisenberg manifold. We then express these asymptotics in a geometric fashion. We first proceed with the Kohn Laplacian and the horizontal sublaplacian on a CR manifold. Then we deal with the horizontal sublaplacian and the contact Laplacian on a contact manifold.

Finally, two appendices are included. In Appendix A we give a version in Heisenberg coordinates of the proof of the invariance by Heisenberg diffeomorphisms of the Heisenberg calculus, which is used in the intrinsic definition of the principal symbol of a $\Psi_H\mathrm{DO}$. In Appendix B we similarly give a version in Heisenberg coordinates of the proof that the transpose of a $\Psi_H\mathrm{DO}$ is again a $\Psi_H\mathrm{DO}$. Both proofs will also be useful for generalizing the aforementionned results to the setting of holomorphic families of $\Psi_H\mathrm{DOs}$.

ACKNOWLEDGEMENTS. I am grateful to Alain Connes, Charles Epstein, Colin Guillermou, Bernard Helffer, Henri Moscovici, Michel Rumin and Elias Stein for helpful and stimulating discussions. I would like also to thank for their hospitality the mathematics departments of Princeton University, Harvard University and University of California at Berkeley where the memoir was finally completed.

On the other hand, some of the results of this memoir were announced in [**Po2**] and presented as part of the author's PhD thesis at University of Paris-Sud (Orsay, France) made under the supervision of Professor Alain Connes.

CHAPTER 2

Heisenberg manifolds and their main differential operators

In this chapter we recall the main definitions and properties of Heisenberg manifolds and we review the construction of the main examples of differential operators on such manifolds.

2.1. Heisenberg manifolds

In this section we gather the main facts about Heisenberg manifolds and their tangent Lie group bundles.

DEFINITION 2.1.1. *1) A Heisenberg manifold is a smooth manifold M equipped with a distinguished hyperplane bundle $H \subset TM$.*

2) A Heisenberg diffeomorphism ϕ from a Heisenberg manifold (M, H) onto another Heisenberg manifold (M, H') is a diffeomorphism $\phi : M \to M'$ such that $\phi^* H = H'$.

DEFINITION 2.1.2. *Let (M^{d+1}, H) be a Heisenberg manifold. Then:*

1) A (local) H-frame for TM is a (local) frame $X_0, X_1, \ldots, X_d$ of TM so that $X_1, \ldots, X_d$ span H.

2) A local Heisenberg chart is a local chart with a local H-frame of TM over its domain.

The main examples of Heisenberg manifolds are the following.

a) Heisenberg group. The $(2n+1)$-dimensional Heisenberg group $\mathbb{H}^{2n+1}$ is $\mathbb{R}^{2n+1} = \mathbb{R} \times \mathbb{R}^{2n}$ equipped with the group law,

$$(2.1.1) \qquad x.y = (x_0 + y_0 + \sum_{1 \leq j \leq n}(x_{n+j}y_j - x_j y_{n+j}), x_1 + y_1, \ldots, x_{2n} + y_{2n}).$$

A left-invariant basis for its Lie algebra $\mathfrak{h}^{2n+1}$ is provided by the vector-fields,

$$(2.1.2) \qquad X_0 = \frac{\partial}{\partial x_0}, \quad X_j = \frac{\partial}{\partial x_j} + x_{n+j}\frac{\partial}{\partial x_0}, \quad X_{n+j} = \frac{\partial}{\partial x_{n+j}} - x_j \frac{\partial}{\partial x_0},$$

with $j = 1, \ldots, n$. For $j, k = 1, \ldots, n$ and $k \neq j$ we have the relations,

$$(2.1.3) \qquad [X_j, X_{n+k}] = -2\delta_{jk}X_0, \quad [X_0, X_j] = [X_j, X_k] = [X_{n+j}, X_{n+k}] = 0.$$

In particular, the subbundle spanned by the vector fields $X_1, \ldots, X_{2n}$ defines a left-invariant Heisenberg structure on $\mathbb{H}^{2n+1}$.

(b) Codimension 1 foliations. These are the Heisenberg manifolds (M, H) such that H is integrable in Fröbenius' sense, i.e., $C^\infty(M, H)$ is closed under the Lie bracket of vector fields.

(c) Contact manifolds. A contact manifold is a Heisenberg manifold (M^{2n+1}, H) such that near any point of M there exists a contact form anihilating H, i.e., a 1-form θ such that $d\theta_{|H}$ is non-degenerate. When M is orientable it is equivalent to require the existence of a globally defined contact form on M anihilating H. More specific examples of contact manifolds include the Heisenberg group $\mathbb{H}^{2n+1}$, boundaries of strictly pseudoconvex domains $D \subset \mathbb{C}^{2n+1}$, like the sphere S^{2n+1}, or even the cosphere bundle S^*M of a Riemannian manifold M^{n+1}.

d) Confoliations. The confoliations of Elyashberg and Thurston in [**ET**] interpolate between contact manifolds and foliations. They can be seen as oriented Heisenberg manifolds (M^{2n+1}, H) together with a non-vanishing 1-form θ on M anihilating H and such that $(d\theta)^n \wedge \theta \geq 0$.

e) CR manifolds. If $D \subset \mathbb{C}^{n+1}$ a bounded domain with boundary ∂D then the maximal complex structure, or CR structure, of $T(\partial D)$ is given by $T_{1,0} = T(\partial D) \cap T_{1,0}\mathbb{C}^{n+1}$, where $T_{1,0}$ denotes the holomorphic tangent bundle of $\mathbb{C}^{n+1}$. More generally, a CR structure on an orientable manifold M^{2n+1} is given by a complex rank n integrable subbundle $T_{1,0} \subset T_{\mathbb{C}}M$ such that $T_{1,0} \cap \overline{T_{1,0}} = \{0\}$. Besides on boundaries of complex domains, and more generally such structures naturally appear on real hypersurfaces in $\mathbb{C}^{n+1}$, quotients of the Heisenberg group $\mathbb{H}^{2n+1}$ by cocompact lattices, boundaries of complex hyperbolic spaces, and circle bundles over complex manifolds.

A real hypersurface $M = \{r = 0\} \subset \mathbb{C}^{n+1}$ is strictly pseudoconvex when the Hessian $\partial\bar{\partial}r$ is positive definite. In general, to a CR manifold M we can associate a Levi form $L_\theta(Z, W) = -id\theta(Z, \overline{W})$ on the CR tangent bundle $T_{1,0}$ by picking a non-vanishing real 1-form θ anihilating $T_{1,0} \oplus T_{0,1}$. We then say that M is strictly pseudoconvex (resp. κ-strictly pseudoconvex) when we can choose θ so that L_θ is positive definite (resp. is nondegenerate with κ negative eigenvalues) at every point. In particular, when this happens θ is non-degenerate on $H = \Re(T_{1,0} \oplus T_{0,1})$ and so (M, H) is a contact manifold.

2.1.1. Tangent Lie group bundle of a Heisenberg manifold. A simple description of the tangent Lie group bundle of a Heisenberg manifold (M^{d+1}, H) is given as follows.

LEMMA 2.1.3 ([**Po6**]). *The Lie bracket of vector fields induces on H a 2-form with values in TM/H,*

(2.1.4) $$\mathcal{L} : H \times H \longrightarrow TM/H,$$

so that for any sections X and Y of H near a point $a \in M$ we have

(2.1.5) $$\mathcal{L}_a(X(a), Y(a)) = [X, Y](a) \mod H_a.$$

DEFINITION 2.1.4. *The 2-form $\mathcal{L}$ is called the Levi form of (M, H).*

The Levi form $\mathcal{L}$ allows us to define a bundle $\mathfrak{g}M$ of graded Lie algebras by endowing the vector bundle $(TM/H) \oplus H$ with the smooth fields of Lie brackets and gradings such that, for sections X_0, Y_0 of TM/H and X', Y' of H and for $t \in \mathbb{R}$, we have

(2.1.6) $$[X_0 + X', Y_0 + Y']_a = \mathcal{L}_a(X', Y'), \qquad t.(X_0 + X') = t^2 X_0 + tX'.$$

As we can easily check $\mathfrak{g}M$ is a bundle of 2-step nilpotent Lie algebras which contains the normal bundle TM/H in its center. Therefore, its associated graded

Lie group bundle GM can be described as follows. As a bundle GM is $(TM/H) \oplus H$ and the exponential map is merely the identity. In particular, the grading of GM is as in (2.1.6). Moreover, since $\mathfrak{g}M$ is 2-step nilpotent the Campbell-Hausdorff formula shows that, for sections X, Y of $\mathfrak{g}M$, we have

(2.1.7) $$(\exp X)(\exp Y) = \exp(X + Y + \frac{1}{2}[X,Y]).$$

From this we deduce that the product on GM is such that

(2.1.8) $$(X_0 + X').(Y_0 + X') = X_0 + Y_0 + \frac{1}{2}\mathcal{L}(X', Y') + X' + Y',$$

for sections X_0, Y_0 of TM/H and sections X', Y' of H.

DEFINITION 2.1.5. *The bundles $\mathfrak{g}M$ and GM are respectively called the tangent Lie group bundle and the tangent Lie group of M.*

In fact, the fibers of GM are classified by the Levi form $\mathcal{L}$ as follows.

PROPOSITION 2.1.6 ([**Po6**]). *1) Let $a \in M$. Then $\mathcal{L}_a$ has rank $2n$ if, and only if, as a graded Lie group G_aM is isomorphic to $\mathbb{H}^{2n+1} \times \mathbb{R}^{d-2n}$.*

2) The Levi form $\mathcal{L}$ has constant rank $2n$ if, and only if, GM is a fiber bundle with typical fiber $\mathbb{H}^{2n+1} \times \mathbb{R}^{d-2n}$.

Now, let $\phi : (M, H) \to (M', H')$ be a Heisenberg diffeomorphism from (M, H) onto another Heisenberg manifold (M', H'). Since $\phi_* H = H'$ we see that ϕ' induces a smooth vector bundle isomorphism $\overline{\phi} : TM/H \to TM'/H'$.

DEFINITION 2.1.7. *We let $\phi'_H : (TM/H) \oplus H \to (TM'/H') \oplus H'$ denote the vector bundle isomorphism such that*

(2.1.9) $$\phi'_H(a)(X_0 + X') = \overline{\phi}'(a)X_0 + \phi'(a)X',$$

for any $a \in M$ and any $X_0 \in T_a/H_a$ and $X' \in H_a$.

PROPOSITION 2.1.8 ([**Po6**]). *The vector bundle isomorphism ϕ'_H is an isomorphism of graded Lie group bundles from GM onto GM'. In particular, the Lie group bundle isomorphism class of GM depends only on the Heisenberg diffeomorphism class of (M, H).*

2.1.2. Heisenberg coordinates and nilpotent approximation of vector fields. It is interesting to relate the intrinsic description of GM above with the more extrinsic description of [**BG**] (see also [**Be**], [**EM**], [**EMM**], [**FS1**], [**Gro**], [**Ro2**]) in terms of the Lie group associated to a nilpotent Lie algebra of model vector fields.

First, let $a \in M$ and let us describe $\mathfrak{g}_a M$ as the graded Lie algebra of left-invariant vector fields on G_aM by identifying any $X \in \mathfrak{g}_a M$ with the left-invariant vector fields L_X on G_aM given by

(2.1.10) $$L_X f(x) = \frac{d}{dt}f[x.(t \exp X)]_{|t=0} = \frac{d}{dt}f[x.(tX)]_{|t=0}, \qquad f \in C^\infty(G_aM).$$

This allows us to associate to any vector fields X near a a unique left-invariant vector fields X^a on G_aM such that

(2.1.11) $$X^a = \begin{cases} L_{X_0(a)} & \text{if } X(a) \notin H_a, \\ L_{X(a)} & \text{otherwise}, \end{cases}$$

where $X_0(a)$ denotes the class of $X(a)$ modulo H_a.

DEFINITION 2.1.9. *The left-invariant vector field X^a is called the model vector field of X at a.*

Let us look at the above construction in terms of a H-frame $X_0, \ldots, X_d$ near a, i.e., of a local trivialization of the vector bundle $(TM/H) \oplus H$. For $j, k = 1, \ldots, d$ we let

(2.1.12) $$\mathcal{L}(X_j, X_k) = [X_j, X_k]X_0 = L_{jk}X_0 \quad \mod H.$$

With respect to the coordinate system $(x_0, \ldots, x_d) \to x_0 X_0(a) + \ldots + x_d X_d(a)$ we can write the product law of $G_a M$ as

(2.1.13) $$x.y = (x_0 + \frac{1}{2}\sum_{j,k=1}^{d} L_{jk} x_j x_k, x_1, \ldots, x_d).$$

Then the vector fields X_j^a, $j = 1, \ldots, d$, in (2.1.11) are just the left-invariant vector fields corresponding to the vector e_j of the canonical basis of $\mathbb{R}^{d+1}$, that is, we have

(2.1.14) $$X_0^a = \frac{\partial}{\partial x_0} \quad \text{and} \quad X_j^a = \frac{\partial}{\partial x_j} - \frac{1}{2}\sum_{k=1}^{d} L_{jk} x_k \frac{\partial}{\partial x_0}, \quad 1 \leq j \leq d.$$

In particular, for $j, k = 1, \ldots, d$ we have the relations,

(2.1.15) $$[X_j^a, X_k^a] = L_{jk}(a) X_0^a, \qquad [X_j^a, X_0^a] = 0.$$

Now, let $\kappa : \text{dom}\,\kappa \to U$ be a Heisenberg chart near $a = \kappa^{-1}(u)$ and let $X_0, \ldots, X_d$ be the associated H-frame of TU. Then there is a unique affine coordinate change $x \to \psi_u(x)$ such that $\psi_u(u) = 0$ and $\psi_{u*} X_j(0) = \frac{\partial}{\partial x_j}$ for $j = 0, 1, \ldots, d$. Indeed, if for $j = 1, \ldots, d$ we set $X_j(x) = \sum_{k=0}^{d} B_{jk}(x) \frac{\partial}{\partial x_k}$ then we have

(2.1.16) $$\psi_u(x) = A(u)(x - u), \qquad A(u) = (B(u)^t)^{-1}.$$

DEFINITION 2.1.10. *1) The coordinates provided by ψ_u are called the privileged coordinates at u with respect to the H-frame $X_0, \ldots, X_d$.*

2) The map ψ_u is called the privileged-coordinate map with respect to the H-frame $X_0, \ldots, X_d$.

REMARK 2.1.11. The privileged coordinates at u are called u-coordinates in [**BG**], but they correspond to the privileged coordinates of [**Be**] and [**Gro**] in the special case of a Heisenberg manifold.

Next, on $\mathbb{R}^{d+1}$ we consider the dilations,

(2.1.17) $$\delta_t(x) = t.x = (t^2 x_0, t x_1, \ldots, t x_d), \qquad t \in \mathbb{R},$$

with respect to which $\frac{\partial}{\partial x_0}$ is homogeneous of degree -2 and $\frac{\partial}{\partial x_1}, \ldots, \frac{\partial}{\partial x_d}$ is homogeneous of degree -1.

Since in the privileged coordinates at u we have $X_j(0) = \frac{\partial}{\partial x_j}$, we can write

(2.1.18) $$X_j = \frac{\partial}{\partial x_j} + \sum_{k=0}^{d} a_{jk}(x) \frac{\partial}{\partial x_k}, \qquad j = 0, 1, \ldots d,$$

where the a_{jk}'s are smooth functions such that $a_{jk}(0)=0$. Thus, we can let

$$(2.1.19) \qquad X_0^{(u)} = \lim_{t\to 0} t^2 \delta_t^* X_0 = \frac{\partial}{\partial x_0},$$

$$(2.1.20) \qquad X_j^{(u)} = \lim_{t\to 0} t^{-1} \delta_t^* X_j = \frac{\partial}{\partial x_j} + \sum_{k=1}^d b_{jk} x_k \frac{\partial}{\partial x_0}, \quad j=1,\ldots,d,$$

where for $j,k=1,\ldots,d$ we have set $b_{jk} = \partial_{x_k} a_{j0}(0)$.

Observe that $X_0^{(u)}$ is homogeneous of degree -2 and $X_1^{(u)},\ldots,X_d^{(u)}$ are homogeneous of degree -1. Moreover, for $j,k=1,\ldots,d$ we have

$$(2.1.21) \qquad [X_j^{(u)}, X_0^{(u)}] = 0 \quad \text{and} \quad [X_j^{(u)}, X_0^{(u)}] = (b_{kj} - b_{jk}) X_0^{(u)}.$$

Thus, the linear space spanned by $X_0^{(u)}, X_1^{(u)}, \ldots, X_d^{(u)}$ is a graded 2-step nilpotent Lie algebra $\mathfrak{g}^{(u)}$. In particular, $\mathfrak{g}^{(u)}$ is the Lie algebra of left-invariant vector fields over the graded Lie group $G^{(u)}$ consisting of $\mathbb{R}^{d+1}$ equipped with the grading (2.1.17) and the group law,

$$(2.1.22) \qquad x.y = (x_0 + \sum_{j,k=1}^d b_{kj} x_j x_k, x_1, \ldots, x_d).$$

Now, if near a we let $\mathcal{L}(X_j, X_k) = [X_j, X_k] = L_{jk}(x) X_0 \bmod H$ then we have

$$(2.1.23) \quad [X_j^{(u)}, X_k^{(u)}] = \lim_{t\to 0}[t\delta_t^* X_j, t\delta_t^* X_k] = \lim_{t\to 0} t^2 \delta_t^* (L_{jk}(\circ \kappa^{-1}(x)) X_0)$$

$$= L_{jk}(a) X_0^{(u)}.$$

Comparing this with (2.1.15) and (2.1.21) then shows that $\mathfrak{g}^{(u)}$ has the same the constant structures as those of $\mathfrak{g}_a M$, hence is isomorphic to $\mathfrak{g}_a M$. Consequently, the Lie groups $G^{(u)}$ and $G_a M$ are isomorphic. In fact, as it follows from [**BG**] and [**Po6**] an explicit isomorphism is given by

$$(2.1.24) \qquad \phi_u(x_0, \ldots, x_d) = (x_0 - \frac{1}{4} \sum_{j,k=1}^d (b_{jk} + b_{kj}) x_j x_k, x_1, \ldots, x_d).$$

DEFINITION 2.1.12. *Let $\varepsilon_u = \phi_u \circ \psi_u$. Then:*

1) The new coordinates provided by ε_u are called Heisenberg coordinates at u with respect to the H-frame $X_0,\ldots,X_d$.

2) The map ε_u is called the u-Heisenberg coordinate map.

REMARK 2.1.13. The Heisenberg coordinates at u have been also considered in [**BG**] as a technical tool for inverting the principal symbol of a hypoelliptic sublaplacian.

Next, as it follows from [**Po6**, Lem. 1.17] we also have

$$(2.1.25) \qquad \phi_* X_0^{(u)} = \frac{\partial}{\partial x_0} = X_0^a,$$

$$(2.1.26) \qquad \phi_* X_j^{(u)} = \frac{\partial}{\partial x_j} - \frac{1}{2} \sum_{k=1}^d L_{jk} x_k \frac{\partial}{\partial x_0} = X_j^a, \quad j=1,\ldots,d.$$

Since ϕ_u commutes with the dilations (2.1.17) using (2.1.19)–(2.1.20) we get

$$(2.1.27) \quad \lim_{t\to 0} t^2 \delta_t^* \phi_{u*} X_0^{(u)} = X_0^a \quad \text{and} \quad \lim_{t\to 0} t \delta_t^* \phi_{u*} X_j^{(u)} = X_j^a, \quad j=1,\ldots,d.$$

In fact, as shown in [**Po6**] for any vector fields X near a, as $t \to 0$ and in Heisenberg coordinates at a, we have

$$(2.1.28) \quad \delta_t^* X = \begin{cases} t^{-2} X^a + \mathrm{O}(t^{-1}) & \text{if } X(a) \in H_a, \\ t^{-1} X^a + \mathrm{O}(1) & \text{otherwise.} \end{cases}$$

Therefore, we obtain:

PROPOSITION 2.1.14 ([**Po6**]). *In the Heisenberg coordinates centered at $a = \kappa^{-1}(u)$ the tangent Lie group $G_a M$ coincides with $G^{(u)}$ and for any vector fields X the model vector fields X^a approximates X near a in the sense of (2.1.28).*

One consequence of the equivalence between the two approaches to GM is a tangent approximation for Heisenberg diffeomorphisms as follows.

Let $\phi : (M,H) \to (M',H')$ be a Heisenberg diffeomorphism from (M,H) to another Heisenberg manifold (M',H'). We also endow $\mathbb{R}^{d+1}$ with the pseudo-norm,

$$(2.1.29) \quad \|x\| = (x_0^2 + (x_1^2 + \ldots + x_d^2)^2)^{1/4}, \quad x \in \mathbb{R}^{d+1},$$

so that for any $x \in \mathbb{R}^{d+1}$ and any $t \in \mathbb{R}$ we have

$$(2.1.30) \quad \|t.x\| = |t|\,\|x\|.$$

PROPOSITION 2.1.15 ([**Po6**, Prop. 2.21]). *Let $a \in M$ and set $a' = \phi(a)$. Then, in Heisenberg coordinates at a and at a' the diffeomorphism $\phi(x)$ has a behavior near $x = 0$ of the form*

$$(2.1.31) \quad \phi(x) = \phi_H'(0)x + (\mathrm{O}(\|x\|^3), \mathrm{O}(\|x\|^2), \ldots, \mathrm{O}(\|x\|^2)).$$

In particular, there is no term of the form $x_j x_k$, $1 \le j,k \le d$, in the Taylor expansion of $\phi_0(x)$ at $x = 0$.

REMARK 2.1.16. An asymptotics similar to (2.1.31) is given in [**Be**, Prop. 5.20] in privileged coordinates at u and $u' = \kappa_1(a')$, but the leading term there is only a Lie algebra isomorphism from $\mathfrak{g}^{(u)}$ onto $\mathfrak{g}^{(u')}$. This is only in Heisenberg coordinates that we recover the Lie group isomorphism $\phi_H'(a)$ as the leading term of the asymptotics.

REMARK 2.1.17. An interesting application of Proposition 2.1.15 in [**Po6**] is the construction of the tangent groupoid $\mathcal{G}_H M$ of (M,H) as the differentiable groupoid encoding the smooth deformation of $M \times M$ to GM. This groupoid is the analogue in the Heisenberg setting of Connes' tangent groupoid (see [**Co**, II.5], [**HS**]) and it shows that GM is tangent to M in a differentiable fashion (compare [**Be**], [**Gro**]).

2.2. Main differential operators on Heisenberg manifolds

In this section we recall the definitions of the most common operators on a Heisenberg manifold. With the exception of the contact Laplacian, all these operators are sublaplacians or are product of such operators up to lower order terms.

A sublaplacian on a Heisenberg manifold (M^{d+1}, H) acting on the sections of a vector bundle $\mathcal{E}$ over M is a differential operator $\Delta : C^\infty(M, \mathcal{E}) \to C^\infty(M, \mathcal{E})$

2.2. MAIN DIFFERENTIAL OPERATORS ON HEISENBERG MANIFOLDS

such that, near any $a \in M$, there exists a H-frame $X_0, X_1, \ldots, X_d$ of TM so that Δ takes the form

$$\Delta = -\sum_{j=1}^{d} X_j^2 + \sum_{j=1}^{d} a_j(x) X_j + c(x), \tag{2.2.1}$$

for some local sections $a_1(x), \ldots, a_d(x)$ and $c(x)$ of $\operatorname{End} \mathcal{E}$.

2.2.1. Hörmander's sum of squares. Let $X_1, \ldots, X_m$ be (real) vector fields on a manifold M^{d+1} and consider the sum of squares,

$$\Delta = -(X_1^2 + \ldots + X_m^2). \tag{2.2.2}$$

By a celebrated theorem of Hörmander [**Hö2**] the operator Δ is hypoelliptic provided that the following bracket condition is satisfied: the vector fields $X_0, \ldots, X_m$ together with their successive Lie brackets $[X_{j_1}, [X_{j_2}, \ldots, X_{j_l}] \ldots]]$ span the tangent bundle TM at every point.

When $X_1, \ldots, X_m$ span a hyperplane bundle H the operator Δ is a sublaplacian with *real* coefficients and the bracket condition reduces to $H + [H, H] = TM$ or, equivalently, to the nonvanishing of the Levi form of (M, H).

In fact, given a vector bundle $\mathcal{E}$, the theorem of Hörmander holds more generally for sublaplacians $\Delta : C^\infty(M, \mathcal{E}) \to C^\infty(M, \mathcal{E})$ of the form

$$\Delta = -(\nabla_{X_1}^2 + \ldots + \nabla_{X_m}^2) + L, \tag{2.2.3}$$

where ∇ is a connection on $\mathcal{E}$ and L is a first order differential operator with *real* coefficients. In particular, if M is endowed with a smooth positive density and $\mathcal{E}$ with a Hermitian metric, this includes the selfadjoint sum of squares,

$$\Delta = \nabla_{X_1}^* \nabla_{X_1} + \ldots + \nabla_{X_m}^* \nabla_{X_m}. \tag{2.2.4}$$

2.2.2. Kohn Laplacian. Let M^{2n+1} be an orientable CR manifold with CR tangent bundle $T_{1,0} \subset T_\mathbb{C} M$, let θ be a non-vanishing real 1-form annihilating the hyperplane bundle $H = \Re(T_{1,0} \oplus T_{0,1})$ and let L_θ be its associated Levi form.

Let $\mathcal{N}$ be a supplement of H in TM. This is an orientable line bundle which gives rise to the splitting,

$$T_\mathbb{C} M = T_{1,0} \oplus T_{0,1} \oplus (\mathcal{N} \otimes \mathbb{C}). \tag{2.2.5}$$

For $p, q = 0, \ldots, n$ let $\Lambda^{p,q} = (\Lambda^{1,0})^p \wedge (\Lambda^{0,1})^q$ be the bundle of (p, q)-forms, where $\Lambda^{1,0}$ and $\Lambda^{0,1}$) denote the annihilators in $T_\mathbb{C}^* M$ of $T_{0,1} \oplus (\mathcal{N} \otimes \mathbb{C})$ and $T_{1,0} \oplus (\mathcal{N} \otimes \mathbb{C})$ respectively. Then we have the splitting,

$$\Lambda^* T_\mathbb{C}^* M = (\bigoplus_{p,q=0}^{n} \Lambda^{p,q}) \oplus (\theta \wedge \Lambda^* T_\mathbb{C}^* M). \tag{2.2.6}$$

Notice that this decomposition does not depend on the choice of θ, but it does depend on that of $\mathcal{N}$.

The complex $\overline{\partial}_b : C^\infty(M, \Lambda^{p,*}) \to C^\infty(M, \Lambda^{p,*+1})$ of Kohn-Rossi ([**KR**], [**Koh1**]) is defined as follows. For any $\eta \in C^\infty(M, \Lambda^{p,q})$ we can uniquely decompose $d\eta$ as

$$d\eta = \overline{\partial}_{b;p,q}\eta + \partial_{b;p,q}\eta + \theta \wedge \mathcal{L}_{X_0}\eta, \tag{2.2.7}$$

where $\overline{\partial}_{b;p,q}\eta$ and $\partial_{b;p,q}\eta$ are sections of $\Lambda^{p,q+1}$ and $\Lambda^{p+1,q}$ respectively and X_0 is the section of $\mathcal{N}$ such that $\theta(X_0) = 1$. Thanks to the integrability of $T_{1,0}$ we have $\overline{\partial}_{b;p,q+1} \circ \overline{\partial}_{b;p,q} = 0$, so we really get a chain complex. This complex depends only on the CR structure of M and on the choice of $\mathcal{N}$, but the latter dependence is only up to the intertwinning by vector bundle isomorphisms (see, e.g., [**Po9**, Lem. 4.1]).

Next, assume that $T_{\mathbb{C}}M$ is endowed with a Hermitian metric compatible with the CR structure in the sense that it commutes with complex conjugation and the splitting (2.2.5) becomes orthogonal. Let $\overline{\partial}_{b;p,q}^*$ be the formal adjoint of $\overline{\partial}_{b;p,q}$. Then the Kohn Laplacian $\square_{b;p,q} : C^\infty(M, \Lambda^{p,q}) \to C^\infty(M, \Lambda^{p,q})$ is

$$(2.2.8) \qquad \square_{b;p,q} = \overline{\partial}_{b;p,q}^* \overline{\partial}_{b;p,q} + \overline{\partial}_{b;p,q-1} \overline{\partial}_{b;p,q-1}^*.$$

The Kohn Laplacian is a sublaplacian (see, e.g., [**FS1**, Sect. 13], [**BG**, Sect. 20]), so is not elliptic. Nevertheless, Kohn [**Koh1**] proved that under a geometric condition on the Levi form L_θ, the so-called condition $Y(q)$, the operator $\square_{b;p,q}$ is hypoelliptic with gain of one derivative, i.e., for any compact $K \subset M$ we have

$$(2.2.9) \qquad \|u\|_{s+1} \leq C_{Ks}(\|\square_{b;p,q} u\|_s + \|u\|_0) \qquad \forall u \in C_K^\infty(M, \Lambda^{p,q}),$$

where $\|.\|_s$ denotes the norm of the Sobolev space $L^2_s(M, \Lambda^{p,q})$.

The condition $Y(q)$ at point $x \in M$ means that if we let $(r(x) - \kappa(x), \kappa(x), n - r(x))$ be the signature of L_θ at x, so that $r(x)$ is the rank of L_θ and $\kappa(x)$ the number of its negative eigenvalues, then we must have

$$(2.2.10) \qquad q \notin \{\kappa(x), \ldots, \kappa(x) + n - r(x)\} \cup \{r(x) - \kappa(x), \ldots, n - \kappa(x)\}.$$

For instance, when M is κ-strictly pseudoconvex, the $Y(q)$-condition exactly means that we must have $q \neq \kappa$ and $q \neq n - \kappa$.

In general this condition is equivalent to the existence of a parametrix within the Heisenberg calculus (see [**BG**] for the case of a smoothly diagonalizable Levi form and Section 3.4 for the general case; see also [**Bo1**], [**FS1**]), from which we recover the hypoellipticity of $\square_{b;p,q}$.

Finally, the condition $Y(q)$ is only a sufficient condition for the hypoellipticity of the Kohn Laplacian, for the latter may be hypoelliptic even when the condition $Y(q)$ fails (see, e.g., [**Koh2**], [**Ko**], [**Ni**]).

2.2.3. Horizontal sublaplacian. Let (M^{d+1}, H) be a Heisenberg manifold endowed with a Riemannian metric. Identifying H^* with the subbundle of T^*M annihilating the orthogonal supplement $H^\perp$, we define the horizontal sublaplacian as the differential operator, $\Delta_{b;k} : C^\infty(M, \Lambda_{\mathbb{C}}^k H^*) \to C^\infty(M, \Lambda_{\mathbb{C}}^{k+1} H^*)$ such that

$$(2.2.11) \qquad \Delta_{b;k} = d_{b;k}^* d_{b;k} + d_{b;k-1} d_{b;k-1}^*, \qquad d_{b;k}\alpha = \pi_{b;k+1}(d\alpha),$$

where $\pi_{b;k+1}$ denotes the orthogonal projection onto $\Lambda_{\mathbb{C}}^{k+1} H^*$.

This operator was first introduced by Tanaka [**Ta**] in the CR setting, but versions of this operator acting on functions were independently defined by Greenleaf [**Gr**] and Lee [**Le**]. Moreover, it can be shown that $d_b^2 = 0$ if, and only if, the subbundle H is integrable, so in general Δ_b is not the Laplacian of a chain complex.

On functions $\Delta_{b;0}$ is a sum of squares modulo a lower order term, hence is hypoelliptic by Hörmander's theorem. On horizontal forms of higher degree, that is, on sections of $\Lambda_{\mathbb{C}}^k H^*$ with $k \geq 1$, it is shown in [**Ta**] and [**Ru**], in the contact case, and in Section 3.5, in the general case, that $\Delta_{b;k}$ is hypoelliptic when some condition, called condition $X(k)$, holds everywhere. More precisely, the condition $X(k)$ is satisfied at a point $x \in M$ when we have

$$(2.2.12) \qquad k \notin \{\frac{1}{2}r(x), \frac{1}{2}r(x) + 1, \ldots, d - \frac{1}{2}r(x)\},$$

where $r(x)$ denotes the rank of the Levi form $\mathcal{L}$ at x. For instance, if M^{2n+1} is a contact manifold or a nondegenerate CR manifold then the Levi form is everywhere nondegenerate, so $r(x) = 2n$ and the $X(k)$-condition becomes $k \neq n$.

Assume now that M is an orientable CR manifold of dimension $2n+1$ with Heisenberg structure $H = \Re(T_{1,0} \oplus T_{0,1})$ and let θ be a global nonvanishing section of TM/H with associated Levi form L_θ. Assume in addition that $T_\mathbb{C}M$ is endowed with a Hermitian metric compatible with its CR structure.

Under these assumptions we have $d_b = \overline{\partial}_b + \partial_b$, where ∂_b denotes the conjugate of $\overline{\partial}_b$, that is, the operator such that $\partial_b \omega = \overline{\overline{\partial}_b \overline{\omega}}$ for any $\omega \in C^\infty(M, \Lambda_\mathbb{C}^* H^*)$. Moreover, one can check that $\overline{\partial}_b \partial_b^* + \partial_b^* \overline{\partial}_b = \overline{\partial}_b^* \partial_b + \partial_b \overline{\partial}_b^* = 0$, from which we get

$$(2.2.13) \qquad \Delta_b = \Box_b + \overline{\Box}_b,$$

where $\overline{\Box}_b$ is the conjugate of $\Box_b$. In particular, this shows that the horizontal sublaplacian preserves the bidegree, i.e., it acts on (p,q)-forms.

Next, as shown in Section 3.5, the operator $\Delta_{b;p,q}$ acting on (p,q)-forms is hypoelliptic when at any point $x \in M$ the condition $X(p,q)$ is satisfied. The latter requires to have

$$(2.2.14) \quad \{(p,q),(q,p)\} \cap \{(\kappa(x)+j, r(x)-\kappa(x)+k);\ \max(j,k) \leq n - r(x)\} = \emptyset,$$

where $r(x)$ denotes the rank of the Levi form L_θ at x and $\kappa(x)$ its number of negative eigenvalues. For instance, when M is κ-strictly pseudoconvex the condition $X(p,q)$ means that we must have $(p,q) \neq (\kappa, n - \kappa)$ and $(p,q) \neq (n-\kappa, \kappa)$.

2.2.4. Gover-Graham operators. Let M^{2n+1} be a strictly pseudoconvex CR manifold. Let $T_{1,0} \subset T_\mathbb{C}M$ be the CR tangent bundle of M and set $T_{0,1} = \overline{T_{1,0}}$ and $H = \Re(T_{1,0} \oplus T_{0,1})$. Let θ be a pseudohermitian contact form, i.e., θ is a 1-form annihilating H such that the associated Levi form L_θ is positive definite on $T_{1,0}$. Thus θ defines a pseudohermitian structure on M in the sense of Webster [**We**].

We extend the Levi form L_θ into a Hermitian metric h_θ on $T_\mathbb{C}M$ such that $T_{1,0}$ and $T_{0,1}$ are orthogonal subspaces, complex conjugation is an (antilinear) isometry and $h_{\theta|_{H^\perp}} = \theta^2$. Then as shown by Tanaka [**Ta**] and Webster [**We**] there is a unique unitary connection on $T_\mathbb{C}M$ preserving the pseudohermitian structure. Note that the contact form θ is unique up to a conformal change $f \to e^{2f}\theta$, $f \in C^\infty(M, \mathbb{R})$.

In order to study the analogue in CR geometry of the Yamabe problem Jerison-Lee [**JL1**] (see also [**JL2**], [**JL3**]) introduced a conformal version of the horizontal sublaplacian acting on functions as the operator $\Box_\theta : C^\infty(M) \to C^\infty(M)$ such that

$$(2.2.15) \qquad \Box_\theta = \Delta_{b;0} + \frac{n}{n+2} R_n,$$

where R_n denotes the scalar curvature of the Tanaka-Webster connection. This is a conformal operator in the sense that we have

$$(2.2.16) \qquad \Box_{e^{2f}\theta} = e^{-(n+2)f} \Box_\theta\, e^{nf} \qquad \forall f \in C^\infty(M, \mathbb{R}).$$

The construction of Jerison-Lee has been generalized by Gover-Graham [**GG**], who produced CR analogues of the conformal operators of [**GJMS**]. For $k = 1, \ldots, n+1$ and for $k = n+2, n+4, \ldots$ their constructions yield a selfadjoint differential operator $\Box_\theta^{(k)} : C^\infty(M) \to C^\infty(M)$ such that

$$(2.2.17) \qquad \Box_{e^{2f}\theta}^{(k)} = e^{-(n+1+k)f} \Box_\theta^{(k)} e^{(n+1-k)f}, \qquad \forall f \in C^\infty(M, \mathbb{R}).$$

We make the convention that for $k = 1, 2, \ldots, n+1$ the operator $\square_\theta^{(k)}$ corresponds to the operator $P_{w,w}$ of [**GG**] with $w = \frac{k-1-n}{2}$ under the canonical trivializations of the density bundles $\mathcal{E}(w,w) = |\Lambda^{n,n}|^w$ coming from the trivialization of $\Lambda^{n,n}$ provided by $d\theta^n$. For $k = n+2, n+4, \ldots$ the operator $\square_\theta^{(k)}$ similarly corresponds to the operator $\mathcal{P}_{w,w}$ of [**GG**] with $w = \frac{k-1-n}{2}$.

The operator $P_{w,w}$ is obtained by pushing down to M the GJMS operator of order k on the associated Fefferman bundle, while is $\mathcal{P}_{w,w}$ is contructed by making use of a CR geometric version of the tractor calculus of [**BEG**]. In particular, for $k = 1$ the operator $\square_\theta^{(1)}$ agrees with the conformal sublaplacian of Jerison-Lee. In general, if we let X_0 denote the Reeb vector field of θ, so that $\imath_{X_0}\theta = 1$ and $\imath_{X_0}d\theta = 0$, then $\square_\theta^{(k)}$ has same principal part (in the Heisenberg sense) as

$$(2.2.18) \qquad (\Delta_{b;0} + i(k-1)X_0)(\Delta_{b;0} + i(k-3)X_0)\cdots(\Delta_{b;0} - i(k-1)X_0).$$

In particular, unless for the value $k = n+1$ the operator $\square_\theta^{(k)}$ is hypoelliptic (see Proposition 3.5).

Finally, let Q_θ denote the CR Q curvature as defined by Fefferman-Hirachi [**FH**]. This the CR analogue of Branson's Q curvature in conformal Riemannian geometry and as with the GJMS operators we have

$$(2.2.19) \qquad e^{2(n+1)f}Q_{e^{2f}\theta} = Q_\theta + \square_\theta^{n+1} \qquad \forall f \in C^\infty(M, \mathbb{R}).$$

2.2.5. Contact complex and contact Laplacian. Let (M^{2n+1}, H) be an orientable contact manifold, let θ be a contact form on H and let $J \in C^\infty(M, \text{End } H)$, $J^2 = -1$, be an almost complex structure on H which is calibrated, i.e., $d\theta(X, JX) = -d\theta(JX, X) > 0$ for any section X of H. Then we can endow M with the Riemannian metric,

$$(2.2.20) \qquad g_{\theta, J} = d\theta(., J.) + \theta^2.$$

The contact complex of Rumin [**Ru**] can be seen as an attempt to get on M a complex of horizontal differential forms by forcing up the equalities $d_b^2 = 0$ and $(d_b^*)^2 = 0$ as follows.

Let X_0 be the Reeb field associated to θ, that is, so that $\imath_{X_0}\theta = 1$ and $\imath_{X_0}d\theta = 0$. Then we have

$$(2.2.21) \qquad d_b^2 = -\mathcal{L}_{X_0}\varepsilon(d\theta) = -\varepsilon(d\theta)\mathcal{L}_{X_0},$$

where $\varepsilon(d\theta)$ denotes the exterior multiplication by $d\theta$.

There are two natural ways of modifying the space $\Lambda_\mathbb{C}^* H^*$ of horizontal forms to get a complex. The first one is to force the equality $d_b^2 = 0$ by restricting the operator d_b to $\Lambda_2^* := \ker \varepsilon(d\theta) \cap \Lambda_\mathbb{C}^* H^*$ since this bundle is closed under d_b and annihilates d_b^2.

The second way is to similarly force the equality $(d_b^*)^2 = 0$ by restricting d_b^* to $\Lambda_1^* := \ker \iota(d\theta) \cap \Lambda_\mathbb{C}^* H^* = (\text{im }\varepsilon(d\theta))^\perp \cap \Lambda_\mathbb{C}^* H^*$, where $\iota(d\theta)$ denotes the interior product with $d\theta$. This amounts to replace d_b by the operator $d_b' = \pi_1 \circ d_b$, where π_1 is the orthogonal projection onto Λ_1^*.

In fact, as $d\theta_{|_H}$ is nondegenerate the operator $\varepsilon(d\theta): \Lambda_\mathbb{C}^k H^* \to \Lambda_\mathbb{C}^{k+2} H^*$ is injective for $k \leq n-1$ and surjective for $k \geq n-1$. This implies that $\Lambda_2^k = 0$ for $k \leq n-1$ and $\Lambda_1^k = 0$ for $k \geq n+1$. Therefore, we only have two halves of complexes, but we can get a whole complex by connecting the two halves to each other as follows.

2.2. MAIN DIFFERENTIAL OPERATORS ON HEISENBERG MANIFOLDS

Consider the differential operator $D_{R;n} : C^\infty(M, \Lambda_\mathbb{C}^n H^*) \to C^\infty(M, \Lambda_\mathbb{C}^n H^*)$ such that

(2.2.22) $$D_{R;n} = \mathcal{L}_{X_0} + d_{b;n-1}\varepsilon(d\theta)^{-1}d_{b;n},$$

where $\varepsilon(d\theta)^{-1}$ is the inverse of $\varepsilon(d\theta) : \Lambda_\mathbb{C}^{n-1} H^* \to \Lambda_\mathbb{C}^{n+1} H^*$. Notice that $D_{R,n}$ has order 2. Then we have:

LEMMA 2.2.1 ([**Ru**]). *The operator $D_{R;n}$ maps to sections of Λ_2^n and satisfies $d_{b;n}D_{R;n} = 0$ and $D_{R;n}d'_{b;n} = 0$.*

PROOF. Let $\alpha \in C^\infty(M, \Lambda_\mathbb{C}^n H^*)$ and set $\beta = \varepsilon(d\theta)^{-1}d_{b;n}\alpha$, so that $d_{b;n}\alpha = d\theta \wedge \beta$. Then we have

(2.2.23) $\theta \wedge D_{R;n}\alpha = \theta \wedge \mathcal{L}_{X_0}\alpha + \theta \wedge d_{b;n-1}\beta = d\alpha - d_{b;n}\alpha + \theta \wedge d\beta$
$= d\alpha - d\theta \wedge \beta + \theta \wedge d\beta = d(\alpha - \theta \wedge \beta).$

Hence $0 = d(\theta \wedge D_{R;n}\alpha) = d\theta \wedge D_{R;n}\alpha - \theta \wedge d_{b;n}D_{R;n}\alpha$, which gives $d_{b;n}D_{R;n}\alpha = 0$ and $d\theta \wedge D_{R;n}\alpha = 0$. In particular $D_{R;n}\alpha$ is a section of $\ker \varepsilon(d\theta) \cap \Lambda_\mathbb{C}^n H^* = \Lambda_2^n$.

Next, let $\alpha \in C^\infty(M, \Lambda_\mathbb{C}^{n-1} H^*)$. Since $d'_{b;n}\alpha$ is the orthogonal projection of $d_{b;n}\alpha$ onto $\ker \iota(d\theta) = \ker \varepsilon(d\theta)^\perp$ we can write $d'_{b;n}\alpha = d_{b;n}\alpha - d\theta \wedge \gamma$ for some section γ of $\Lambda_\mathbb{C}^{n-2} H^*$. Then using (2.2.21) we get $d_{b;n+1}d'_{b;n}\alpha = d_b^2\alpha - d_b(\theta \wedge \gamma) = -d\theta \wedge \mathcal{L}_{X_0}\alpha - d\theta \wedge d_{b;n-2}\gamma$. Thus,

(2.2.24) $$d_{b;n-1}\varepsilon(d\theta)^{-1}d_{b;n+1}d'_{b;n}\alpha = -d_{b;n}\mathcal{L}_{X_0}\alpha - d_b^2\alpha = -d_{b;n}\mathcal{L}_{X_0}\alpha + d\theta \wedge \mathcal{L}_{X_0}\gamma.$$

On the other hand, we have

(2.2.25) $$\mathcal{L}_{X_0}d'_{b;n}\alpha = \mathcal{L}_{X_0}d_{b;n}\alpha - \mathcal{L}_{X_0}(d\theta \wedge \gamma) = \mathcal{L}_{X_0}d_{b;n}\alpha - d\theta \wedge \mathcal{L}_{X_0}\gamma.$$

Furthemore, we see that $\mathcal{L}_{X_0}d_{b;n}\alpha$ is equal to

(2.2.26) $$\iota_{X_0}d[d\alpha - \theta \wedge \mathcal{L}_{X_0}\alpha] = -\iota_{X_0}(d\theta \wedge \mathcal{L}_{X_0}\alpha - \theta \wedge d_{b;n}\mathcal{L}_{X_0}\alpha) = d_{b;n}\mathcal{L}_{X_0}\alpha.$$

Combining this with (2.2.24) then gives

(2.2.27) $$D_{R;n}d'_{b;n}\alpha = \mathcal{L}_{X_0}d'_{b;n}\alpha + d_{b;n-1}\varepsilon(d\theta)^{-1}d_{b;n+1}d'_{b;n}\alpha = 0.$$

The lemma is thus proved. □

All this shows that we have the complex,

(2.2.28) $$C^\infty(M) \xrightarrow{d_{R;0}} \ldots C^\infty(M, \Lambda^n) \xrightarrow{d_{R;n}} C^\infty(M, \Lambda^n) \ldots \xrightarrow{d_{R;2n-1}} C^\infty(M, \Lambda^{2n}),$$

where $d_{R;k}$ agrees with $\pi_1 \circ d_{b;k}$ for $k = 0, \ldots, n-1$ and with $d_{b;k}$ otherwise. This complex is called the contact complex.

The above definition depends on the choice of the contact form θ and of the calibrated almost complex structure J. As shown by Rumin [**Ru**] there is an alternative description of the contact complex depending only on $H = \ker \theta$. For $k = 0, 1, \ldots, 2n+1$ consider the vector bundles,

(2.2.29) $$\Lambda_{H,1}^k = \Lambda_\mathbb{C}^k T^*M/[(\operatorname{im} \varepsilon(\theta) + \operatorname{im} \varepsilon(d\theta)) \cap \Lambda_\mathbb{C}^k T^*M],$$

(2.2.30) $$\Lambda_{H,2}^k = \ker \varepsilon(\theta) \cap \ker \varepsilon(d\theta) \cap \Lambda_\mathbb{C}^k T^*M.$$

Notice that these bundles depend only on H and that $\Lambda_{H,1}^k = \{0\}$ for $k \geq n+1$ and $\Lambda_{H,2}^k = \{0\}$ for $k \leq n$. Furthermore, there are natural vector bundle isomorphisms $\phi_k : \Lambda_1^k \to \Lambda_{H,1}^k$ and $\psi_k : \Lambda_2^k :\to \Lambda_{H,2}^{k+1}$. The former is obtained by restricting to

Λ_1^k the canonical projection from $\Lambda_{\mathbb{C}}^k T^*M$ onto $\Lambda_{H,1}^k$ and the latter is given by the restriction to Λ_2^k of $\varepsilon(\theta)$, whose inverse is the restriction of ι_{X_0} to $\Lambda_{H,2}^{k+1}$.

Next, the de Rham differential maps induces differential operators,

(2.2.31) $\qquad d_k : C^\infty(M, \Lambda_{H,1}^k) \to C^\infty(M, \Lambda_{H,1}^{k+1}), \quad k = 0, 1, \ldots, n-1,$

(2.2.32) $\qquad d_k : C^\infty(M, \Lambda_{H,2}^k) \to C^\infty(M, \Lambda_{H,2}^{k+1}), \quad k = n, \ldots, 2n,$

which give rise to two halves of complex. As shown by Rumin these two halves can be connected by means of a differential operator $D_n : C^\infty(M, \Lambda_{H,1}^n) \to C^\infty(M, \Lambda_{H,2}^{n+1})$ in such way that we get a full complex,

(2.2.33) $\quad C^\infty(M) \xrightarrow{d_0} \ldots C^\infty(M, \Lambda_{H,1}^n) \xrightarrow{D_n} C^\infty(M, \Lambda_{H,2}^{n+1}) \ldots \xrightarrow{d_{2n}} C^\infty(M, \Lambda_{H,2}^{2n+1}).$

In addition, the isomorphisms ϕ_k and ψ_k above intertwine this complex with the complex (2.2.28), that is, we have:

- $\phi_{k+1} d_{R;k} = d_k \phi_k$ for $k = 0, 1, \ldots, n-1$;
- $\psi_n d_{R;n} = D_n \phi_n$;
- $\psi_{k+1} d_{R;k} = d_{k+1} \psi_k$ for $k = n, \ldots, 2n$.

Let θ' be another contact form annihilating H and let J' be an almost complex structure on H which is calibrated with respect to θ'. Then, assigning the superscript $'$ to objects associated to θ' and J', we see from the above discussion that there exist vector bundle isomorphisms $\phi_k : \Lambda_1^k \to \Lambda_1^{k'}$, $k = 0, \ldots, n$, and $\psi_k : \Lambda_2^k \to \Lambda_2^{k'}$, $k = n, \ldots, 2n$, intertwining the contact complexes associated to (θ, J) and (θ', J'). Thus, up to the intertwining by vector bundle isomorphisms, the contact complex depend only on $H = \ker \theta$.

The contact Laplacian is defined as follows. In degree $k \neq n$ this is the differential operator $\Delta_{R;k} : C^\infty(M, \Lambda^k) \to C^\infty(M, \Lambda^k)$ such that

(2.2.34)
$$\Delta_{R;k} = \begin{cases} (n-k)d_{R;k-1}d_{R;k}^* + (n-k+1)d_{R;k+1}^*d_{R;k}, & k = 0, \ldots, n-1, \\ (k-n-1)d_{R;k-1}d_{R;k}^* + (k-n)d_{R;k+1}^*d_{R;k}, & k = n+1, \ldots, 2n. \end{cases}$$

For $k = n$ we have the differential operators $\Delta_{R;nj} : C^\infty(M, \Lambda_j^n) \to C^\infty(M, \Lambda_j^n)$, $j = 1, 2$, given by the formulas,

(2.2.35) $\quad \Delta_{R;n1} = (d_{R;n-1}d_{R;n}^*)^2 + d_{R;n}^*d_{R;n}, \quad \Delta_{R;n2} = d_{R;n}d_{R;n}^* + (d_{R;n+1}^*d_{R;n}).$

Observe that outside middle degree $k \neq n$ the operator $\Delta_{R;k}$ is a differential operator order 2, whereas Δ_{Rn1} and Δ_{Rn2} are differential operators of order 4. Moreover, Rumin [**Ru**] proved that in every degree the contact Laplacian is maximal hypoelliptic in the sense of [**HN3**].

Alternatively, we can show that in every degree the contact Laplacian admits a parametrix in the Heisenberg calculus and recover its hypoellipticity from this fact (see [**JK**] and Section 3.5).

CHAPTER 3

Intrinsic Approach to the Heisenberg Calculus

This chapter is organized as follows. In Section 3.1 we give a detailed review of the main definitions and properties of the Heisenberg calculus, following mostly the point of view of [**BG**].

In Section 3.2 we define intrinsic notions of principal symbol and model operator for the Heisenberg calculus and check their main properties.

In Section 3.3 we establish an invertibility criterion for the principal symbol in terms of the so called Rockland condition.

After this we explain in more details the invertibility criterion in some specific cases. First, we deal with general sublaplacians in Section 3.4, for which the results of [**BG**] even yield an explicit formula for the inverse of the principal symbol. Second, we deal more specifically with the main differential operators on Heisenberg manifolds in Section 3.5.

3.1. Heisenberg calculus

The Heisenberg calculus is the relevant pseudodifferential tool to study hypoelliptic operators on Heisenberg manifolds. It was independently invented by Beals-Greiner [**BG**] and Taylor [**Tay**], extending previous works of Boutet de Monvel [**Bo1**], Folland-Stein [**FS1**] and Dynin ([**Dy1**], [**Dy2**]) (see also [**BGH**], [**CGGP**], [**EM**], [**Gri**], [**Hö3**], [**RS**]).

The idea in the Heisenberg calculus, which is due to Elias Stein, is to have a pseudodifferential calculus on a Heisenberg manifold (M, H) which is modeled at any point $a \in M$ by the calculus of left-invariant pseudodifferential operators on the tangent group $G_a M$.

3.1.1. Left-invariant pseudodifferential operators.
Let (M^{d+1}, H) be a Heisenberg manifold and let $G = G_a M$ be the tangent group of M at a point $a \in M$. We shall now recall the main facts about left-invariant pseudodifferential operators on G (see also [**BG**], [**CGGP**], [**Tay**]).

Recall that for any finite dimensional real vector space E the Schwartz class $\mathcal{S}(E)$ is a Fréchet space and the Fourier transform is the continuous isomorphism of $\mathcal{S}(E)$ onto $\mathcal{S}(E^*)$ given by

$$(3.1.1) \qquad \hat{f}(\xi) = \int_E e^{i\langle \xi, x \rangle} f(x) dx, \qquad f \in \mathcal{S}(E), \quad \xi \in E^*,$$

where dx denotes the Lebesgue measure of E.

DEFINITION 3.1.1. $\mathcal{S}_0(E)$ is the closed subspace of $\mathcal{S}(E)$ consisting of $f \in \mathcal{S}(E)$ such that for any differential operator P on E^* we have $(P\hat{f})(0) = 0$.

Since G has the same underlying set as that of its Lie algebra $\mathfrak{g} = \mathfrak{g}_x M$ we can let $\mathcal{S}(G)$ and $\mathcal{S}_0(G)$ denote the Fréchet spaces $\mathcal{S}(E)$ and $\mathcal{S}_0(E)$ associated to the underlying linear space E of $\mathfrak{g}$ (notice that the Lebesgue measure of E coincides with the Haar measure of G since G is nilpotent).

Next, for $\lambda \in \mathbb{R}$ and $\xi = \xi_0 + \xi'$ in $\mathfrak{g}^* = (T_a^*M/H_a^*) \oplus H_a$ we let

$$(3.1.2) \qquad \lambda.\xi = \lambda.(\xi_0 + \xi') = \lambda^2 \xi_0 + \lambda \xi'.$$

DEFINITION 3.1.2. $S_m(\mathfrak{g}^*)$, $m \in \mathbb{C}$, is the space of functions $p \in C^\infty(\mathfrak{g}^* \setminus 0)$ which are homogeneous of degree m in the sense that, for any $\lambda > 0$, we have

$$(3.1.3) \qquad p(\lambda.\xi) = \lambda^m p(\xi) \qquad \forall \xi \in \mathfrak{g}^* \setminus 0.$$

In addition $S_m(\mathfrak{g}^*)$ is endowed with the Fréchet space topology coming from that of $C^\infty(\mathfrak{g}^* \setminus 0)$.

Note that the image $\hat{\mathcal{S}}_0(G)$ of $\mathcal{S}(G)$ under the Fourier transform consists of functions $v \in \mathcal{S}(\mathfrak{g}^*)$ such that, given any norm $|.|$ on G, near $\xi = 0$ we have $|g(\xi)| = O(|\xi|^N)$ for any integer $N \geq 0$. Thus, any $p \in S_m(\mathfrak{g}^*)$ defines an element of $\hat{\mathcal{S}}_0(\mathfrak{g}^*)'$ by letting

$$(3.1.4) \qquad \langle p, g \rangle = \int_{\mathfrak{g}^*} p(\xi) g(\xi) d\xi, \qquad g \in \hat{\mathcal{S}}_0(\mathfrak{g}^*).$$

This allows us to define the inverse Fourier transform of p as the element $\check{p} \in \mathcal{S}_0(G)'$ such that

$$(3.1.5) \qquad \langle \check{p}, f \rangle = \langle p, \check{f} \rangle \qquad \forall f \in \mathcal{S}_0(G).$$

PROPOSITION 3.1.3 ([**BG**], [**CGGP**]). *1) For any $p \in S_m(\mathfrak{g}^*)$ the left-convolution operator by $\check{p}$, i.e.,*

$$(3.1.6) \qquad \check{p} * f(x) := \langle \check{p}(y), f(x.y^{-1}) \rangle, \qquad f \in \mathcal{S}_0(G),$$

gives rise to a continuous endomorphism of $\mathcal{S}_0(G)$.

2) There is a continuous bilinear product,

$$(3.1.7) \qquad * : S_{m_1}(\mathfrak{g}^*) \times S_{m_2}(\mathfrak{g}^*) \longrightarrow S_{m_1+m_2}(\mathfrak{g}^*),$$

such that, for any $p_1 \in S_{m_1}(\mathfrak{g}^*)$ and any $p_2 \in S_{m_2}(\mathfrak{g}^*)$, the composition of the left-convolution operators by $\check{p}_1$ and $\check{p}_2$ is the left-convolution operator by $(p_1 * p_2)^\vee$, that is, we have

$$(3.1.8) \qquad \check{p}_1 * (\check{p}_2 * f) = (p_1 * p_2)^\vee * f \qquad \forall f \in \mathcal{S}_0(G).$$

Let us also mention that if $p \in S_m(\mathfrak{g}^*)$ then the convolution operator $Pu = \check{p} * f$ is a pseudodifferential operator. Indeed, let $X_0(a), \ldots, X_d(a)$ be a (linear) basis of $\mathfrak{g}$ so that $X_0(a)$ is in T_aM/H_a and $X_1(a), \ldots, X_d(a)$ span H_a. For $j = 0, \ldots, d$ let X_j^a be the left-invariant vector fields on G such that $X_{j|x=0}^w = X_j(a)$. The basis $X_0(a), \ldots, X_d(a)$ yields a linear isomorphism $\mathfrak{g} \simeq \mathbb{R}^{d+1}$, hence a global chart of G. In this chart p is a homogeneous symbol on $\mathbb{R}^{d+1} \setminus 0$ with respect to the dilations

$$(3.1.9) \qquad \lambda.x = (\lambda^2 x_0, \lambda x_1, \ldots, \lambda x_d), \qquad x \in \mathbb{R}^{d+1}, \quad \lambda > 0.$$

Similarly, each vector field $\frac{1}{i}X_j^a$, $j = 0, \ldots, d$, corresponds to a vector field on $\mathbb{R}^{d+1}$ whose symbol is denoted $\sigma_j^a(x, \xi)$. Then, setting $\sigma = (\sigma_0, \ldots, \sigma_d)$, it can be shown that in the above chart the operator P is given by

$$(3.1.10) \qquad Pf(x) = \int_{\mathbb{R}^{d+1}} e^{ix.\xi} p(\sigma^a(x, \xi)) \hat{f}(\xi), \qquad f \in \mathcal{S}_0(\mathbb{R}^{d+1}).$$

In other words P is the pseudodifferential operator $p(-iX^a) := p(\sigma^a(x,D))$ acting on $\mathcal{S}_0(\mathbb{R}^{d+1})$.

3.1.2. Ψ_HDOs on an open subset of $\mathbb{R}^{d+1}$. Let U be an open subset of $\mathbb{R}^{d+1}$ together with a hyperplane bundle $H \subset TU$ and a H-frame $X_0, X_1, \ldots, X_d$ of TU. Then the class of Ψ_HDOs on U is a class of pseudodifferential operators modelled on that of homogeneous convolution operators on the fibers of GU.

DEFINITION 3.1.4. $S_m(U \times \mathbb{R}^{d+1})$, $m \in \mathbb{C}$, is the space of functions $p(x,\xi)$ in $C^\infty(U \times \mathbb{R}^{d+1}\backslash 0)$ which are homogeneous of degree m with respect to ξ, that is,

$$(3.1.11) \qquad p(x, \lambda.\xi) = \lambda^m p(x,\xi) \qquad \text{for any } \lambda > 0,$$

where $\lambda.\xi$ is defined as in (3.1.9).

Observe that the homogeneity of $p \in S_m(U \times \mathbb{R}^{d+1})$ implies that, for any compact $K \subset U$, we have the estimates

$$(3.1.12) \qquad |\partial_x^\alpha \partial_\xi^\beta p(x,\xi)| \leq C_{K\alpha\beta} \|\xi\|^{\Re m - \langle \beta \rangle}, \qquad x \in K, \ \xi \neq 0,$$

where $\|\xi\| = (|\xi_0|^2 + |\xi_1|^4 + \ldots + |\xi_d|^4)^{1/4}$ and $\langle \alpha \rangle = 2\alpha_0 + \alpha_1 + \ldots + \alpha_d$.

DEFINITION 3.1.5. $S^m(U \times \mathbb{R}^{d+1})$, $m \in \mathbb{C}$, consists of symbols $p \in C^\infty(U \times \mathbb{R}^{d+1})$ with an asymptotic expansion $p \sim \sum_{j \geq 0} p_{m-j}$, $p_k \in S_k(U \times \mathbb{R}^{d+1})$, in the sense that, for any integer N and for any compact $K \subset U$, we have

$$(3.1.13) \qquad |\partial_x^\alpha \partial_\xi^\beta (p - \sum_{j<N} p_{m-j})(x,\xi)| \leq C_{\alpha\beta NK}\|\xi\|^{\Re m - \langle \beta \rangle - N}, \qquad x \in K, \ \|\xi\| \geq 1.$$

Next, for $j = 0, \ldots, d$ let $\sigma_j(x,\xi)$ denote the symbol of $\frac{1}{i}X_j$ (in the classical sense) and set $\sigma = (\sigma_0, \ldots, \sigma_d)$. For any $p \in S^m(U \times \mathbb{R}^{d+1})$ it can be shown that the symbol $p_\sigma(x,\xi) := p(x, \sigma(x,\xi))$ is in the Hörmander class of symbols of type $(\frac{1}{2}, \frac{1}{2})$ (see [**BG**, Prop. 10.22]). Therefore, we define a continuous linear operator from $C_c^\infty(U)$ to $C^\infty(U)$ by letting

$$(3.1.14) \quad p(x, -iX)f(x) = (2\pi)^{-(d+1)} \int e^{ix.\xi} p(x, \sigma(x,\xi)) \hat{f}(\xi) d\xi, \qquad f \in C_c^\infty(U).$$

In the sequel we let $\Psi^{-\infty}(U)$ denotes the class of smoothing operators, i.e., the class of operators $P : C_c^\infty(U) \to C^\infty(U)$ given by a smooth kernel.

DEFINITION 3.1.6. $\Psi_H^m(U)$, $m \in \mathbb{C}$, consists of operators $P : C_c^\infty(U) \to C^\infty(U)$ of the form

$$(3.1.15) \qquad P = p(x, -iX) + R,$$

with p in $S^m(U \times \mathbb{R}^{d+1})$, called the symbol of P, and R in $\Psi^{-\infty}(U)$.

The above definition of the symbol of P differs from that of [**BG**], since there the authors defined it to be $p_\sigma(x,\xi) = p(x, \sigma(x,\xi))$. Note also that p is unique modulo $S^{-\infty}(U \times \mathbb{R}^{d+1})$.

LEMMA 3.1.7. For $j = 0, 1, \ldots$ let $p_{m-j} \in S_{m-j}(U \times \mathbb{R}^{d+1})$. Then there exists $P \in \Psi_H^m(U)$ with symbol $p \sim \sum_{j \geq 0} p_{m-j}$. Moreover, the operator P is unique modulo smoothing operators.

The class $\Psi_H^m(U)$ does not depend on the choice of the H-frame $X_0, \ldots, X_d$ (see [**BG**, Prop. 10.46]). Moreover, since it is contained in the class of ΨDOs of type $(\frac{1}{2}, \frac{1}{2})$ we obtain:

PROPOSITION 3.1.8. *Let P be a $\Psi_H DO$ of order m on U.*

1) The operator P extends to a continuous linear mapping from $\mathcal{E}'(U)$ to $\mathcal{D}'(U)$ and has a distribution kernel which is smooth off the diagonal.

3) Let $k = \Re m$ if $\Re m \geq 0$ and $k = \frac{1}{2}\Re m$ otherwise. Then P extends to a continuous mapping from $L^2_{s,\mathrm{comp}}(U)$ to $L^2_{s-k,\mathrm{loc}}(U)$ for any s in $\mathbb{R}$.

3.1.3. Composition of Ψ_HDOs. Recall that there is no symbolic calculus for ΨDOs of type $(\frac{1}{2}, \frac{1}{2})$ since the product of two such ΨDOs needs not be again a ΨDO of type $(\frac{1}{2}, \frac{1}{2})$. However, the fact that the Ψ_HDOs are modelled on left-invariant pseudodifferential operators allows us to construct a symbolic calculus for Ψ_HDOs.

First, for $j = 0, \ldots, d$ let $X_j^{(x)}$ be the leading homogeneous part of X_j in privileged coordinates centered at x defined according to (2.1.19) and (2.1.20). These vectors span a nilpotent Lie algebra of left-invariant vector fields on a nilpotent graded Lie group G^x which corresponds to $G_x U$ by pulling back the latter from the Heisenberg coordinates at x to the privileged coordinates at x.

As alluded to above the product law of $G^{(x)}$ defines a convolution product for symbols,

$$(3.1.16) \qquad *^{(x)} : S_{m_1}(\mathbb{R}^{d+1}) \times S_{m_2}(\mathbb{R}^{d+1}) \longrightarrow S_{m_1+m_2}(\mathbb{R}^{d+1}).$$

such that, with the notations of (3.1.10), on $\mathcal{L}(\mathcal{S}_0(\mathbb{R}^{d+1}))$ we have

$$(3.1.17) \qquad p_1(-iX^{(x)})p_2(-iX^{(x)}) = (p_1 *^{(x)} p_2)(-iX^{(x)}) \qquad \forall p_j \in S_{m_j}(\mathbb{R}^{d+1}).$$

As it turns out the product $*^{(x)}$ depends smoothly on x (see [**BG**, Prop. 13.33]). Therefore, we get a continuous bilinear product,

$$(3.1.18) \qquad * : S_{m_1}(U \times \mathbb{R}^{d+1}) \times S_{m_2}(U \times \mathbb{R}^{d+1}) \to S_{m_1+m_2}(U \times \mathbb{R}^{d+1}),$$

$$(3.1.19) \qquad p_1 * p_2(x, \xi) = (p_1(x,.) *^{(x)} p_2(x,.))(\xi), \qquad p_j \in S_{m_j}(U \times \mathbb{R}^{d+1}).$$

PROPOSITION 3.1.9 ([**BG**, Thm. 14.7]). *For $j = 1,2$ let $P_j \in \Psi_H^{m_j}(U)$ have symbol $p_j \sim \sum_{k \geq 0} p_{j,m_j-k}$ and assume that one of these operators is properly supported. Then the operator $P = P_1 P_2$ is a $\Psi_H DO$ of order $m_1 + m_2$ and has symbol $p \sim \sum_{k \geq 0} p_{m_1+m_2-k}$, with*

$$(3.1.20) \quad p_{m_1+m_2-k} = \sum_{k_1+k_2 \leq k} \sum_{\alpha,\beta,\gamma,\delta}^{(k-k_1-k_2)} h_{\alpha\beta\gamma\delta}(D_\xi^\delta p_{1,m_1-k_1}) * (\xi^\gamma \partial_x^\alpha \partial_\xi^\beta p_{2,m_2-k_2}),$$

where $\sum\limits_{\alpha\beta\gamma\delta}^{(l)}$ denotes the sum over all the indices such that $|\alpha|+|\beta| \leq \langle \beta \rangle - \langle \gamma \rangle + \langle \delta \rangle = l$ and $|\beta| = |\gamma|$, and the functions $h_{\alpha\beta\gamma\delta}(x)$'s are polynomials in the derivatives of the coefficients of the vector fields $X_0, \ldots, X_d$.

3.1.4. The distribution kernels of Ψ_HDOs. An important fact about ΨDOs is their characterization in terms of their distribution kernels.

First, we extend the notion of homogeneity of functions to distributions. For $K \in \mathcal{S}'(\mathbb{R}^{d+1})$ and for $\lambda > 0$ we let K_λ denote the element of $\mathcal{S}'(\mathbb{R}^{d+1})$ such that

$$(3.1.21) \qquad \langle K_\lambda, f \rangle = \lambda^{-(d+2)} \langle K(x), f(\lambda^{-1}.x) \rangle \qquad \forall f \in \mathcal{S}(\mathbb{R}^{d+1}).$$

In the sequel we will also use the notation $K(\lambda.x)$ for denoting $K_\lambda(x)$. We then say that K is homogeneous of degree m, $m \in \mathbb{C}$, when $K_\lambda = \lambda^m K$ for any $\lambda > 0$.

DEFINITION 3.1.10. $\mathcal{S}'_{reg}(\mathbb{R}^{d+1})$ consists of tempered distributions on $\mathbb{R}^{d+1}$ which are smooth outside the origin. We equip it with the weakest topology such that the inclusions of $\mathcal{S}'_{reg}(\mathbb{R}^{d+1})$ into $\mathcal{S}'(\mathbb{R}^{d+1})$ and $C^\infty(\mathbb{R}^{d+1}\setminus 0)$ are continuous.

DEFINITION 3.1.11. $\mathcal{K}_m(U \times \mathbb{R}^{d+1})$, $m \in \mathbb{C}$, consists of distributions $K(x,y)$ in $C^\infty(U)\hat{\otimes}\mathcal{S}'_{reg}(\mathbb{R}^{d+1})$ so that for some functions $c_\alpha(x) \in C^\infty(U)$, $\langle\alpha\rangle = m$, we have

(3.1.22) $K(x, \lambda.y) = \lambda^m K(x,y) + \lambda^m \log \lambda \sum_{\langle\alpha\rangle=m} c_\alpha(x) y^\alpha$ for any $\lambda > 0$.

The interest of considering the distribution class $\mathcal{K}_m(U \times \mathbb{R}^{d+1})$ stems from:

LEMMA 3.1.12 ([**BG**, Prop. 15.24], [**CM**, Lem. I.4]). *1) Any $p \in S_m(U \times \mathbb{R}^{d+1})$ agrees on $U \times (\mathbb{R}^{d+1}\setminus 0)$ with a distribution $\tau(x,\xi) \in C^\infty(U)\hat{\otimes}\mathcal{S}'(\mathbb{R}^{d+1})$ such that $\check{\tau}_{\xi\to y}$ is in $\mathcal{K}_{\hat{m}}(U \times \mathbb{R}^{d+1})$, $\hat{m} = -(m+d+2)$.*

2) If $K(x,y)$ belongs to $\mathcal{K}_{\hat{m}}(U \times \mathbb{R}^{d+1})$ then the restriction of $\hat{K}_{y\to\xi}(x,\xi)$ to $U \times (\mathbb{R}^{d+1}\setminus 0)$ belongs to $S_m(U \times \mathbb{R}^{d+1})$.

This result is a consequence of the solution to the problem of extending a homogeneous function $p \in C^\infty(\mathbb{R}^{d+1}\setminus 0)$ into a homogeneous distribution on $\mathbb{R}^{d+1}$ and of the fact that for $\tau \in \mathcal{S}'(\mathbb{R}^{d+1})$ we have

(3.1.23) $(\hat{\tau})_\lambda = |\lambda|^{-(d+2)}(\tau_{\lambda^{-1}})^\wedge$ $\forall \lambda \in \mathbb{R}\setminus 0$.

In particular, if τ is homogeneous of degree m then $\hat{\tau}$ is homogeneous of degree $-(m+d+2)$.

DEFINITION 3.1.13. $\mathcal{K}^m(U \times \mathbb{R}^{d+1})$, $m \in \mathbb{C}$, consists of distributions K in $\mathcal{D}'(U \times \mathbb{R}^{d+1})$ with an asymptotic expansion $K \sim \sum_{j\geq 0} K_{m+j}$, $K_l \in \mathcal{K}_l(U \times \mathbb{R}^{d+1})$, in the sense that, for any integer N, as soon as J is large enough we have

(3.1.24) $K - \sum_{j\leq J} K_{m+j} \in C^N(U \times \mathbb{R}^{d+1})$.

Since under the Fourier transform the asymptotic expansion (3.1.13) for symbols corresponds to that for distributions in (3.1.24), using Lemma 3.1.12 we get:

LEMMA 3.1.14 ([**BG**, pp. 133–134]). *Let $K \in \mathcal{D}'(U \times \mathbb{R}^{d+1})$. Then the following are equivalent:*

(i) The distribution K belongs to $\mathcal{K}^m(U \times \mathbb{R}^{d+1})$;

(ii) We can put K into the form

(3.1.25) $K(x,y) = \check{p}_{\xi\to y}(x,y) + R(x,y)$,

for some $p \in S^{\hat{m}}(U \times \mathbb{R}^{d+1})$, $\hat{m} = -(m+d+2)$, and some $R \in C^\infty(U \times \mathbb{R}^{d+1})$.

Moreover, if (i) and (ii) holds and $K \sim \sum_{j\geq 0} K_{m+j}$, $K_l \in \mathcal{K}_l(U \times \mathbb{R}^{d+1})$, then we have $p \sim \sum_{j\geq 0} p_{\hat{m}-j}$ where $p_{\hat{m}-j} \in S_{\hat{m}-j}(U \times \mathbb{R}^{d+1})$ is the restriction to $U \times (\mathbb{R}^{d+1}\setminus 0)$ of $(K_{m+j})^\wedge_{y\to\xi}$.

Next, for $x \in U$ let ψ_x denote the affine change to the privileged coordinates at x and let us write $(A_x^t)^{-1}\xi = \sigma(x,\xi)$ with $A_x \in \mathrm{GL}_{d+1}(\mathbb{R})$. Since $\psi_x(x) = 0$ and $\psi_{x*} X_j = \frac{\partial}{\partial y_j}$ at $y = 0$ for $j = 0, \ldots, d$, one checks that $\psi_x(y) = A_x(y - x)$.

Let $p \in S^m(U \times \mathbb{R}^{d+1})$. As by definition we have $p(x, -iX) = p_\sigma(x, D_x)$ with $p_\sigma(x, \xi) = p(x, \sigma(x, \xi)) = p(x, (A_x^t)^{-1}\xi)$, the distribution kernel $k_{p(x,-iX)}(x, y)$ of $p(x, -iX)$ is represented by the oscillating integrals

$$(3.1.26) \quad (2\pi)^{-(d+1)} \int e^{i(x-y)\cdot\xi} p(x, (A_x^t)^{-1}\xi) d\xi = \frac{|A_x|}{(2\pi)^{d+1}} \int e^{iA_x(x-y)\cdot\xi} p(x, \xi) d\xi.$$

Since $\psi_x(y) = A_x(y - x)$ we deduce that

$$(3.1.27) \qquad k_{p(x,-iX)}(x, y) = |\psi_x'| \check{p}_{\xi \to y}(x, -\psi_x(y)).$$

Combining this with Lemma 3.1.14 then gives:

PROPOSITION 3.1.15 ([**BG**, Thms. 15.39, 15.49]). *Let $P : C_c^\infty(U) \to C^\infty(U)$ be a continuous linear operator with distribution kernel $k_P(x, y)$. Then the following are equivalent:*

(i) P is a $\Psi_H DO$ of order m, $m \in \mathbb{C}$.

(ii) We can write $k_P(x, y)$ in the form,

$$(3.1.28) \qquad k_P(x, y) = |\psi_x'| K(x, -\psi_x(y)) + R(x, y),$$

with $K \in \mathcal{K}^{\hat{m}}(U \times \mathbb{R}^{d+1})$, $\hat{m} = -(m + d + 2)$, and $R \in C^\infty(U \times U)$.

Furthermore, if (i) and (ii) hold and $K \sim \sum_{j \geq 0} K_{\hat{m}+j}$, $K_l \in \mathcal{K}_l(U \times \mathbb{R}^{d+1})$, then P has symbol $p \sim \sum_{j \geq 0} p_{m-j}$, $p_l \in S_l(U \times \mathbb{R}^{d+1})$, where p_{m-j} is the restriction to $U \times (\mathbb{R}^{d+1} \backslash 0)$ of $(K_{m+j})_{y \to \xi}^\wedge$.

In the sequel we will need a version of Proposition 3.1.15 in Heisenberg coordinates. To this end let ε_x denote the coordinate change to the Heisenberg coordinates at x and set $\phi_x = \varepsilon_x \circ \psi_x^{-1}$. Recall that ϕ_x is a Lie group isomorphism from $G^{(x)}$ to $G_x U$ such that $\phi_x(\lambda.y) = \lambda.\phi_x(y)$ for any $\lambda \in \mathbb{R}$. Moreover, using (2.1.24) one can check that $|\phi_x'| = 1$ and $\phi_x^{-1}(y) = -\phi_x(-y)$. Therefore, from (3.1.27) we see that we can put $k_{p(x,-iX)}(x, y)$ into the form

$$(3.1.29) \qquad k_{p(x,-iX)}(x, y) = |\varepsilon_x'| K_P(x, -\varepsilon_x(y)),$$

where we have let

$$(3.1.30) \qquad K_P(x, y) = \check{p}_{\xi \to y}(x, -\phi_x(-y)) = \check{p}_{\xi \to y}(x, \phi_x^{-1}(y)).$$

In fact, the coordinate changes ϕ_x, $x \in U$, give rise to an action on distributions on $U \times \mathbb{R}^{d+1}$ given by

$$(3.1.31) \qquad K(x, y) \longrightarrow \phi_x^* K(x, y), \qquad \phi_x^* K(x, y) = K(x, \phi_x^{-1}(y)).$$

Since ϕ_x depends smoothly on x, this action induces a continuous linear isomorphisms of $C^N(U \times \mathbb{R}^{d+1})$, $N \geq 0$, and $C^\infty(U \times \mathbb{R}^{d+1})$ onto themselves. As $\phi_x(y)$ is polynomial in y in such way that $\phi_x(0) = 0$ and $\phi_x(\lambda.y) = \lambda.\phi_x(y)$ for every $\lambda \in \mathbb{R}$, we deduce that the above action also yields a continuous linear isomorphism of $C^\infty(U) \hat{\otimes} \mathcal{S}_{\text{reg}}'(\mathbb{R}^{d+1})$ onto itself and, for every $\lambda > 0$, we have

$$(3.1.32) \qquad (\phi_x^* K)(x, \lambda.y) = \phi_x^*[K(x, \lambda.y)], \qquad K \in \mathcal{D}'(U \times \mathbb{R}^{d+1}).$$

Furthermore, as $\phi_x(y)$ is polynomial in y we see that for every $\alpha \in \mathbb{N}^{d+1}$ we can write $\phi_x(y)^\alpha$ in the form $\phi_x(y)^\alpha = \sum_{\langle \beta \rangle = \langle \alpha \rangle} d_{\alpha\beta}(x) y^\beta$ with $d_{\alpha\beta} \in C^\infty(U \times \mathbb{R}^{d+1})$. It then follows that, for every $m \in \mathbb{C}$, the map $K(x, y) \to \phi_x^* K(x, y)$ induces a linear isomorphisms of $\mathcal{K}_m(U \times \mathbb{R}^{d+1})$ and $\mathcal{K}^m(U \times \mathbb{R}^{d+1})$ onto themselves. Combining this with (3.1.29) and Proposition 3.1.15 then gives:

PROPOSITION 3.1.16. *Let* $P : C_c^\infty(U) \to C^\infty(U)$ *be a continuous linear operator with distribution kernel* $k_P(x,y)$. *Then the following are equivalent:*

(i) P *is a* $\Psi_H DO$ *of order* m, $m \in \mathbb{C}$.

(ii) *We can write* $k_P(x,y)$ *in the form,*

$$(3.1.33) \qquad k_P(x,y) = |\varepsilon'_x| K_P(x, -\varepsilon_x(y)) + R(x,y),$$

with $K_P \in \mathcal{K}^{\hat{m}}(U \times \mathbb{R}^{d+1})$, $\hat{m} = -(m+d+2)$, *and* $R \in C^\infty(U \times U)$.

Furthermore, if (i) and (ii) hold and $K_P \sim \sum_{j \geq 0} K_{P,\hat{m}+j}$, $K_l \in \mathcal{K}_l(U \times \mathbb{R}^{d+1})$, *then* P *has symbol* $p \sim \sum_{j \geq 0} p_{m-j}$, $p_l \in S_l(U \times \mathbb{R}^{d+1})$, *where* p_{m-j} *is the restriction to* $U \times (\mathbb{R}^{d+1} \setminus 0)$ *of* $[K_{P,\hat{m}+j}(x, \phi_x^{-1}(y))]^\wedge_{y \to \xi}$.

REMARK 3.1.17. Let $a \in U$. Then (3.1.33) shows that the distribution kernel of $\tilde{P} = (\varepsilon_a)_* P$ at $x = 0$ is

$$(3.1.34) \qquad k_{\tilde{P}}(0,y) = |\varepsilon'_a|^{-1} k_P(\varepsilon_a^{-1}(0), \varepsilon_a^{-1}(y)) = K_P(a, -y).$$

Moreover, as we are in Heisenberg coordinates already, we have $\psi_0 = \varepsilon_0 = \phi_0 = \text{id}$. Thus, in the form (3.1.33) for $\tilde{P}$ we have $K_{\tilde{P}}(0, y) = K_P(a, y)$. Therefore, if we let $p_m(x, \xi)$ denote the principal symbol of P and let $K_{P,\hat{m}} \in \mathcal{K}_{\hat{m}}(U \times \mathbb{R}^{d+1})$ denote the leading kernel of K_P, then by Proposition 3.1.16 we have

$$(3.1.35) \qquad p_m(0, \xi) = [K_{P,\hat{m}}]^\wedge_{y \to \xi}(a, \xi).$$

This shows that $[K_{P,\hat{m}}]^\wedge_{y \to \xi}(a, \xi)$ is the principal symbol of P at $x = 0$ in Heisenberg coordinates centered at a.

3.1.5. Ψ_HDOs on a general Heisenberg manifold. Let (M^{d+1}, H) be a Heisenberg manifold. As alluded to before the Ψ_HDOs on an subset of $\mathbb{R}^{d+1}$ are ΨDOs of type $(\frac{1}{2}, \frac{1}{2})$. However, the latter don't make sense on a general manifold, for their class is not preserved by an arbitrary change of chart. Nevertheless, when dealing with Ψ_HDOs this issue is resolved if we restrict ourselves to changes of Heisenberg charts. Indeed, we have:

PROPOSITION 3.1.18. *Let* U *(resp.* $\tilde{U}$*) be an open subset of* $\mathbb{R}^{d+1}$ *together with a hyperplane bundle* $H \subset TU$ *(resp.* $\tilde{H} \subset T\tilde{U}$*) and a* H*-frame of* TU *(resp. a* $\tilde{H}$*-frame of* $T\tilde{U}$*). Let* $\phi : (U, H) \to (\tilde{U}, \tilde{H})$ *be a Heisenberg diffeomorphism and let* $\tilde{P} \in \Psi^m_{\tilde{H}}(\tilde{U})$.

1) *The operator* $P = \phi^* \tilde{P}$ *is a* $\Psi_H DO$ *of order* m *on* U.

2) *If we write the distribution kernel of* $\tilde{P}$ *in the form (3.1.33) with* $K_{\tilde{P}}(\tilde{x}, \tilde{y})$ *in* $\mathcal{K}^{\hat{m}}(\tilde{U} \times \mathbb{R}^{d+1})$, *then the distribution kernel of* P *can be written in the form (3.1.33) with* $K_P(x, y) \in \mathcal{K}^{\hat{m}}(U \times \mathbb{R}^{d+1})$ *such that*

$$(3.1.36) \qquad K_P(x, y) \sim \sum_{\langle \beta \rangle \geq \frac{3}{2}\langle \alpha \rangle} \frac{1}{\alpha!\beta!} a_{\alpha\beta}(x) y^\beta (\partial^\alpha_{\tilde{y}} K_{\tilde{P}})(\phi(x), \phi'_H(x)y),$$

where we have let $a_{\alpha\beta}(x) = \partial^\beta_y [|\partial_y(\tilde{\varepsilon}_{\phi(x)} \circ \phi \circ \varepsilon_x^{-1})(y)| (\tilde{\varepsilon}_{\phi(x)} \circ \phi \circ \varepsilon_x^{-1}(y) - \phi'_H(x)y)^\alpha]|_{y=0}$ *and* $\tilde{\varepsilon}_{\tilde{x}}$ *denote the change to the Heisenberg coordinates at* $\tilde{x} \in \tilde{U}$. *In particular,*

$$(3.1.37) \qquad K_P(x, y) = |\phi'_H(x)| K_{\tilde{P}}(\phi(x), \phi'_H(x)y) \qquad \mod \mathcal{K}^{\hat{m}+1}(U \times \mathbb{R}^{d+1}).$$

REMARK 3.1.19. The version of the above statement in [**BG**] does not contain the asymptotics (3.1.37), which will be crucial for giving a global definition of the principal symbol of a Ψ_HDO in the next section. For this reason a detailed proof of the above version is given in Appendix A. This proof will also be useful in Chapter 4 and [**Po12**] for generalizing Proposition 3.1.18 to holomorphic families of Ψ_HDOs and to Ψ_HDOs with parameter.

As a consequence of Proposition 3.1.18 we can define Ψ_HDOs on M acting on the sections of a vector bundle $\mathcal{E}$ over M.

DEFINITION 3.1.20. $\Psi_H^m(M, \mathcal{E})$, $m \in \mathbb{C}$, consists of continuous operators P from $C_c^\infty(M, \mathcal{E})$ to $C^\infty(M, \mathcal{E})$ such that:

(i) The distribution kernel of P is smooth off the diagonal;

(ii) For any trivialization $\tau : \mathcal{E}_{|U} \to U \times \mathbb{C}^r$ over a local Heisenberg chart $\kappa : U \to V \subset \mathbb{R}^{d+1}$ the operator $\kappa_* \tau_*(P_{|U})$ belongs to $\Psi_H^m(V, \mathbb{C}^r) := \Psi_H^m(V) \otimes \mathrm{End}\,\mathbb{C}^r$.

All the previous properties of ΨDOs on an open subset of $\mathbb{R}^{d+1}$ hold *mutatis mutandis* for Ψ_HDOs on M acting on sections of $\mathcal{E}$.

3.1.6. Transposes and adjoints of Ψ_HDOs. Let us now look at the transpose and adjoints of Ψ_HDOs. First, given a Heisenberg chart $U \subset \mathbb{R}^{d+1}$ we have:

PROPOSITION 3.1.21. *Let* $P \in \Psi_H^m(U)$. *Then:*

1) The transpose operator P^t is a Ψ_HDO of order m on U.

2) If we write the distribution kernel of P in the form (3.1.33) with $K_P(x, y)$ in $\mathcal{K}^{\hat{m}}(U \times \mathbb{R}^{d+1})$ then P^t can be written in the form (3.1.33) with $K_{P^t} \in \mathcal{K}^{\hat{m}}(U \times \mathbb{R}^{d+1})$ such that

$$(3.1.38) \quad K_{P^t}(x, y) \sim \sum_{\frac{3}{2}\langle\alpha\rangle \leq \langle\beta\rangle} \sum_{|\gamma| \leq |\delta| \leq 2|\gamma|} a_{\alpha\beta\gamma\delta}(x) y^{\beta+\delta} (\partial_x^\gamma \partial_y^\alpha K_P)(x, -y),$$

where $a_{\alpha\beta\gamma\delta}(x) = \frac{|\varepsilon_x^{-1}|}{\alpha!\beta!\gamma!\delta!}[\partial_y^\beta(|\varepsilon'_{\varepsilon_x^{-1}(-y)}|(y - \varepsilon_{\varepsilon_x^{-1}(y)}(x))^\alpha)\partial_y^\delta(\varepsilon_x^{-1}(-y) - x)^\gamma](x, 0)$. *In particular, we have*

$$(3.1.39) \quad K_{P^t}(x, y) = K_P(x, -y) \mod \mathcal{K}^{\hat{m}+1}(U \times \mathbb{R}^{d+1}).$$

REMARK 3.1.22. The asymptotic expansion (3.1.38) is not stated in [**BG**], but we need it in order to determine the global principal symbol of the transpose of a Ψ_HDO (see next section). A detailed proof of Proposition 3.1.21 can be found in Appendix B.

Using this result, or its version in [**BG**], we obtain:

PROPOSITION 3.1.23 ([**BG**, Thm. 17.4]). *Let* $P : C^\infty(M, \mathcal{E}) \to C^\infty(M, \mathcal{E})$ *be a Ψ_HDO of order m. Then:*

1) The transpose operator $P^t : \mathcal{E}'(M, \mathcal{E}^*) \to \mathcal{D}'(M, \mathcal{E}^*)$ is a Ψ_HDO of order m;

2) If M is endowed with a smooth positive density and $\mathcal{E}$ with a Hermitian metric then the adjoint $P^* : C^\infty(M, \mathcal{E}) \to C^\infty(M, \mathcal{E})$ is a Ψ_HDO of order $\overline{m}$.

3.2. Principal symbol and model operators.

3.2.1. Principal symbol and model operators.
In this section we define the principal symbols and model operators of Ψ_HDOs and check their main properties. Let $\mathfrak{g}^*M$ be the dual bundle of $\mathfrak{g}M$ with canonical projection $\pi : \mathfrak{g}^*M \to M$.

DEFINITION 3.2.1. *For $m \in \mathbb{C}$ the space $S_m(\mathfrak{g}^*M, \mathcal{E})$ consists of sections $p(x, \xi)$ in $C^\infty(\mathfrak{g}^*M \setminus 0, \operatorname{End} \pi^*\mathcal{E})$ which are homogeneous of degree m, i.e., we have*

$$(3.2.1) \qquad p(x, \lambda.\xi) = \lambda^m p(x, \xi) \qquad \forall \text{for any } \lambda > 0,$$

where $\lambda.\xi$ is defined as in (3.1.2).

Let $P \in \Psi_H^m(M)$ and for $j = 1,2$ let κ_j be a Heisenberg chart with domain $V_j \subset M$ and let $\phi : U_1 \to U_2$ be the corresponding transition map, where we have let $U_j = \kappa_j(V_1 \cap V_2) \subset \mathbb{R}^{d+1}$.

Let us first assume that $\mathcal{E}$ is the trivial line bundle, so that P is a scalar operator. For $j = 1, 2$ we let $P_j := \kappa_{j*}(P_{|V_1 \cap V_2})$, so that $P_1 = \phi^* P_2$. Since P_j belongs to $\Psi_H^m(U_j)$ its distribution kernel is of the form (3.1.33) with $K_{P_j} \in \mathcal{K}^{\hat{m}}(U_j \times \mathbb{R}^{d+1})$. Moreover, by Proposition 3.1.18 we have

$$(3.2.2) \qquad K_{P_1}(x, y) = |\phi_H'(x)| K_{P_2}(\phi(x), \phi_H'(x)y) \qquad \mod \mathcal{K}^{\hat{m}+1}(U_1 \times \mathbb{R}^{d+1}).$$

Therefore, if $K_{P_j, \hat{m}} \in \mathcal{K}_{\hat{m}}(U_j \times \mathbb{R}^{d+1})$ is the leading kernel of K_{P_j} then we get

$$(3.2.3) \qquad K_{P_1, \hat{m}}(x, y) = |\phi_H'(x)| K_{P_2, \hat{m}}(\phi(x), \phi_H'(x)y).$$

Next, for $j = 1, 2$ we define

$$(3.2.4) \qquad p_{j,m}(x, \xi) = [K_{P_j, \hat{m}}]^\wedge_{y \to \xi}(x, \xi), \qquad (x, \xi) \in U_j \times \mathbb{R}^{d+1} \setminus 0.$$

By Remark 3.1.17 for any $a \in U_j$ the symbol $p_j(a, .)$ yields in Heisenberg coordinates centered at a the principal symbol of P_j at $x = 0$. Moreover, since $\phi_H'(a)$ is a linear map, from (3.2.3) we get

$$(3.2.5) \qquad p_{1,m}(x, \xi) = p_{2,m}(\phi(x), [\phi_H'(x)^{-1}]^t \xi).$$

This shows that $p_m := \kappa_1^* p_{1,m}$ is an element of $S_m(\mathfrak{g}^*(V_1 \cap V_1))$ which is independent of the choice of the chart κ_1. Since $S_m(\mathfrak{g}^*M)$ is a sheaf this gives rise this uniquely defines a symbol $p_m(x, \xi)$ in $S_m(\mathfrak{g}^*M)$.

When $\mathcal{E}$ is a general vector bundle, the above construction can be carried out similarly, so that we obtain:

THEOREM 3.2.2. *For any $P \in \Psi_H^m(M, \mathcal{E})$ there is a unique symbol $\sigma_m(P)(x, \xi)$ in $S_m(\mathfrak{g}^*M, \mathcal{E})$ such that, if in a local trivializing Heisenberg chart $U \subset \mathbb{R}^{d+1}$ we let $K_{P, \hat{m}}(x, y) \in \mathcal{K}_{\hat{m}}(U \times \mathbb{R}^{d+1})$ be the leading kernel for the kernel $K_P(x, y)$ in the form (3.1.33) for P, then we have*

$$(3.2.6) \qquad \sigma_m(P)(x, \xi) = [K_{P, \hat{m}}]^\wedge_{y \to \xi}(x, \xi), \qquad (x, \xi) \in U \times \mathbb{R}^{d+1} \setminus 0.$$

Equivalently, for any $x_0 \in M$ the symbol $\sigma_m(P)(x_0, .)$ agrees in trivializing Heisenberg coordinates centered at x_0 with the principal symbol of P at $x = 0$.

DEFINITION 3.2.3. *For $P \in \Psi_H^m(M, \mathcal{E})$ the symbol $\sigma_m(P) \in S_m(\mathfrak{g}^*M, \mathcal{E})$ provided by Theorem 3.2.2 is called the principal symbol of P.*

REMARK 3.2.4. Since we have two notions of principal symbol we shall distinguish between them by saying that $\sigma_m(P)$ is the global principal symbol of P and that in a local trivializing chart the principal symbol p_m of P in the sense of (3.1.13) is the local principal symbol of P in this chart.

In a local Heisenberg chart $U \subset \mathbb{R}^{d+1}$ the global symbol $\sigma_m(P)$ and the local principal symbol p_m of $P \in \Psi_H^m(U)$ can be easily related to each other. Indeed, by Proposition 3.1.16 we have

$$p_m(x,\xi) = [K_{P,\hat{m}}(x,\phi_x^{-1}(y))]_{y\to\xi}^{\wedge}(x,\xi), \tag{3.2.7}$$

where $K_{P,\hat{m}}$ denotes the leading kernel for the kernel K_P in the form (3.1.33) for P. By combining this with the definition (3.2.4) of $\sigma_m(P)$ we thus get

$$p_m(x,\xi) = (\hat{\phi}_x^* \sigma_m(P))(x,\xi), \tag{3.2.8}$$

$$(\hat{\phi}_x^* \sigma_m(P)) = [[\sigma_m(P)]_{\xi\to y}^{\vee}(x,\phi_x^{-1}(y))]_{y\to\xi}^{\wedge} = [\phi_x^*[\sigma_m(P)]_{\xi\to y}^{\vee}]_{y\to\xi}^{\wedge}, \tag{3.2.9}$$

where ϕ_x^* is the isomorphism map (3.1.31). In particular, since the latter is a linear isomorphism of $\mathcal{K}_m(U \times \mathbb{R}^{d+1})$ onto itself, we see that the map $p \to \hat{\phi}_x^* p$ is a linear isomorphism of $S_m(U \times \mathbb{R}^{d+1})$ onto itself.

EXAMPLE 3.2.5. Let $X_0, \ldots, X_d$ be a local H-frame of TM near a point $a \in M$. In any Heisenberg chart associated with this frame the Heisenberg symbol of X_j is $\frac{1}{i}\xi_j$. In particular, this is true in Heisenberg coordinates centered at a. Thus the (global) principal symbol of X_j is equal to $\frac{1}{i}\xi_j$ in the local trivialization of $\mathfrak{g}^*M \setminus 0$ defined by the frame $X_0, \ldots, X_d$. More generally, for any differential $P = \sum_{\langle\alpha\rangle \leq m} a_\alpha(x) X^\alpha$ on M we have

$$\sigma_m(P)(x,\xi) = \sum_{\langle\alpha\rangle \leq m} a_\alpha(x) i^{-|\alpha|} \xi^\alpha. \tag{3.2.10}$$

Thus, for differential operators the global and local principal symbols agree in suitable coordinates. Alternatively, this result follows from the fact that the isomorphism (3.1.31) induces the identity map on distributions supported in $U \times \{y = 0\}$.

PROPOSITION 3.2.6. *The principal symbol* $\sigma_m : \Psi_H^m(M, \mathcal{E}) \to S_m(\mathfrak{g}^*M, \mathcal{E})$ *gives rise to a linear isomorphism* $\Psi_H^m(M, \mathcal{E})/\Psi_H^{m-1}(M, \mathcal{E}) \xrightarrow{\sim} S_m(\mathfrak{g}^*M, \mathcal{E})$.

PROOF. Since the principal symbol of $P \in \Psi_H^m(M, \mathcal{E})$ vanishes everywhere if, and only if, P has order $\leq m - 1$, we see that the principal symbol map σ_m induces an injective linear map from $\Psi_H^m(M, \mathcal{E})/\Psi_H^{m-1}(M, \mathcal{E})$ to $S_m(\mathfrak{g}^*M, \mathcal{E})$.

It remains to show that σ_m is surjective. To this end consider a symbol $p_m(x,\xi) \in S_m(\mathfrak{g}^*M, \mathcal{E})$ and let $(\varphi_i)_{i \in I}$ be a partition of the unity subordinated to an open covering $(U_i)_{i \in I}$ of M by domains of Heisenberg charts $\kappa_i : U_i \to V_i$ over which there are trivializations $\tau_i : \mathcal{E}_{|U_i} \to U_i \times \mathbb{C}^r$. For each index i let $\psi_i \in C^\infty(U_i)$ be such that $\psi_i = 1$ near $\operatorname{supp} \varphi_i$ and set

$$p_m^{(i)}(x,\xi) = (1 - \chi(\xi))(\hat{\phi}_{i,x}^* \kappa_{i*} \tau_{i*} p_{m|\mathfrak{g}^*U_i \setminus 0})(x,\xi) \in S_m(V_i) \otimes \operatorname{End} \mathbb{C}^r, \tag{3.2.11}$$

where $\chi \in C^\infty(\mathbb{R}^{d+1})$ is such that $\chi = 1$ near the origin and $\hat{\phi}_{i,x}^*$ denotes the isomorphism (3.2.9) with respect to the chart V_i. Then we define a a Ψ_HDO of order m by letting

$$P = \sum \varphi_i [\tau_i^* \kappa_i^* p_m^{(i)}(x, -iX)] \psi_i. \tag{3.2.12}$$

For for every index i the local principal symbol of $\varphi_i [\tau_i^* \kappa_i^* p_m^{(i)}(x, -iX)] \psi_i$ in the chart V_i is $\varphi_i \circ \kappa_i^{-1}(\hat{\phi}_{i,x}^* \kappa_{i*} \tau_{i*} p_{m|\mathfrak{g}^*U_i \setminus 0})$, so by (3.2.8) its global principal is $\varphi_i \circ \kappa_i^{-1}(\kappa_{i*} \tau_{i*} p_{m|\mathfrak{g}^*U_i \setminus 0})$, which pulls back to $\varphi_i p_m$ on U_i. It follows that the global principal symbol of P is $\sigma_m(P) = \sum_i \varphi_i p_m = p_m$. This proves the surjectivity of the map σ_m, so the proof is now complete. $\square$

Next, granted the above definition of the principal symbol, we can define the model operator at a point as follows.

DEFINITION 3.2.7. *Let $P \in \Psi_H^m(M, \mathcal{E})$ have (global) principal symbol $\sigma_m(P)$. Then the model operator of P at $a \in M$ is the left-invariant $\Psi_H DO$-operator P^a from $\mathcal{S}_0(G_a M, \mathcal{E}_a)$ to itself with symbol $\sigma_m(P)^{\vee}_{\xi \to y}(a, .)$, i.e., we have*

$$(3.2.13) \qquad P^a f(x) = \langle \sigma_m(P)^{\vee}_{\xi \to y}(a, y), f(x.y^{-1}) \rangle, \qquad f \in \mathcal{S}_0(G_a M, \mathcal{E}_a).$$

Consider a local trivializing chart $U \subset \mathbb{R}^{d+1}$ near a and let us relate the model operator P^a on $G_a M$ to the operator $P^{(a)} = \tilde{p}_m^a(-iX^{(a)})$ on $G^{(a)}$ defined using the local principal symbol $\tilde{p}_m(x, \xi)$ of P in this chart. Using (3.1.6) and (3.2.8) for $f \in \mathcal{S}_0(\mathbb{R}^{d+1})$ we get

$$(3.2.14) \quad P^{(a)} f(y) = \langle (p_m^a)^{\vee}(z), f(y.z^{-1}) \rangle = \langle (\sigma_m(P)^{\vee}_{\xi \to y}(x, \phi_a^{-1}(y)), f(y.z^{-1}) \rangle.$$

Since $|\phi_a'| = 1$ and ϕ_a is a Lie group isomorphism from $G^{(a)}$ onto $G_a M$ we obtain

$$(3.2.15) \qquad P^{(a)} f(y) = \langle (\sigma_m(P)^{\vee}_{\xi \to y}(x, y), f \circ \phi_a^{-1}(y.\phi_a(x)^{-1}) \rangle = (\phi_a^* P^a) f(y).$$

In particular, we have

$$(3.2.16) \qquad\qquad P^a = (\phi_a)_* p_m^a(-iX^{(a)}).$$

3.2.2. Composition of principal symbols and model operators. Let us now look at the composition of principal symbols. To this end for $a \in M$ we let $*^a : S_{m_1}(\mathbb{R}^{d+1}) \times S_{m_2}(\mathbb{R}^{d+1}) \to S_{m_1+m_2}(\mathbb{R}^{d+1})$ be the convolution product for symbols defined by the product law of $G_a M$ under the identification $G_a M \simeq \mathbb{R}^{d+1}$ provided by a H-frame $X_0, \ldots, X_d$ of TM near a, that is,

$$(3.2.17) \ (p_{m_1} *^a p_{m_j})(-iX^a) = p_{m_1}(-iX^a) \circ p_{m_2}(-iX^a), \qquad p_{m_j} \in S_{m_j}(\mathbb{R}^{d+1}).$$

Let $U \subset \mathbb{R}^{d+1}$ be a local trivializing Heisenberg chart chart near a and for $j = 1, 2$ let $P_j \in \Psi_H^{m_j}(U)$ have (global) principal symbol $\sigma_{m_j}(P_j)$. Under the trivialization of GU provided by the H-frame $X_0, \ldots, X_d$ we have $P_j^a = \sigma(P_j)(x, -iX^a)$, so we obtain

$$(3.2.18) \qquad [\sigma_{m_j}(P_j)(x,,) *^a \sigma_{m_j}(P_j)(x,.)](-iX^a) = P_1^a P_2^a.$$

On the other hand, using (3.2.8) and (3.2.16) we see that $\hat{\phi}_a^*[p_{m_1} *^a p_{m_2}](-iX^a)$ is equal to

$$(3.2.19) \quad \phi_a^*[p_{m_1}(-iX^a) \circ p_{m_2}(-iX^a)] = \phi_a^*[p_{m_1}(-iX^a)] \circ \phi_a^*[p_{m_2}(-iX^a)]$$
$$= (\hat{\phi}_a^* p_{m_1})(-iX^{(a)}) \circ (\hat{\phi}_a^* p_{m_2})(-iX^{(a)}) = [(\hat{\phi}_a^* p_{m_1}) *^{(a)} (\hat{\phi}_a^* p_{m_2})](-iX^{(a)}).$$

Hence we have

$$(3.2.20) \qquad p_{m_1} *^a p_{m_2} = (\hat{\phi}_a)_* [(\hat{\phi}_a^* p_{m_1}) *^{(a)} (\hat{\phi}_a^* p_{m_2})] \qquad \forall p_{m_j} \in S_{m_j}(\mathbb{R}^{d+1}),$$

where $(\hat{\phi}_a)_*$ denotes the inverse of $\hat{\phi}_a^*$. Since $\hat{\phi}_a^*$, its inverse and $*^{(a)}$ depend smoothly on a, we deduce that that so does $*^a$. Therefore, we get:

PROPOSITION 3.2.8. *The group laws on the fibers of GM give rise to a convolution product,*

$$* : S_{m_1}(\mathfrak{g}^*M, \mathcal{E}) \times S_{m_2}(\mathfrak{g}^*M, \mathcal{E}) \longrightarrow S_{m_1+m_2}(\mathfrak{g}^*M, \mathcal{E}), \tag{3.2.21}$$

$$p_{m_1} * p_{m_2}(x, \xi) = [p_{m_1}(x, .) *^x p_{m_2}(x, .)](\xi), \qquad p_{m_j} \in S_{m_j}(\mathfrak{g}^*M, \mathcal{E}), \tag{3.2.22}$$

where $*^x$ denote the convolution product for symbols on G_xM.

Notice that (3.2.20) shows that, under the relation (3.2.8) between local and global principal symbols, the convolution product (3.2.17) for global principal symbols corresponds to the convolution product (3.1.19) for local principal symbols. Since by Proposition 3.1.9 the latter yields the *local* principal symbol of the product of two Ψ_HDOs in a local chart, we deduce that the convolution product (3.2.17) yields the *global* principal symbol of the product two Ψ_HDOs.

Moreover, by (3.2.18) the global convolution product (3.2.17) corresponds to the product of model operators, so the model operator of a product of two Ψ_HDOs is equal to the product of the model operators. We have thus proved:

PROPOSITION 3.2.9. *For $j = 1,2$ let $P_j \in \Psi_H^{m_j}(M, \mathcal{E})$ and assume that P_1 or P_2 is properly supported.*

1) We have $\sigma_{m_1+m_2}(P_1 P_2) = \sigma_{m_1}(P) * \sigma_{m_2}(P)$.

2) At every $a \in M$ the model operator of $P_1 P_2$ is $(P_1 P_2)^a = P_1^a P_2^a$.

Finally, we look at the continuity of the above product for homogeneous symbols. To this end for each $m \in \mathbb{C}$ we endow $S_m(\mathfrak{g}^*M, \mathcal{E})$ with the Fréchet space topology inherited from that of $C^\infty(\mathfrak{g}^*M \setminus 0, \text{End}\,\mathcal{E})$.

PROPOSITION 3.2.10. *The product $*$ for homogeneous symbols gives rise to a continuous bilinear map from $S_{m_1}(\mathfrak{g}^*M, \mathcal{E}) \times S_{m_2}(\mathfrak{g}^*M, \mathcal{E})$ to $S_{m_1+m_2}(\mathfrak{g}^*M, \mathcal{E})$.*

PROOF. Consider a sequence $(p_k, q_k)_{k \geq 0}$ in $S_{m_1}(\mathfrak{g}^*M, \mathcal{E}) \times S_{m_2}(\mathfrak{g}^*M, \mathcal{E})$ converging to (p, q) and such that $p_k * q_k$ converges to r in $S_{m_1+m_2}(\mathfrak{g}^*M, \mathcal{E})$. Let $a \in M$. Then $(p_k(a, .), q_k(a, .))$ converges to $(p(a, .), q(a, .))$ in $S_{m_1}(\mathfrak{g}_a^*M, \mathcal{E}_a) \times S_{m_2}(\mathfrak{g}_a^*M, \mathcal{E}_a)$ and $(p_k * q_k)(a, .)$ converges to $r(a, .)$ in $S_{m_1+m_2}(\mathfrak{g}_a^*M, \mathcal{E}_a)$.

On the other hand, by Proposition 3.1.3 the product $*^a$ gives rise to a continuous bilinear map from $S_{m_1}(\mathfrak{g}_a^*M, \mathcal{E}_a) \times S_{m_2}(\mathfrak{g}_a^*M, \mathcal{E}_a)$ to $S_{m_1+m_2}(\mathfrak{g}_a^*M, \mathcal{E}_a)$, so $(p_k * q_k)(a, .) = p_k(a, .) *^a q_k(a, .)$ also converges to $p(a, .) *^a q(a, .) = p * q(a, .)$. Hence $p * q(a, .) = r(a, .)$ for any $a \in M$, that is, the symbols $p * q$ and r agree. It then follows from the closed graph theorem that $*$ gives rise a continuous bilinear map from $S_{m_1}(\mathfrak{g}^*M, \mathcal{E}) \times S_{m_2}(\mathfrak{g}^*M, \mathcal{E})$ to $S_{m_1+m_2}(\mathfrak{g}^*M, \mathcal{E})$. □

3.2.3. Principal symbol of transposes and adjoints. In this subsection we shall determine the principal symbols and the model operators of transposes and adjoints of Ψ_HDOs.

Recall that by Proposition 3.1.23 if $P \in \Psi_H^m(M, \mathcal{E})$ then its transpose operator $P^t : C_c^\infty(M, \mathcal{E}^*) \to C^\infty(M, \mathcal{E})$ is a Ψ_HDO of order m and its adjoint $P^* : C_c^\infty(M, \mathcal{E}) \to C^\infty(M, \mathcal{E})$ is a Ψ_HDO of order $\overline{m}$ (assuming M endowed with a positive density and $\mathcal{E}$ with a Hermitian metric in order to define the adjoint).

PROPOSITION 3.2.11. *Let $P \in \Psi_H^m(M, \mathcal{E})$ have principal symbol $\sigma_m(P)$. Then:*

1) The principal symbol of P^t is $\sigma_m(P^t)(x, \xi) = \sigma_m(x, -\xi)^t \in S_m(\mathfrak{g}^*M, \mathcal{E}^*)$;

2) If P^a is the model operator of P at a then the model operator of P^t at a is the transpose operator $(P^a)^t : \mathcal{S}_0(G_xM, \mathcal{E}_x^*) \to \mathcal{S}_0(G_xM, \mathcal{E}_x^*)$.

3.2. PRINCIPAL SYMBOL AND MODEL OPERATORS.

PROOF. Let us first assume that $\mathcal{E}$ is the trivial line bundle and that P is a scalar operator. In a local Heisenberg chart $U \subset \mathbb{R}^{d+1}$ we can write the distribution kernels of P and P^t in the form (3.1.33) with distributions K_P and K_{P^t} in $\mathcal{K}^{\hat{m}}(U \times \mathbb{R}^{d+1})$. Let $K_{P,\hat{m}}$ and $K^t_{P^t,\hat{m}}$ denote the principal parts of K_P and K_{P^t} respectively. Then the principal symbols of P and P^t are $\sigma_m(P)(x,\xi) = (K_{P,\hat{m}})^{\wedge}_{y\to\xi}(x,\xi)$ and $\sigma_m(P^t)(x,\xi) = (K_{P^t,\hat{m}})^{\wedge}_{y\to\xi}(x,\xi)$ respectively. Since (3.1.39) implies that $K_{P^t,\hat{m}}(x,y) = K_{P,\hat{m}}(x,-y)$ and the Fourier transform commutes with the multiplication by -1 we get

$$(3.2.23) \qquad \sigma_m(P^t)(x,\xi) = \sigma_m(P)(x,-\xi).$$

Next, for $a \in U$ let $p \in S_m(G_a)$ and let P be the left-invariant $\Psi_H\text{DO}$ with symbol p. Then the transpose P^t is such that, for f and g in $S_0(G_aU)$, we have

$$(3.2.24) \quad \langle P^tf, g\rangle = \langle f, Pv\rangle = \langle 1, f(x)(Pg)(x)\rangle = \langle 1, f(x)\langle \check{p}(y), g(x.y^{-1})\rangle\rangle$$
$$= \langle 1 \otimes \check{p}(x,y), f(x)g(x.y^{-1})\rangle.$$

Therefore, using the change of variable $(x,y) \to (x.y^{-1}, y^{-1})$ and the fact that $y^{-1} = -y$ we get

$$(3.2.25) \quad \langle P^tf, g\rangle = \langle 1 \otimes \check{p}(x,-y), f(x)g(x.y^{-1})\rangle = \langle 1, f(x)\langle \check{p}(-y), g(x.y^{-1})\rangle\rangle.$$

Since $\check{p}(-y) = \check{p}^t(y)$ with $p^t(\xi) = p(-\xi)$, we obtain

$$(3.2.26) \qquad \langle P^tf, g\rangle = \langle 1, f(x)\langle \check{p}^t(y), g(x.y^{-1})\rangle\rangle = \langle (p^t * f)(x), g(x)\rangle.$$

Thus P^t is the left-convolution operator with symbol $p^t(\xi) = p(-\xi)$.

Now, since the model operator $(P^t)^a$ is the left-invariant $\Psi_H\text{DO}$ with symbol $\sigma_m(P^t)(a,\xi) = \sigma_m(P)(a,-\xi)$ we see that it agrees with the transpose $(P^a)^t$.

In the general case, when $\mathcal{E}$ is not the trivial bundle, we can similarly show that P^t is a $\Psi_H\text{DO}$ of order m with principal symbol $\sigma_m(P^t)(x,\xi) = \sigma_m(P)(x,-\xi)^t$ and such that at every point $a \in M$ its model operator at a is the transpose $(P^a)^t$. □

Assume now that M is endowed with a positive density and $\mathcal{E}$ with a Hermitian metric respectively and let $L^2(M, \mathcal{E})$ be the associated L^2-Hilbert space.

PROPOSITION 3.2.12. *Let $P \in \Psi_H^m(M, \mathcal{E})$ have principal symbol $\sigma_m(P)$. Then:*

1) The principal symbol of P^* is $\sigma_{\bar{m}}(P^*)(x,\xi) = \sigma_m(P)(x,\xi)^*$.

2) If P^x denotes the model operator of P at $x \in M$ then the model operator of P^* at x is the adjoint $(P^x)^*$ of P^x.

PROOF. Let us first assume that $\mathcal{E}$ is the trivial line bundle, so that P is a scalar operator. Moreover, since the above statements are local ones, it is enough to prove them in a local Heisenberg chart $U \subset \mathbb{R}^{d+1}$ and we may assume that P is a $\Psi_H\text{DO}$ on U.

Let $\overline{P} : C_c^\infty(U) \to C^\infty(U)$ be the conjugate operator of P, so that $\overline{P}u = \overline{P(\bar{f})}$ for any $f \in C_c^\infty(U)$. By Proposition 3.1.16 the distribution kernel of P of the form (3.1.33) with $K_P(x,y)$ in $\mathcal{K}_{\hat{m}}(U \times \mathbb{R}^{d+1})$, so the kernel of $\overline{P}$ takes the form

$$(3.2.27) \qquad k_{\overline{P}}(x,y) = \overline{k_P(x,y)} = |\varepsilon'_x|\overline{K_P(x,y)} \mod C^\infty(U \times U).$$

Since the conjugation of distribution $K(x,y) \to \overline{K(x,y)}$ induces an anti-linear isomorphism from $\mathcal{K}^{\hat{m}}(U \times \mathbb{R}^{d+1})$ onto $\mathcal{K}^{\hat{m}}(U \times \mathbb{R}^{d+1})$, it follows from Proposition 3.1.16 that $\overline{P}$ is a $\Psi_H\text{DO}$ of order $\hat{m}$ and its kernel can be put into the form (3.1.33) with $K_{\overline{P}}(x,y) = \overline{K_P(x,y)}$. In particular, if $K_{P,\hat{m}} \in \mathcal{K}_{\hat{m}}(U \times \mathbb{R}^{d+1})$

denotes the leading kernel of K_P then the leading kernel of $K_{\overline{P}}$ is $\overline{K_{P,\hat{m}}}$. Thus $\overline{P}$ has principal symbol

$$(3.2.28) \qquad \sigma_{\bar{m}}(\overline{P})(x,\xi) = [\overline{K_{P,\hat{m}}}]^{\wedge}_{\xi \to y}(x,\xi) = \overline{[(K_{P,\hat{m}})^{\wedge}_{\xi \to y}(x,-\xi)} = \overline{\sigma_m(x,-\xi)}.$$

Moreover, we have $\sigma_{\bar{m}}(\overline{P})^{\vee}_{\xi \to y}(x,y) = \overline{\sigma_m(P)^{\vee}_{\xi \to y}(x,y)}$. Therefore, for any f in $\mathcal{S}_0(G_aU)$ the function $(\overline{P})^a f(x)$ is equal to

$$(3.2.29) \qquad \langle \overline{\sigma_m(P)^{\vee}_{\xi \to y}(x,y)}, f(x.y^{-1}) \rangle = \overline{\langle \sigma_m(P)^{\vee}_{\xi \to y}(x,y), \overline{f(x.y^{-1})} \rangle} = \overline{P^a} f(x).$$

Hence $(\overline{P})^a$ agrees with $\overline{P^a}$.

Combining all this with Proposition 3.1.21 and Proposition 3.2.11 we see that $\overline{P}^t$ is a Ψ_HDO of order $\overline{m}$ such that:

- If we put the kernel of $\overline{P}^t$ into the form (3.1.33) with a distribution $K_{\overline{P}^t}(x,y)$ in $\mathcal{K}_{\hat{m}}(U \times \mathbb{R}^{d+1})$, then the leading kernel of $K_{\overline{P}^t}$ is $K_{\overline{P}^t,\hat{m}} = \overline{K_{P,\hat{m}}(x,-y)}$;

- The global principal symbol of $\overline{P}^t$ is $\sigma_{\bar{m}}(\overline{P}^t) = \overline{\sigma_m(P^t)(x,\xi)} = \overline{\sigma_{\bar{m}}(P)(x,\xi)}$;

- The model operator at $a \in U$ of $\overline{P}^t$ is $(\overline{P}^t)^a = \overline{P^{a^t}} = (P^a)^*$.

Now, let $d\rho(x) = \rho(x)dx$ be the smooth positive density on U coming from that of M. Then the adjoint $P^* : C_c^{\infty}(U) \to C^{\infty}(U)$ of P with respect to $d\rho$ is such that

$$(3.2.30) \qquad \int_f \overline{Pf(x)}g(x)\rho(x)dx = \int_U \overline{f(x)}P^*g(x)\rho(x)dx, \qquad f,g \in C_c^{\infty}(U).$$

Thus $P^* = \rho^{-1}\overline{P}^t\rho$, which shows that P^* is a Ψ_HDO of order $\bar{m}$. Moreover, as in the proof of Proposition 3.1.21 in Appendix B, we can prove that the kernel of P^* can be put into the form (3.1.33) with $K_{P^*}(x,y) \in \mathcal{K}^{\bar{m}}(U \times \mathbb{R}^{d+1})$ equal to

$$(3.2.31) \qquad \rho(x)^{-1}K_{\overline{P}^t}(x,y)\rho(\varepsilon_x^{-1}(y)) \sim \sum_\alpha \frac{1}{\alpha!}\rho(x)^{-1}\partial_y^\alpha(\rho(\varepsilon_x^{-1}(y))_{|y=0}K_{\overline{P}^t}(x,y).$$

In particular, $K_{P^*}(x,y)$ and $K_{\overline{P}^t}(x,y)$ agree modulo $\mathcal{K}^{\hat{m}+1}(U \times \mathbb{R}^{d+1})$, hence have same leading kernels. It then follows that P^* and $\overline{P}^t$ have same principal symbol and same model operator at a point $a \in U$, that is, $\sigma_{\bar{m}}(P^*)(a,\xi) = \overline{\sigma_m(P)(x,\xi)}$ and $(P^*)^a = (P^a)^*$.

Finally, assume that $\mathcal{E}$ is an arbitrary vector bundle of rank r, so that the restriction of P to U is given by a matrix $P = (P_{ij})$ of Ψ_HDOs of order r. Let $h(x) \in C^{\infty}(U, GL_r(\mathbb{C}))$, $h(x)^* = h(x)$, be the Hermitian metric on $U \times \mathbb{C}^r$ coming from that of $\mathcal{E}$. Then the adjoint of P with respect to this Hermitian metric is $P^* = \rho^{-1}h^{-1}\overline{P}^t h\rho$. Therefore, in the same way as in the scalar case we can prove that P^* has principal symbol $h(x)^{-1}\overline{\sigma_m(P)(x,\xi)}^t h(x) = \sigma_m(P)(x,\xi)^*$ and its model operator at any point $a \in U$ is $h(a)^{-1}\overline{P^a}^t h(a) = (P^a)^*$. $\square$

3.3. Hypoellipticity and Rockland condition

In this section we define a Rockland condition for Ψ_HDOs and relate it to the invertibility of the principal symbol to get hypoellipticity criterions.

3.3.1. Principal symbol and Parametrices.

By [**BG**, Sect. 18] in a local Heisenberg chart the invertibility of the local principal symbol of a Ψ_HDO is equivalent to the existence of a Ψ_HDO-parametrix. Using the global principal symbol we can give a global reformulation of this result as follows.

PROPOSITION 3.3.1. *Let* $P : C_c^\infty(M, \mathcal{E}) \to C^\infty(M, \mathcal{E})$ *be a* $\Psi_H DO$ *of order* m. *The following are equivalent:*

1) The principal symbol $\sigma_m(P)$ *of* P *is invertible with respect to the convolution product for homogeneous symbols;*

2) P admits a parametrix Q *in* $\Psi_H^{-m}(M, \mathcal{E})$, *so that* $PQ = QP = 1 \mod \Psi^{-\infty}(M, \mathcal{E})$.

PROOF. First, it follows from Proposition 3.2.9 that 2) implies 1). Conversely, the formula (3.2.20) shows that in a local trivializing Heisenberg chart the invertibility of the global principal $\sigma_m(P)$ is equivalent to that of the local principal symbol. Once the latter is granted then, as shown in [**BG**, p. 142], Lemma 3.1.7 and Proposition 3.1.9 allows us to carry out in a local trivializing Heisenberg chart the standard parametrix construction to obtain a parametrix for P as a Ψ_HDO of order $-m$. A standard partition of the unity argument then allows us to construct a parametrix for P in $\Psi_H^{-m}(M, \mathcal{E})$. □

When a Ψ_HDO has an invertible principal symbol the Sobolev regularity properties of its parametrices allows us to get:

PROPOSITION 3.3.2 ([**BG**, p. 142]). *Let* $P : C_c^\infty(M, \mathcal{E}) \to C^\infty(M, \mathcal{E})$ *be a* $\Psi_H DO$ *of order* m *with* $\Re m \geq m$ *and such that its principal symbol is invertible. Then* P *is hypoelliptic with gain of* $\frac{1}{2}\Re m$ *derivatives, i.e., setting* $k = \frac{1}{2}\Re m$, *for any* $a \in M$, *any* $u \in \mathcal{E}'(M, \mathcal{E})$ *and any* $s \in \mathbb{R}$, *we have*

(3.3.1) $\qquad Pu$ *is* L_s^2 *near* $a \implies u$ *is* L_{s+k}^2 *near* a.

In particular, if M *is compact then, for any reals* s *and* s', *we have the hypoelliptic estimate,*

(3.3.2) $\qquad \|f\|_{L_{s+k}^2} \leq C_s(\|Pf\|_{L_s^2} + \|f\|_{L_{s'}^2}), \qquad f \in C^\infty(M, \mathcal{E})$.

REMARK 3.3.3. It used to be customary to call the above property hypoellipticity with *gain* of k derivatives (see, e.g., [**BG**]). We have followed here the recent terminology of [**Koh3**], where is constructed an example of sum of squares which is hypoelliptic but, instead of gaining of derivatives as in [**Hö2**], it gains derivatives.

REMARK 3.3.4. We can give sharper regularity results for Ψ_HDOs in terms of suitably weighted Sobolev spaces (see [**FS1**] and Section 5.5). When P is a differential operator and the Levi form is non-vanishing these results correspond to the maximal hypoellipticity of P as in [**HN3**].

REMARK 3.3.5. As it follows from the proof in [**BG**, p. 142] in order to have the hypoelliptic properties (3.3.1) and (3.3.2) it is enough to have a left-parametrix for Q rather than a two-sided parametrix. Therefore, Proposition remains valid when we assume the principal symbol of P to be left-invertible only.

3.3.2. Rockland condition. Assume that M is endowed with a positive density and $\mathcal{E}$ with a Hermitian metric and let $P : C_c^\infty(M, \mathcal{E}) \to C^\infty(M, \mathcal{E})$ be a $\Psi_H\text{DO}$ of order m. Let P^a be the model operator of P at a point $a \in M$ and let $\pi : G \to \mathcal{H}_\pi$ be a (nontrivial) unitary representation of $G = G_a M$. We define the symbol π_{P^a} as follows (see also [**Ro1**], [**Gł1**], [**CGGP**]).

Let $\mathcal{H}_\pi^0(\mathcal{E}_a)$ be the subspace of $\mathcal{H}_\pi(\mathcal{E}_a) := \mathcal{H}_\pi \otimes \mathcal{E}_a$ spanned by the vectors,

$$(3.3.3) \qquad \pi_f \xi = \int_G (\pi_x \otimes 1_{\mathcal{E}_a})(\xi \otimes f(x)) dx,$$

where ξ ranges over $\mathcal{H}_\pi$ and f over $\mathcal{S}_0(G, \mathcal{E}_a) = \mathcal{S}_0(G) \otimes \mathcal{E}_a$. Then π_{P^a} is the (unbounded) operator of $\mathcal{H}_\pi(\mathcal{E}_a)$ with domain $\mathcal{H}_\pi^0(\mathcal{E}_a)$ such that

$$(3.3.4) \qquad \pi_{P^a}(\pi_f \xi) = \pi_{P^a f} \xi \qquad \forall f \in \mathcal{S}_0(G, \mathcal{E}_a) \quad \forall \xi \in \mathcal{H}_\pi.$$

One can check that $\pi_{P^{a*}}$ is the adjoint of π_{P^a} on $\mathcal{H}_\pi^0$, hence is densely defined. Thus π_{P^a} is closeable and we can let $\overline{\pi_{P^a}}$ denote its closure.

In the sequel we let $C_\pi^\infty(\mathcal{E}_a) = C_\pi^\infty \otimes \mathcal{E}_a$, where $C_\pi^\infty \subset \mathcal{H}_\pi$ denotes the space of smooth vectors of π, i.e., the subspace of vectors $\xi \in \mathcal{H}_\pi$ so that $x \to \pi(x)\xi$ is smooth from G to $\mathcal{H}_\pi$.

PROPOSITION 3.3.6 ([**CGGP**]). *1) The domain of $\overline{\pi_{P^a}}$ always contains $C_\pi^\infty(\mathcal{E}_a)$.*
2) If $\Re m \leq 0$ then the operator $\overline{\pi_{P^a}}$ is bounded.
3) We have $\overline{(\pi_{P^a})^*} = (\overline{\pi_{P^a}})^*$.
4) If P_1 and P_2 are ΨDOs on M then $\overline{\pi_{(P_1 P_2)^a}} = \overline{\pi_{P_1^a}} \, \overline{\pi_{P_2^a}}$.

REMARK 3.3.7. If $\mathcal{E}_a = \mathbb{C}$ and P^a is a differentiable operator then, as it is left-invariant, P^a belongs to the enveloping algebra $\mathcal{U}(\mathfrak{g})$ of the Lie algebra $\mathfrak{g} = \mathfrak{g}_a M$ of G. In this case $\overline{\pi_{P^a}}$ coincides on C_π^∞ with the operator $d\pi(P^a)$, where $d\pi$ is the representation of $\mathcal{U}(\mathfrak{g})$ induced by π.

DEFINITION 3.3.8. *We say that P satisfies the Rockland condition at a if for any nontrivial unitary irreducible representation π of $G_a M$ the operator $\overline{\pi_{P^a}}$ is injective on $C_\pi^\infty(\mathcal{E}_a)$.*

Set $2n = \text{rk}\,\mathcal{L}_a$. Under the identification $G = G_a M \simeq \mathbb{H}^{2n+1} \times \mathbb{R}^{d-2n}$ given by Proposition 2.1.6 there are left-invariant vector fields $X_0, \ldots, X_d$ on G such that $X_0, \ldots, X_{2n}$ are given by (2.1.2) and $X_k = \frac{\partial}{\partial x_k}$ for $k \geq 2n+1$. Then, up to unitary equivalence, the nontrivial irreducible representations of G are of two types:

(i) Infinite dimensional representations $\pi^{\lambda,\xi} : G \to L^2(\mathbb{R}^n)$ parametrized by $\lambda \in \mathbb{R} \setminus 0$ and $\xi = (\xi_{2n+1}, \ldots, \xi_{2n})$ such that

$$(3.3.5) \qquad d\pi^{\lambda,\xi}(X_0) = i\lambda|\lambda|, \qquad d\pi^{\lambda,\xi}(X_k) = i\lambda\xi_k, \quad k = 2n+1, \ldots, d,$$

$$(3.3.6) \qquad d\pi^{\lambda,\xi}(X_j) = |\lambda|\frac{\partial}{\partial \xi_j}, \qquad d\pi^{\lambda,\xi}(X_{n+j}) = i\lambda\xi_j, \quad j = 1, \ldots, n.$$

Moreover, in this case we have $C^\infty(\pi^{\pm,\xi}) = \mathcal{S}(\mathbb{R}^n)$.

(ii) One dimensional representations $\pi^\xi : G \to \mathbb{C}$ indexed by $\xi = (\xi_1, \ldots, \xi_d)$ in $\mathbb{R}^d \setminus 0$ such that

$$(3.3.7) \qquad d\pi^\xi(X_0) = 0, \qquad d\pi^\xi(X_j) = i\xi_j, \qquad j = 1, \ldots, d.$$

In particular, if $P = p_m(-iX)$ with $p \in S_m(G)$ then the homogeneity of the symbol p implies that we have $\overline{\pi_P^{\lambda,\xi}} = |\lambda|^m \overline{\pi_P^{\pm,\xi}}$, where $\overline{\pi_P^{\pm,\xi}} = \overline{\pi_P^{\pm 1,\xi}}$ accordingly with the sign of λ, while for the representations in (ii) we have $\overline{\pi_P^\xi} = \pi_P^\xi = p_m(0, \xi)$. Therefore, we obtain:

PROPOSITION 3.3.9. *Let $p_m \in S_m(G_a M)$. Then the Rockland condition for $P = p_m(-iX^a)$ is satisfied if, and only if, the following two conditions hold:*

(i) The operators $\overline{\pi_P^{\pm,\xi}}$, $\xi \in \mathbb{R}^{d-2n}$, are injective on $\mathcal{S}(\mathbb{R}^n)$;

(ii) The restriction of p_m to $\{0\} \times (\mathbb{R}^d \setminus 0) \simeq H_a^* \setminus 0$ is pointwise invertible.

3.3.3. Parametrices and Rockland condition. The aim of this subsection is to show that the Rockland condition is enough to insure us the invertibility of the principal symbol and the existence of a parametrix in the Heisenberg calculus.

First, we deal with zero'th order Ψ_HDOs. In this case, we have:

THEOREM 3.3.10. *Let $P : C_c^\infty(M, \mathcal{E}) \to C^\infty(M, \mathcal{E})$ be a zero'th order $\Psi_H DO$. Then the following are equivalent:*

(i) The principal symbol of P is invertible;

(ii) For every point $a \in M$ the model operator P^a is invertible on $L^2(G_a M, \mathcal{E}_a)$.

(iii) P and P^t satisfy the Rockland condition at every point $a \in M$.

Furthermore, if M is endowed with smooth density > 0 and $\mathcal{E}$ with a Hermitian metric, then in (iii) we can replace the Rockland condition for P^t by that for P^*.

PROOF. When GM is a (trivial) fiber bundle of Lie groups, i.e., the Levi form of (M, H) has constant rank, the theorem can be deduced from the results of [**CGGP**, Sect. 5], which are based on an idea due to Christ [**Ch2**] (see [**Po8**]). As we shall now see elaborating on the same idea allows us to deal with the general case as well.

First, assume that the principal symbol of P is invertible, so that P admits a parametrix $Q \in \Psi_H^{-m}(M, \mathcal{E})$, which without any loss of generality may be assumed to be properly property. Then, for any $a \in M$ the operators $Q^a P^a$ and $(Q^a)^t(P^a)^t$ are equal to 1 on $\mathcal{S}_0(G_a M, \mathcal{E}_a)$ and $\mathcal{S}_0(G_a M, \mathcal{E}_a^*)$ respectively. Therefore, it follows from Proposition 3.3.6 that for any nontrivial irreducible unitary representation π of G_a the operators $\overline{\pi_{P^a}}$ and $\overline{\pi_{(P^a)^t}}$ are injective on $C^\infty(\pi)$, i.e., P and P^t satisfy the Rockland condition at every point of M. Hence (i) implies (iii).

Second, by a result of Głowacki [**Gł2**, Thm. 4.3] for any $a \in M$ the Rockland condition for P^a is equivalent to the left-invertibility of P^a on $L^2(G_a M, \mathcal{E}_a)$. The same is true for $(P^t)^a = (P^a)^t$, so P^a has a two-sided inverse on $L^2(G_a M, \mathcal{E}_a)$ if, and only if, P^a and $(P^t)^a$ satisfy the Rockland condition. Therefore, we see that the conditions (ii) and (iii) are equivalent.

Notice also that if M is endowed with smooth density > 0 and $\mathcal{E}$ is endowed with a Hermitian metric then the above arguments can be carried through without any changes if we replace the transpose of P by the adjoint P^*. Therefore, in (iii) we may replace the Rockland condition for P^t by that for P^*.

It remains now to prove that (ii) implies (i). Observe that it is enough to proceed locally in a trivializing Heisenberg chart $U \subset \mathbb{R}^{d+1}$ with H-frame $X_0, \ldots, X_d$. In fact, we may further assume that $\mathcal{E}$ is a trivial line bundle since, as we sill see later, the arguments below can be carried out *verbatim* for systems of Ψ_HDOs. Therefore, we are reduced to prove:

PROPOSITION 3.3.11. *Let $p \in S_0(U \times \mathbb{R}^{d+1})$ be such that for any $a \in U$ the operator $P^a = p(a, -iX^a)$ is invertible on $L^2(\mathbb{R}^{d+1})$. Then p is invertible in $S_0(U \times \mathbb{R}^{d+1})$.*

The proof will follow from a series of lemmas. In the sequel we let $S = \{x \in \mathbb{R}^{d+1}; \|x\| = 1\}$ and we endow its with its induced Riemannian metric. Let $K \in \mathcal{S}'(\mathbb{R}^{d+1})$ be homogeneous of degree $-(d+2)$ and such that $K_{|S^1}$ is in $L^2(S)$. Then K is of the form,

$$(3.3.8) \qquad K = \mathrm{pv}(K) + c(K)\delta_0, \qquad c(K) \in \mathbb{C},$$

where $c(K)$ is a complex constant and $\mathrm{pv}(K)$ is the principal value distribution,

$$(3.3.9) \qquad \langle \mathrm{pv}(K), f \rangle = \lim_{\epsilon \to 0^+} \int_{|x| > \epsilon} K(x) f(x) dx, \qquad f \in \mathcal{S}(\mathbb{R}^{d+1}).$$

The decomposition (3.3.8) is unique and, in particular, K is uniquely determined by its restriction to S and the constant $c(K)$ (see [**FS2**, Prop. 6.13], [**Ch1**, Lem. 2.4]).

DEFINITION 3.3.12. *$\mathcal{A}_k$, $k \in \mathbb{N}$, consists of distributions $K \in \mathcal{S}'(\mathbb{R}^{d+1})$ that are homogeneous of degree $-(d+2)$ and such that $(\partial^\alpha K)_{|S} \in L^2(S)$ for $\langle \alpha \rangle \leq k$.*

We turn $\mathcal{A}_k$ into a Banach space by endowing it with the norm,

$$(3.3.10) \qquad |K|_k = |c(K)| + \sum_{\langle \alpha \rangle \leq k} \|(\partial^\alpha K)_{|S}\|_{L^2(S)}, \qquad K \in \mathcal{A}_k.$$

Notice that $\mathcal{K}_{-(d+2)}(\mathbb{R}^{d+1})$ is contained in all the spaces $\mathcal{A}_k$, $k \in \mathbb{N}$, and the norms $|.|_k$'s give rise to a system of semi-norms on $\mathcal{K}_{-(d+2)}(\mathbb{R}^{d+1})$ whose corresponding topology is the weakest topology making the maps $K \to c(K)$ and $K \to K_{|S}$ be continuous from $\mathcal{K}_{-(d+2)}(\mathbb{R}^{d+1})$ to $\mathbb{C}$ and $C^\infty(S)$ respectively.

In the sequel for $a \in U$ and $K \in \mathcal{A}_k$ we let P_K^a denote the convolution operator $P_K^a f = K *^a f$, $f \in \mathcal{S}_0(\mathbb{R}^{d+1})$. Then we have:

LEMMA 3.3.13 ([**KS**, Thm. 1], [**FS2**, Thm. 6.19], [**Ch2**]). *Let k be an integer greater than or equal to $d + 3$. Then:*

(i) For any $a \in U$ and any $K \in \mathcal{A}_k$ the operator P_K^a extends to a bounded operator on $L^2(\mathbb{R}^{d+1})$.

(ii) For any compact $L \subset U$ there exists $C_{Lk} > 0$ such that for any $a \in L$ and any $K \in \mathcal{A}_k$ we have

$$(3.3.11) \qquad \|P_K^a\| \leq C_{Lk} |K|_k.$$

PROOF. Let $\mathcal{B}_1$ be the space of distributions $K \in \mathcal{S}'(\mathbb{R}^{d+1})$ that are homogeneous of degree $-(d+2)$ and such that $K_{|S}$ is C^1. This becomes a Banach space when endowed with the norm,

$$(3.3.12) \qquad |K|_{\mathcal{B}_1} = |c(K)| + \|K_{|S}\|_1, \qquad K \in \mathcal{B}_1,$$

where $\|K_{|S}\|_1$ is some Banach norm on $C^1(S)$. Then it follows from a theorem of Knapp-Stein [**KS**, Thm. 1] (see also [**FS2**, Thm. 6.19]) that, given $a \in U$, for any $K \in \mathcal{B}_1$ the operator $P_{K^a}^a$ extends to a bounded operator on $L^2(\mathbb{R}^{d+1})$ and there exists a constant $C_a > 0$ such that for any $K \in \mathcal{B}_1$ we have

$$(3.3.13) \qquad \|P_K^a\| \leq C(a) |K|_{\mathcal{B}_1}.$$

Furthermore, it follows from the proof in [**FS2**] that the constant C_a can be chosen independently of a provided that a remains in a compact set of U.

3.3. HYPOELLIPTICITY AND ROCKLAND CONDITION

On the other hand, if $K \in \mathcal{A}_k$ with $k \geq d+3$ then $(\partial^\alpha K)_{|S}$ is in $L^2(S)$ for $|\alpha| \leq \frac{d+3}{2}$. Therefore, it follows from the Sobolev embedding theorem that $\mathcal{A}_k$ embeds continuously into $\mathcal{B}_1$. Combining this with the first part of the proof then gives the lemma. □

REMARK 3.3.14. The L^2-boundedness of P_K^a above is actually true for any $K \in \mathcal{A}_0$ (see [**Ch2**]), but the uniform dependence with respect to a is more difficult to keep track in [**Ch2**] than in [**FS2**]. This is not really relevant in the sequel, because in order to obtain Theorem 3.3.10 it is enough to prove Lemma 3.3.13 for k large enough.

LEMMA 3.3.15 ([**CGGP**, Lem. 5.7]). *Let k be an integer $\geq d+3$. Then:*

(i) For any $a \in U$ the convolution $*^a$ induces a bilinear product on $\mathcal{A}_k$.

(ii) For any compact $L \subset U$ there exists $C_{Lk} > 0$ such that, for any $a \in L$ and any K_1 and K_2 in $\mathcal{A}_k$, we have

$$(3.3.14) \qquad |K_1 *^a K_2|_k \leq C_{Lk}(\|P_{K_1}^a\| |K_2|_k + |K_1|_k \|P_{K_2}^a\|).$$

REMARK 3.3.16. The fact that the constant in (3.3.14) can be chosen independently of a when a remains in a compact set of U is not explicitly stated in [**CGGP**], but this follows from its proof, noticing that the proof of the Lemma 2.10 of [**Ch1**] shows that in the Lemma 5.8 of [**CGGP**] the constant C can be chosen independently of a when a stays in a compact set of U.

As we will see below it is essential that the estimate (3.3.14) involves also the operator norm and not just the norm of $\mathcal{A}_k$. This trick is initially due to Christ [**Ch2**]. Note also that Lemma 3.3.15 allows us to endow $L^\infty_{\text{loc}}(U, \mathcal{A}_k)$ with the convolution product,

$$(3.3.15) \qquad (K_1^a)_{a \in U} * (K_2^a)_{a \in U} = (K_1^a *^a K_2^a)_{a \in U}, \qquad (K_j^a)_{a \in U} \in L^\infty_{\text{loc}}(U, \mathcal{A}_k).$$

In particular, the constant family $(\delta_0)_{a \in U}$ is a unit for this product.

LEMMA 3.3.17. *Let $k \in \mathbb{N}$, $k \geq d+3$, let $L \subset U$ be compact and consider a family $K = (K^a)_{a \in L}$ in $L^\infty(L, \mathcal{A}_k)$ such that $\sup_{a \in L} \|P_{K^a}^a\| < 1$. Then $\delta_0 - K$ is invertible in $L^\infty(L, \mathcal{A}_k)$.*

PROOF. Let C_{Lk} be the sharpest constant in the estimate (3.3.14) and let us endow $L^\infty(L, \mathcal{A}_k)$ with the Banach norm,

$$(3.3.16) \qquad |K|_{k,L} = \frac{1}{C_{Lk}} \sup_{a \in L} |K^a|_k, \qquad K = (K^a)_{a \in U} \in L^\infty(L, \mathcal{A}_k).$$

Then for any K_1 and K_2 in $L^\infty(L, \mathcal{A}_k)$ we have

$$(3.3.17) \qquad |K_1 * K_2|_{k,L} \leq (|K_1|_{k,L} \|P_{K_2}\|_L + \|P_{K_1}\|_L |K_2|_{k,L}).$$

where we have let $\|P_{K_j}\| = \sup_{a \in L} \|P_{K_j^a}^a\|$.

For $j = 0, 1, 2, \ldots$ let $K^{(j)}$ be the j'th power of K with respect to the product $*$ on $L^\infty(L, \mathcal{A}_k)$. From (3.3.17) we deduce by induction that for $j = 1, 2, \ldots$ we have

$$(3.3.18) \qquad |K^{(j)}|_{k,L} \leq (j-1)|K|_{k,L} \|P_K\|_L^{j-1}.$$

It follows that $\limsup |K^{(j)}|_{k,L}^{1/j} = \|P_K\|_L < 1$, so that the series $\sum_{j=0}^\infty K^{(j)}$ converges normally in $L^\infty(L, \mathcal{A}_k)$ to the inverse of $\delta_0 - K$. Hence the result. □

With all this preparation we are ready to prove Proposition 3.3.11.

PROOF OF PROPOSITION 3.3.11. Let $p \in S_0(U \times \mathbb{R}^{d+1})$ be such that for any $a \in U$ the operator $P^a = p(a, -iX^a)$ is invertible on $L^2(\mathbb{R}^{d+1})$. Then for any $a \in U$ the operators $(P^a)^* P^a$ and $P^a (P^a)^*$ are invertible on $L^2(\mathbb{R}^{d+1})$ and have respective symbols $\overline{p} * p$ and $p * \overline{p}$ in $S_0(U \times \mathbb{R}^{d+1})$. Moreover, if $\overline{p} * p$ and $p * \overline{p}$ admits respective inverses q_1 and q_2 in $S_0(U \times \mathbb{R}^{d+1})$, then $q_1 * \overline{p}$ and $\overline{p} * q_2$ are respectively a left-inverse and a right-inverse for p. Therefore, it is enough to prove that $\overline{p} * p$ and $p * \overline{p}$ are invertible. Incidentally, we may assume that P^a is a positive operator on $L^2(\mathbb{R}^{d+1})$ for any $a \in U$.

On the other hand, recall that for any $a \in U$ the convolution product $*^a$ on $\mathcal{K}_{-(d+2)}(U \times \mathbb{R}^{d+1})$ corresponds under the Fourier transform to the product $*^a$ on $S_0(\mathbb{R}^{d+1})$. Since the latter depends smoothly on a and by [**BG**, Prop. 15.30] the Fourier transform is a topological isomorphism from $\mathcal{K}_{-(d+2)}(\mathbb{R}^{d+1})$ onto $S_0(\mathbb{R}^{d+1})$ it follows that the convolution product $*^a$ depends smoothly on a, i.e., gives rise to smooth family of bilinear maps from $\mathcal{K}_{-(d+2)}(U \times \mathbb{R}^{d+1}) \times \mathcal{K}_{-(d+2)}(U \times \mathbb{R}^{d+1})$ to $\mathcal{K}_{-(d+2)}(\mathbb{R}^{d+1})$. Therefore, we get a convolution product on $\mathcal{K}_{-(d+2)}(U \times \mathbb{R}^{d+1})$ by letting

$$(3.3.19) \qquad K_1 * K_2(x, .) = K_1(x, ,) *^x K_2(x, .), \qquad K_j \in \mathcal{K}_{-(d+2)}.$$

Bearing all this in mind, define $K(x, y) = \check{p}_{\xi \to y}(x, y)$. This is an element of $\mathcal{K}_{-(d+2)}(U \times \mathbb{R}^{d+1})$ which we can write as a smooth family $(K^a)_{a \in U} = (K(a, .))_{a \in U}$ with values in $\mathcal{K}_{(d+2)}(\mathbb{R}^{d+1})$.

CLAIM. *The family $(K^a)_{a \in U}$ is invertible in $L^\infty_{\text{loc}}(U, \mathcal{K}_{-(d+2)}(\mathbb{R}^{d+1}))$.*

PROOF OF THE CLAIM. It is enough to show that for any integer $k \geq d+3$ and any compact $L \subset U$ the family $(K^a)_{a \in L}$ is invertible in $L^\infty_{\text{loc}}(L, \mathcal{A}_k)$.

By assumption for any $a \in L$ the operator $P_{K^a} = P^a$ is an invertible positive operator on $L^2(\mathbb{R}^{d+1})$. Since by Lemma 3.3.13 the family $(P^a_{K^a})_{a \in L}$ is bounded in $\mathcal{L}(L^2(\mathbb{R}^{d+1})$, the same is true for the family $((P^a_{K^a})^{-1})_{a \in U}$.

Furthermore, since $P_{K^a} = P^a$ is positive its spectrum is contained in the interval $[\|(P^a)^{-1}\|^{-1}, \|P^a\|]$. Therefore, there exist constants c_1 and c_2 with $0 < c_1 < c_2$ such that for any $a \in L$ the spectrum of $P^a_{K^a}$ is contained in the interval $[c_1, c_2]$. Without any loss of generality we may assume $c_2 < 1$. Then the spectrum of $1 - P^a_{K^a}$ is contained in $[1 - c_2, 1 - c_1]$ for any $a \in L$, hence we get

$$(3.3.20) \qquad \sup_{a \in L} \|1 - P^a_{K^a}\| \leq 1 - c_1 < 1.$$

Since $1 - P^a_{K^a} = P^a_{\delta_0 - K^a}$ it follows from Lemma 3.3.17 that $(K^a)_{a \in L}$ is invertible in $L^\infty(L, \mathcal{A}_k)$. The proof is therefore complete. □

Let $J = (J^a)_{a \in U} \in L^\infty_{\text{loc}}(U, \mathcal{K}_{-(d+2)}(\mathbb{R}^{d+1}))$ be the inverse of K. The next step is to prove:

CLAIM. *The family $(J^a)_{a \in U}$ belongs to $C^\infty(U, \mathcal{K}_{-(d+2)})$.*

PROOF. For any a and b in U we have the equality

$$(3.3.21) \qquad J^b - J^a = -J^a *^a (K^b - K^a) *^b J^b.$$

We know that $\lim_{b \to a} (K^b - K^a) = 0$ in $\mathcal{K}_{-(d+2)}(\mathbb{R}^{d+1})$ and that the convolution product $*^a$ on $\mathcal{K}_{-(d+2)}$ depends smoothly on a. As near a the family $(J^a)_{a \in U}$ is bounded in $\mathcal{K}_{-(d+2)}(\mathbb{R}^{d+1})$ it follows that we have $\lim_{b \to a} J^b = J^a$ in $\mathcal{K}_{-(d+2)}(\mathbb{R}^{d+1})$. Hence the family $(J_a)_{a \in U}$ depends continuously on a.

Now, if we let $e_1, \ldots, e_{d+1}$ be the canonical basis of $\mathbb{R}^{d+1}$ then we have

(3.3.22) $\displaystyle\lim_{t \to 0} \frac{1}{t}(J^{a+te_j} - J^a) = -\lim_{t \to 0} J^a *^a \frac{1}{t}(K^{a+te_j} - K^a) *^{a+te_j} J^{a+te_j}$
$$= -J^a *^a \partial_{a_j} K^a *^a J^a.$$

Thus $\partial_{a_j} J^a$ exists and is equal to $-J^a *^a \partial_{a_j} K^a *^a J^a$. The latter depends continuously on a, so $(J^a)_{a \in U}$ is a C^1-family with values on $\mathcal{K}_{-(d+2)}(\mathbb{R}^{d+1})$. An induction then shows that it is actually of class C^k for every integer $k \geq 1$, hence is a smooth family. The claim is thus proved. □

We now can complete the proof of Proposition 3.3.11. For $(x, \xi) \in U \times (\mathbb{R}^{d+1}\backslash 0)$ let $q(x, \xi) = \hat{J}_{y \to \xi}(x, \xi)$. Then q belongs to $S_0(U \times \mathbb{R}^{d+1})$ and we have $(q * p)^\vee_{\xi \to y} = J * K = \delta_0$, hence $q * p = 1$. Similarly, we have $p * q = 1$, so q is an inverse for p in $S_0(U \times \mathbb{R}^{d+1})$. The lemma is thus proved. □

All this proves that in Theorem 3.3.10 the condition (ii) implies (i) when $\mathcal{E}$ is the trivial line bundle over M. In fact, all the Lemmas 3.3.13–3.3.17 are true for systems as well. More precisely, for Lemma 3.3.13 this follows from the fact that the proof of [**FS2**, Thm. 6.19] can be carried out *verbatim* for systems, while the extension to systems of Lemmas 3.3.15 and 3.3.17 is immediate. Henceforth, the arguments in the proof of Proposition 3.3.11 remain valid for systems. It then follows that for general bundles too (ii) implies (i). The proof of Theorem 3.3.10 is thus achieved. □

THEOREM 3.3.18. *Let $P : C_c^\infty(M, \mathcal{E}) \to C^\infty(M, \mathcal{E})$ be a $\Psi_H DO$ of integer order $m \geq 1$. Then the following are equivalent:*

(i) The principal symbol of P is invertible;

(ii) P and P^t satisfy the Rockland condition at every point $a \in M$.

Furthermore, when M is endowed with smooth density > 0 and $\mathcal{E}$ with a Hermitian metric then in (ii) we can replace the Rockland condition for P^t by that for P^*. In any case, when (i) or (ii) holds the operator P admits a parametrix in $\Psi_H^{-m}(M, \mathcal{E})$.

PROOF. First, in the same way as in the proof of Theorem 3.3.10 we can show that if the principal symbol of P is invertible then P and P^t satisfy the Rockland condition at every point $a \in M$.

Conversely, assume that P and P^t satisfy the Rockland condition at every point. Without any loss of generality we may assume that P and P^t are properly supported. In order to show that the principal symbol of P is invertible it is enough to proceed locally in a trivializing Heisenberg chart $U \subset \mathbb{R}^{d+1}$ with H-frame $X_0, \ldots, X_d$. As a consequence we may assume that $\mathcal{E}$ is the trivial line bundle and P is a scalar $\Psi_H DO$, since the arguments below can be carried out *verbatim* for systems of $\Psi_H DOs$.

On U consider the sublaplacian $\Delta = -(X_1^2 + \ldots + X_d^2) + X_0$. By [**BG**, Thm. 18.4] the principal symbol of Δ is invertible, so Δ admits a parametrix Q in $\Psi_H^{-2}(U)$, which may be assumed to be properly supported. Then for proving that the principal symbol of P is left-invertible it is enough to check the invertibility of that of $Q^m P^t P$, for if $q_0(x, \xi)$ is the inverse of $\sigma_0(Q^m P^t P)$ then $q * \sigma_{-m}(Q^m P^t)$ is a left-inverse for $\sigma_m(P)$. Similarly, to show that $\sigma_m(P)$ is right-invertible it is sufficient to prove that the principal symbol of $PP^t Q^m$ is invertible.

Next, as the operators P^a, $(P^t)^a = (P^a)^t$ and $(Q^m)^a = (Q^a)^m$ satisfy the Rockland condition at every point $a \in U$, the same is true for the operators $(Q^m P^t P)^a = (Q^m)^a (P^t)^a P^a$ and $(PP^t Q^m)^a = P^a (P^t)^a (Q^m)^a$, that is, $Q^m P^t P$ and $PP^t Q^m$ satisfy the Rockland condition at every point. Furthermore, as Q^t has an invertible principal symbol it satisfies the Rockland condition at every point. Therefore, by arguing as above we see that the operators $(Q^m P^t P) = P^t P (Q^t)^m$ and $(PP^t Q^m)^t = (Q^t)^m PP^t$ satisfy the Rockland condition at every point.

Now, $Q^m P^t P$ and $PP^t Q^m$ both have order 0, so it follows from Theorem 3.3.10 that their principal symbol are invertible. As alluded to above this implies that the principal symbol of P is invertible.

Finally, when M is endowed with a smooth density > 0 and $\mathcal{E}$ with a Hermitian metric, the above arguments remain valid if we replace P^t by the adjoint P^*, so that in (ii) we may replace the Rockland condition for P^t by that for P^*. □

REMARK 3.3.19. In Chapter 5 we will extend Theorem 3.3.18 to Ψ_HDOs with non-integer orders.

Let us now mention few applications of Theorems 3.3.10 and 3.3.18. First, we have the following hypoellipticity criterion:

PROPOSITION 3.3.20. *Let $P : C_c^\infty(M, \mathcal{E}) \to C^\infty(M, \mathcal{E})$ be a $\Psi_H DO$ of integer order $m \geq 0$ such that P satisfies the Rockland condition at every point. Then P is hypoelliptic with gain of $\frac{m}{2}$-derivatives in the sense of (3.3.1).*

PROOF. Without any loss of generality we may assume that M is endowed with smooth density > 0 and $\mathcal{E}$ with a Hermitian metric and we may also assume that P is properly supported. As explained in Remark 3.3.5 a sufficient condition for P to be hypoelliptic with gain of $\frac{m}{2}$-derivatives it is that its principal symbol is left-invertible. To this end it is enough to show that the principal symbol of P^*P is invertible, because if q is an inverse for $\sigma_{2m}(P^*P) = \sigma_m(P)^* * \sigma_m(P)$ then $q * \sigma_m(P)^*$ is a left inverse for $\sigma_m(P)$.

On the other hand, by Theorems 3.3.10 and 3.3.18 the principal symbol of P^*P is invertible if, and only if, P^*P satisfies the Rockland condition at every point. Observe that for any $a \in M$ and any nontrivial irreducible representation of $G_a M$ the operators $\pi_{(P^*P)^a} = \pi_{P^a}^* \pi_{P^a}$ and π_{P^a} have same kernels, so the Rockland conditions at a for P^*P and P are equivalent.

Combining all this, we see that if P satisfies the Rockland condition at every point then P is hypoelliptic with gain of $\frac{m}{2}$-derivatives. □

Next, even though the representation theory of the tangent group may vary from point to point, the Rockland condition and the invertibility of the principal symbol are open properties. Indeed, we have:

PROPOSITION 3.3.21. *Let $P : C_c^\infty(M, \mathcal{E}) \to C^\infty(M, \mathcal{E})$ be a $\Psi_H DO$ of integer order m with principal symbol $p_m(x, \xi)$ and let $a \in M$.*

1) If P satisfies the Rockland condition at a then there exists an open neighborhood V of a such that P satisfies the Rockland condition at every point of V.

2) If $p_m(a, \xi)$ is invertible in $S_m(\mathfrak{g}_a^* M, \mathcal{E}_a)$ then there exists an open neighborhood V of a such that $p_{m|_V}$ is invertible on $S_m(\mathfrak{g}^* V, \mathcal{E})$.

PROOF. Let us first prove 2). To achieve this it is enough to proceed in a trivializing Heisenberg chart U near a and we may assume that $\mathcal{E}$ is the trivial line bundle since the proof for systems follows along similar lines. Furthermore, arguing as in the proof of Theorem 3.3.18 allows us to reduce the proof to the case $m = 0$.

Assume now that P has order 0. Then by Theorem 3.3.13 the invertibility of $p_0(a,.)$ in $S_0(\mathbb{R}^{d+1})$ with respect to the product $*^a$ is equivalent to the invertibility of the model operator P^a on $L^2(\mathbb{R}^{d+1})$. By Lemma 3.3.13 the L^2-extension of P^x depends continuously on x, so there exists an open neighborhood V of a such that P^x is invertible on $L^2(\mathbb{R}^{d+1})$ for any $x \in V$. Then Theorem 3.3.10 insures us that $p_0|_V$ is invertible in $S_0(V \times \mathbb{R}^{d+1})$. Hence the result.

Let us now deduce 1) from 2). Assume that P satisfies the Rockland condition at a and let us endow M with a density > 0 and $\mathcal{E}$ with a Hermitian metric. Then P^*P satisfies the Rockland condition at a, so by Theorem 3.3.18 its principal symbol is invertible at a. Therefore, the principal symbol of P is left-invertible on V and along similar lines as that at the beginning of Theorem 3.3.10 we can show that P satisfies the Rockland condition at every point of V. □

Finally, we look at families of invertible symbols parametrized by a smooth manifold B. As in Proposition 3.2.10 each space $S_m(\mathfrak{g}^*M, \mathcal{E})$ is endowed with the topology inherited from that of $C^\infty(\mathfrak{g}^*M \setminus 0, \text{End}\,\mathcal{E})$. In addition, if $p \in S_m(\mathfrak{g}^*M, \mathcal{E})$ is invertible then for $k \in \mathbb{Z}$ we let $p^{(k)}$ denote the k'th power of p with respect to the product for homogeneous symbols, e.g., $p^{(-1)}$ is the inverse of p.

PROPOSITION 3.3.22. *Let $(p_\nu)_{\nu \in B}$ be a smooth family of invertible symbols in $S_m(\mathfrak{g}^*M, \mathcal{E})$. Then $(p_\nu^{(-1)})_{\nu \in B}$ is a smooth family with values in $S_{-m}(\mathfrak{g}^*M, \mathcal{E})$.*

PROOF. First, as in the proofs of Theorem 3.3.10 and Proposition 3.3.21 it is enough to prove the result in a trivializing Heisenberg chart $U \subset \mathbb{R}^{d+1}$ and we may assume that $\mathcal{E}$ is the trivial line bundle.

Next, we can reduce the proof to the case $m = 0$ as follows. Let $X_0, \ldots, X_d$ be a H-frame on U and let p_2 be the principal symbol in the Heisenberg sense of the sublaplacian $-(X_1^2 + \ldots + X_d^2) - X_0$, so that by [**BG**, Thm. 8.4] p_2 is an invertible symbol. Then we have $p_\nu^{(-1)} = p_2^{(-1)} * p_\nu * (p_2^{(-m)} * p_\nu^{(2)})^{(-1)}$. Since by Proposition 3.2.10 is bilinear continuous, hence smooth, it follows that in order to prove that $p_\nu^{(-1)}$ depends smoothly on ν it is enough to do it for $(p_2^{(-m)} * p_\nu^{(2)})^{(-1)}$. As the latter is the inverse of a symbol in $S_0(U \times \mathbb{R}^{d+1})$ we see that it is sufficient to prove the proposition in the case $m = 0$.

Suppose now that p_ν belongs to $S_0(U \times \mathbb{R}^{d+1})$. We shall use here the same notation as that of the proof of Theorem 3.3.10 and regard $(p_\nu(a,.))_{(\nu,a) \in B \times U}$ as a smooth family with values in $S_0(\mathbb{R}^{d+1})$ parametrized by $B \times U$. For (ν, a) in $B \times U$ let $K_\nu^a = (p_\nu)_{\xi \to y}^\vee(a,.)$. Since by [**BG**, Prop. 15.30] the inverse Fourier transform is a topological isomorphism from $S_0(\mathbb{R}^{d+1})$ onto $\mathcal{K}_{-(d+2)}(\mathbb{R}^{d+1})$ we see that $(K_\nu^a)_{(\nu,a) \in B \times U}$ is a smooth family with values in $\mathcal{K}_{-(d+2)}(U \times \mathbb{R}^{d+1})$.

Since p_ν is an invertible symbol for every ν we see that for every $\nu \in B$ and every $a \in U$ the model operator $P_{K_\nu^a}^a$ is invertible on $L^2(\mathbb{R}^{d+1})$. Therefore, a simple modification of the arguments at the end of the proof of the Theorem 3.3.10 shows that there exists a smooth family $(J_\nu^a)_{(\nu,a) \in B \times U}$ with values in $K_{-(d+2)}(\mathbb{R}^{d+1})$ such that $J_\nu^a *^a K_\nu^a = K_\nu^a *^a J_\nu^a = \delta_0$. Clearly, we have $p_\nu^{(-1)}(a,.) = (J_\nu^a)^\wedge$, so $(p_\nu^{(-1)}(a,.))_{(\nu,a) \in B \times U}$ is a smooth family with values in $S_0(\mathbb{R}^{d+1})$, that is, $(p_\nu^{(-1)})$ is a smooth family with values in $S_0(U \times \mathbb{R}^{d+1})$. The proof is thus achieved. □

REMARK 3.3.23. For $m \in \mathbb{C}$ let $S_m(\mathfrak{g}^*M, \mathcal{E})^\times$ denote the set of invertible elements of $S_m(\mathfrak{g}^*M, \mathcal{E})$. Then with some additional work it is possible to show that $S_m(\mathfrak{g}^*M, \mathcal{E})^\times$ is an open subset of $S_m(\mathfrak{g}^*M, \mathcal{E})$ and that the the map $p \to p^{(-1)}$ is continuous from $S_m(\mathfrak{g}^*M, \mathcal{E})^\times$ to $S_{-m}(\mathfrak{g}^*M, \mathcal{E})^\times$ and, in fact, is infinitely many times differentiable.

3.4. Invertibility criteria for sublaplacians

In this section we focus on sublaplacians, which yield several important examples of operators on Heisenberg manifolds. The scalar case was dealt with in [**BG**], but the results were not extended to sublaplacians acting on sections of vector bundles. These extensions are necessary in order to deal with sublaplacians acting on forms such as the Kohn Laplacian or the horizontal sublaplacian (see next section).

In this section, after having explained the scalar case from the point of view of this memoir, we extend the results to the non-scalar case.

Recall that a differential operator $\Delta : C^\infty(M, \mathcal{E}) \to C^\infty(M, \mathcal{E})$ is a sublaplacian when, near any point $a \in M$, we can put Δ in the form,

$$(3.4.1) \qquad \Delta = -(X_1^2 + \ldots + X_d^2) - i\mu(x)X_0 + \mathrm{O}_H(1),$$

where $X_0, X_1, \ldots, X_d$ is a local H-frame of TM, the coefficient $\mu(x)$ is a local section of $\operatorname{End}\mathcal{E}$ and the notation $\mathrm{O}_H(1)$ means a differential operator of Heisenberg order ≤ 1.

Let us look at the Rockland condition for a sublaplacian $\Delta : C^\infty(M) \to C^\infty(M)$ acting on functions. Let $a \in M$ and let $X_0, X_1, \ldots, X_d$ be a local H-frame of TM so that near a we can write

$$(3.4.2) \qquad \Delta = -\sum_{j=1}^d X_j^2 - i\mu(x)X_0 + \mathrm{O}_H(1),$$

where $\mu(x)$ is a smooth function near a. Using (3.2.10) we see that the principal symbol of Δ is

$$(3.4.3) \qquad \sigma_2(\Delta)(x, \xi) = |\xi'|^2 + \mu(a)\xi_0, \qquad \xi' = (\xi_1, \ldots, \xi_d).$$

In particular we have $\sigma_2(\Delta)(x, 0, \xi') = |\xi'|^2 > 0$ for $\xi' \neq 0$, which shows that the condition (i) of Proposition 3.3.9 is always satisfied.

Let $L(x) = (L_{jk}(x))$ be the matrix of the Levi form $\mathcal{L}$ with respect to the H-frame $X_0, \ldots, X_d$, so that for $j, k = 1, \ldots, d$ we have

$$(3.4.4) \qquad \mathcal{L}(X_j, X_k) = [X_j, X_k] = L_{jk}X_0 \quad \mathrm{mod}\ H.$$

Equivalently, if we let $g(x)$ be the metric on H making orthonormal the frame $X_1, \ldots, X_d$, then for any sections X and Y of H we have

$$(3.4.5) \qquad \mathcal{L}(X, Y) = g(x)(L(x)X, Y)X_0 \quad \mathrm{mod}\ H.$$

The matrix $L(x)$ is antisymmetric, so up to an orthogonal change of frame of H, which does not affect the form (3.4.2), we may assume that $L(a)$ is in the normal form,

$$(3.4.6) \qquad L(a) = \begin{pmatrix} 0 & D & 0 \\ -D & 0 & 0 \\ 0 & 0 & 0 \end{pmatrix}, \qquad D = \operatorname{diag}(\lambda_1, \ldots, \lambda_n), \qquad \lambda_j > 0,$$

so that $\pm i\lambda_1, \ldots, \pm i\lambda_{2n}, 0, \ldots, 0$ are the eigenvalues of $L(a)$ counted with multiplicity. Then the model vector fields $X_0^a, \ldots, X_d^a$ are:

(3.4.7) $$X_0^a = \frac{\partial}{\partial x_0}, \qquad X_k^a = \frac{\partial}{\partial x_k}, \qquad k = 2n+1, \ldots, d,$$

(3.4.8) $$X_j^a = \frac{\partial}{\partial x_j} - \frac{1}{2}\lambda_j x_{n+j}\frac{\partial}{\partial x_0}, \qquad X_{n+j}^a = \frac{\partial}{\partial x_j} + \frac{1}{2}\lambda_j x_j \frac{\partial}{\partial x_0}, \ j = 1, \ldots, n.$$

In terms of these vector fields the model operator of Δ at a is

(3.4.9) $$\Delta^a = -[(X_1^a)^2 + \ldots + (X_1^a)^2] - i\mu(a) X_0^a.$$

Next, under the isomorphism $\phi : \mathbb{H}^{2n+1} \times \mathbb{R}^{d-2n} \to G_a M$ given by

(3.4.10) $$\phi(x_0, \ldots, x_d) = (x_0, \lambda_1^{\frac{1}{2}} x_1, \ldots, \lambda_n^{\frac{1}{2}} x_n, \lambda_1^{\frac{1}{2}} x_{n+1}, \ldots \lambda_n^{\frac{1}{2}} x_{2n}, x_{2n+1}, \ldots, x_d),$$

the representations $\pi^{\pm,\xi} = \pi^{\pm 1, \xi}$, $\xi \in \{0\}^{2n} \times \mathbb{R}^{d-2n}$, become the representations of $G_a M$ such that

(3.4.11) $$d\pi^{\pm,\xi}(X_0) = \pm i, \qquad d\pi^{\pm,\xi}(X_k) = \pm i\xi_k, \qquad k = 2n+1, \ldots, d,$$

(3.4.12) $$d\pi^{\pm,\xi}(X_j) = \lambda_j^{\frac{1}{2}} \frac{\partial}{\partial \xi_j}, \qquad d\pi^{\pm,\xi}(X_{n+j}) = \pm i\lambda_j^{\frac{1}{2}} \xi_j, \qquad j = 1, \ldots, n,$$

(3.4.13) $$\pi_{\Delta^a}^{\pm,\xi} = d\pi^{\pm,\xi}(\Delta^a) = \sum_{j=1}^n \lambda_j(-\partial_{\xi_j}^2 + \xi_j^2) \pm (\xi_{2n+1}^2 + \ldots + \xi_d^2 + \mu(a)).$$

The spectrum of the harmonic oscillator $\sum_{j=1}^n \lambda_j(-\partial_{\xi_j}^2 + \xi_j^2)$ is $\sum_{j=1}^n \lambda_j(1+2\mathbb{N})$ and all its eigenvectors belong to $\mathcal{S}(\mathbb{R}^n)$. Thus, the operator $\pi_{\Delta^a}^{\pm,\xi}$ is injective on $\mathcal{S}(\mathbb{R}^n)$ if, and only if, $\xi_{2n+1}^2 + \ldots + \xi_d^2 + \mu(a)$ is not $\pm\sum_{j=1}^n \lambda_j(1+2\mathbb{N})$. This occurs for any $\xi \in \{0\}^{2n} \times \mathbb{R}^{d-2n}$ if, and only if, the following condition holds

(3.4.14) $\qquad\qquad \mu(a)$ is not in the singular set Λ_a,

(3.4.15) $$\Lambda_a = (-\infty, -\frac{1}{2}\mathrm{Trace}\,|L(a)|] \cup [\frac{1}{2}\mathrm{Trace}\,|L(a)|, \infty) \qquad \text{if } 2n < d,$$

(3.4.16) $$\Lambda_a = \{\pm(\frac{1}{2}\mathrm{Trace}\,|L(a)| + 2\sum_{1\leq j \leq n} \alpha_j |\lambda_j|); \alpha_j \in \mathbb{N}^d\} \qquad \text{if } 2n = d.$$

In particular, the condition (ii) of Proposition 3.3.9 is equivalent to (3.4.14). Since the condition (i) is always satisfied, it follows that the Rockland condition for Δ is equivalent to (3.4.14).

Notice also that, independently of the equivalence with the Rockland condition, the condition (3.4.14) does not depend on the choice of the H-frame, because as Λ_a depends only on the eigenvalues of $L(a)$ which scale in the same way as $\mu(a)$ under a change of H-frame preserving the form (3.4.2).

On the other hand, since the transpose $(\Delta^a)^t = (\Delta^t)^a$ is given by the formula (3.4.9) with $\mu(a)$ replaced by $-\mu(a)$, which has no effect on (3.4.14), we see that the Rockland condition for $(\Delta^t)^a$ too is equivalent to (3.4.14). Therefore, we have obtained:

PROPOSITION 3.4.1. *The Rockland conditions for Δ^t and Δ at a are both equivalent to (3.4.14).*

In particular, we see that if the principal symbol of Δ is invertible then the condition (3.4.14) holds at every point. As shown by Beals-Greiner the converse is true as well. The key result is the following.

PROPOSITION 3.4.2 ([**BG**, Sect. 5]). *Let $U \subset \mathbb{R}^{d+1}$ be a Heisenberg chart near a and define*

(3.4.17) $$\Omega = \{(\mu, x) \in \mathbb{C} \times U;\ \mu \notin \Lambda_x\}.$$

Then Ω is an open set and there exists $q_\mu(x, \xi) \in C^\infty(\Omega, S_{-2}(\mathbb{R}^{d+1}))$ such that:

(i) $q_\mu(x, \xi)$ is analytic with respect to μ;

(ii) For any $(\mu, x) \in \Omega$ the symbol $q_\mu(x, .)$ inverts $|\xi'|^2 + i\mu\xi_0$ on G_xU, that is,

(3.4.18) $$q_\mu(x, .) *^x (|\xi'|^2 + i\mu\xi_0) = (|\xi'|^2 + i\mu\xi_0) *^x q_\mu(x, .) = 1.$$

More precisely, $q_\mu(x, \xi)$ is obtained from the analytic continuation of the function,

$$q_\mu(x, \xi) = \int_0^\infty e^{-t\mu\xi_0} G(x, \xi, t) dt, \qquad |\Re\mu| < \frac{1}{2} \operatorname{Tr}|L(x)|,$$

$$G(x, \xi, t) = \det{}^{-\frac{1}{2}}[\cosh(t|\xi_0||L(x)|)] \exp[-t\langle \frac{\tanh(t|\xi_0||L(x)|)}{t|\xi_0||L(x)|} \xi', \xi' \rangle].$$

This implies that if the condition (3.4.14) is satisfied at every point $x \in U$ then we get an inverse $q_{-2} \in S_{-2}(U \times \mathbb{R}^{d+1})$ for $\sigma_2(\Delta)(x, \xi) = |\xi'|^2 + i\mu(x)\xi_0$ on $U \times \mathbb{R}^{d+1}$ by letting $q_{-2}(x, \xi) = q_{\mu(x)}(x, \xi)$ for $(x, \xi) \in U \times (\mathbb{R}^{d+1} \setminus 0)$. It thus follows that if (3.4.14) holds at every point of M then the principal symbol of Δ is invertible near any point of M, hence admits an inverse in $S_{-2}(\mathfrak{g}^*M)$. Therefore, we get:

PROPOSITION 3.4.3. *A sublaplacian $\Delta : C^\infty(M) \to C^\infty(M)$ has an invertible principal symbol if, and only if, it satisfies the condition (3.4.14) at every point.*

Let us now deal with a sublaplacian $\Delta : C^\infty(M, \mathcal{E}) \to C^\infty(M, \mathcal{E})$ acting on the sections of the vector bundle $\mathcal{E}$.

Let $a \in M$ and let $X_0, \ldots, X_d$ be a local H-frame near a such that

(3.4.19) $$\Delta = -\sum_{j=1}^d X_j^2 - i\mu(x)X_0 + \mathrm{O}_H(1),$$

where $\mu(x)$ is a smooth local section of $\operatorname{End}\mathcal{E}$.

In a suitable basis of $\mathcal{E}_a$ the matrix of $\mu(a)$ is in triangular form,

(3.4.20) $$\mu(a) = \begin{pmatrix} \mu_1(a) & * & * \\ 0 & \ddots & * \\ 0 & 0 & \mu_r(a) \end{pmatrix}.$$

where $\mu_1(a), \ldots, \mu_r(a)$ denote the eigenvalues of $\mu(a)$ counted with multiplicity. Therefore, the model operator of Δ at a is of the form,

(3.4.21) $$\Delta^a = \begin{pmatrix} \Delta_1^a & * & * \\ 0 & \ddots & * \\ 0 & 0 & \Delta_r^a \end{pmatrix}, \qquad \Delta_j^a = -[(X_1^a)^2 + \ldots + (X_1^a)^2] - i\mu_j(a)X_0^a.$$

It follows that Δ^a satisfies the Rockland condition if, and only if, so does each sublaplacian Δ_j^a, $j = 1, \ldots, r$. Using Proposition 3.4.3 we then deduce that the

3.4. INVERTIBILITY CRITERIA FOR SUBLAPLACIANS

Rockland condition Δ^a is equivalent to the condition,

(3.4.22) $$\operatorname{Sp}\mu(a) \cap \Lambda_a = \emptyset.$$

Notice that the same is true for the transpose $(\Delta^a)^t$. Moreover, the condition (3.4.22) is independent of the choice of the basis of $\mathcal{E}_a$ or of the H-frame since the condition involves $\mu(a)$ only though its eigenvalues of $\mu(a)$ and the latter scale in the same way as that of $L(a)$ under a change of H-frame preserving the form (3.4.19).

Next, concerning the invertibility of the principal symbol of Δ the following extension of Proposition 3.4.2 holds.

PROPOSITION 3.4.4. *Let $U \subset \mathbb{R}^{d+1}$ be a trivializing Heisenberg chart near a and define*

(3.4.23) $$\Omega = \{(\mu, x) \in M_r(\mathbb{C}) \times U;\ \mu \notin \Lambda_x\}.$$

Then Ω is an open set and there exists $q_\mu(x, \xi) \in C^\infty(\Omega, S_{-2}(\mathbb{R}^{d+1}, \mathbb{C}^r))$ so that:

(i) *$q_\mu(x, \xi)$ is analytic with respect to μ;*

(ii) *For any $(\mu, x) \in \Omega$ the symbol $q_\mu(x, .)$ inverts $|\xi'|^2 + i\mu\xi_0$ on $G_x U$, that is,*

(3.4.24) $$q_\mu(x, .) *^x (|\xi'|^2 + i\mu\xi_0) = (|\xi'|^2 + i\mu\xi_0) *^x q_\mu(x, .) = 1.$$

PROOF. It is enough to prove that near point $(\mu_0, x_0) \in \Omega$ there exists an open neighborhood Ω' contained in Ω and a function $q_\mu(x, \xi) \in C^\infty(\Omega', S_{-2}(\mathbb{R}^{d+1}, \mathbb{C}^r))$ satisfying the properties (i) and (ii) on Ω'.

To this end observe that since $\operatorname{Sp}\mu \subset \overline{D}(0, \|\mu\|)$ for any $\mu \in M_r(\mathbb{C})$, we see that if we let $K = \overline{B(0, \|\mu_0\| + 1)}$ then any $\mu \in M_r(\mathbb{C})$ close enough to μ_0 has its spectrum contained in K. In addition, let $\delta \in (0, \frac{1}{2}\operatorname{dist}(\operatorname{Sp}\mu_0, \Lambda_{x_0}))$ and set $V_1 = \operatorname{Sp}\mu_0 + D(0, \delta)$ and $V_2 = \Lambda_{x_0} + D(0, \delta)$, so that V_1 and V_2 are disjoint open subsets of $\mathbb{C}$ containing $\operatorname{Sp}\mu_0$ and Λ_{x_0} respectively.

Notice that for any μ close enough to μ_0 we have $\operatorname{Sp}\mu \subset V_1$. Otherwise there exists a sequence $(\mu_k)_{k\geq 1} \subset M_r(\mathbb{C})$ converging to μ_0 and a sequence of eigenvalues $(\lambda_k)_{k\geq 1} \subset K$, $\lambda_k \in \operatorname{Sp}\mu_k$, such that $\lambda_k \notin V_1$ for any $k \geq 1$. Since the sequence $(\lambda_k)_{k\geq 1}$ is bounded, we may assume that it converges to some $\lambda \notin V_1$. Necessarily λ is an eigenvalue of μ_0, which contradicts the fact that $\lambda \notin V_1$. Thus there exists $\eta_1 > 0$ so that for any $\mu \in B(\mu_0, \eta_1)$ we have $\operatorname{Sp}\mu \subset V_1$.

Similarly, there exists $\eta_2 > 0$ so that for any $x \in B(x_0, \eta_2)$ we have $\operatorname{Sp}|L(x)|$ is contained in $\operatorname{Sp}|L(x_0)| + D(0, \delta)$, which implies $\Lambda_x \subset \Lambda_{x_0} + D(0, \delta) = V_2$. Therefore, the open set $\Omega' = B(\mu_0, \eta_1) \times B(x_0, \eta_2)$ is such that for any $(\mu, x) \in \Omega'$ we have $\operatorname{Sp}\mu \cap \Lambda_x \subset V_1 \cap V_2 = \emptyset$, i.e., Ω' is an open neighborhood of (μ_0, x_0) in Ω.

Next, let Γ be a smooth curve of index 1 such that the bounded connected component of $\mathbb{C} \setminus \Gamma$ contains V_1 and its unbounded component contains V_2. Then we define an element of $\operatorname{Hol}(B(\mu_0, \eta_1)) \hat{\otimes} C^\infty(B(x_0, \eta_2) \times \mathbb{R}^{d+1} \setminus 0)$ by letting

(3.4.25) $$q_\mu(x, \xi) = \frac{1}{2i\pi} \int_\Gamma q_\gamma(x, \xi)(\gamma - \mu) d\gamma, \qquad (\mu, x, \xi) \in \Omega' \times \mathbb{R}^{d+1} \setminus 0.$$

This function is homogeneous of degree -2 with respect to ξ and for any $(\mu, x) \in \Omega'$ we have

$$(3.4.26) \quad q_\mu(x,.) *^x (|\xi'|^2 + i\mu\xi_0) = \frac{1}{2i\pi} \int_\Gamma q_\gamma(x,.) *^x (|\xi'|^2 + i\mu\xi_0)(\gamma - \mu)^{-1} d\gamma$$

$$= \frac{1}{2i\pi} \int_\Gamma [(\gamma - \mu)^{-1} - iq_\gamma(x,.) * \xi_0] d\gamma = 1.$$

Similarly, we have $(|\xi'|^2 + i\mu\xi_0) *^x q_\mu(x,.) = 1$. Thus $q_\mu(x, \xi)$ is an element of $C^\infty(\Omega', S_{-2}(\mathbb{R}^{d+1}, \mathbb{C}^r))$ satisfying the properties (i) and (ii) on Ω'. The proof is therefore complete. □

In the same way as Proposition 3.4.2 in the scalar case, Proposition 3.4.4 implies that when the condition (3.4.22) holds everywhere the principal symbol of Δ admits an inverse in $S_{-2}(\mathfrak{g}^*M, \mathcal{E})$. We have thus proved:

PROPOSITION 3.4.5. *1) At every point $a \in M$ the Rockland conditions for Δ and Δ^t are equivalent to (3.4.22).*

2) The principal symbol of Δ is invertible if, and only if, the condition (3.4.22) holds everywhere. Moreover, when the latter occurs Δ admits a parametrix in $\Psi_H^{-2}(M, \mathcal{E})$ and is hypoelliptic with gain of 1 derivative.

3.5. Invertibility criteria for the main differential operators

In this section we explain how the previous results of this monograph can be used to deal with the hypoellipticity for the main geometric operators on Heisenberg manifolds: Hörmander's sum of squares, Kohn Laplacian, horizontal sublaplacian and contact Laplacian. In particular, we complete the treatment in [**BG**] of the Kohn Laplacian and we establish a criterion for the invertibility of the horizontal sublaplacian.

3.5.1. Hörmander's sum of squares.
Let (M^{d+1}, H) be a Heisenberg manifold and let $\Delta : C^\infty(M, \mathcal{E}) \to C^\infty(M, \mathcal{E})$ be a generalized sum of squares of the form (2.2.3), that is,

$$(3.5.1) \qquad \Delta = -(\nabla_{X_1}^2 + \ldots + \nabla_{X_m}^2) + O_H(1),$$

where ∇ is a connection on $\mathcal{E}$ and the vector $X_1, \ldots, X_m$ span H. Then, in the local form (3.4.1) for Δ the matrix $\mu(x)$ vanishes, so (3.4.22) holds at a point $a \in M$ if, and only if, the Levi form does not vanish at a. Combining this with Proposition 3.4.5 then gives:

PROPOSITION 3.5.1. *1) At a point $x \in M$ the operators Δ and Δ^t satisfies the Rockland condition if, and only if, the Levi form $\mathcal{L}$ does not vanish at x.*

2) The principal symbol of Δ is invertible if, and only if, the Levi form is non-vanishing. In particular, when the latter occurs Δ admits a parametrix in $\Psi_H^{-2}(M, \mathcal{E})$ and is hypoelliptic with gain of one derivative.

In particular, since the nonvanishing of the Levi form is equivalent to the bracket condition $H + [H, H] = TM$, we see that, the special case of Heisenberg manifolds, we recover the hypoellipticity result of [**Hö2**] for sums of squares.

3.5. INVERTIBILITY CRITERIA FOR THE MAIN DIFFERENTIAL OPERATORS

3.5.2. Kohn Laplacian. Let M^{2n+1} be an orientable CR manifold with CR tangent bundle $T_{1,0} \subset T_\mathbb{C} M$, so that $H = \Re(T_{1,0} \oplus T_{0,1})$ yields a Heisenberg structure on M, and let $\mathcal{N}$ be a line bundle supplement of H in TM. Assuming that $T_\mathbb{C} M$ endowed with a Hermitian metric commuting with complex conjugation and making orthogonal the splitting $T_\mathbb{C} M = T_{1,0} \oplus T_{0,1} \oplus (\mathcal{N} \otimes \mathbb{C})$, we let $\Box_{b;p,q} : C^\infty(M, \Lambda^{p,q}) \to C^\infty(M, \Lambda^{p,q})$ be the Kohn Laplacian acting on (p,q)-forms.

Let θ be a global nonvanishing real 1-form anihilating H with Levi form L_θ. Recall that the condition $Y(q)$ at a point x requires to have

$$(3.5.2) \qquad q \notin \{\kappa(x), \ldots, \kappa(x) + n - r(x)\} \cup \{r(x) - \kappa(x), \ldots, n - \kappa(x)\},$$

where $\kappa(x)$ denotes the number of negative eigenvalues of L_θ at x and $r(x)$ its rank.

It is shown in [**BG**, Sect. 21] that at every point $a \in M$ the condition $Y(q)$ is equivalent to the condition (3.4.22). Therefore, from Proposition 3.4.5 we get:

PROPOSITION 3.5.2. *1) At a point $x \in M$ the Rockland condition for $\Box_{b;p,q}$ is equivalent to the condition $Y(q)$.*

2) The principal symbol of $\Box_{b;p,q}$ is invertible if, and only if, the condition $Y(q)$ is satisfied at every point. In particular, when the latter occurs $\Box_{b;p,q}$ admits a parametrix in $\Psi_H^{-2}(M, \Lambda^{p,q})$ and is hypoelliptic with gain of one derivative.

The proof of the second part above is not quite complete in [**BG**, Sect. 21]. In fact, Beals-Greiner claimed that diagonalizing the leading part of the Kohn Laplacian allows us make use of the criterion from Proposition 3.4.3 for *scalar* sublaplacians. This fact is definitely true in case of a Levi Metric (see [**FS1**]) or even a smoothly diagonalizable Levi form, but it fails in general. Indeed, for the Kohn Laplacian the eigenvalues of the matrix $\mu(x)$ in (3.4.1) with respect to an orthonormal H-frame of TM are given in terms of eigenvalues of the Levi form (see Eq. (21.31) in [**BG**]), but the latter need not depend smoothly on x.

This shows that in order to deal with the Kohn Laplacian acting on forms, and more generally with sublaplacians acting on sections of a vector bundle, we really need to use Proposition 3.4.5, as we can cannot in general reduce the study to the scalar case.

3.5.3. The horizontal sublaplacian. Let (M^{d+1}, H) be a Heisenberg manifold endowed with a Riemannian metric, let $\Lambda_\mathbb{C}^* H^* = \oplus_{k=0}^d \Lambda_\mathbb{C}^k H^*$ be the (complexified) bundle of horizontal forms and let $\Delta_{b;k} : C^\infty(M, \Lambda_\mathbb{C}^k H^*) \to C^\infty(M, \Lambda_\mathbb{C}^k H^*)$ be the associated horizontal sublaplacian on horizontal forms of degree k as defined in (2.2.11).

We shall now express the condition (3.4.22) in terms of the more geometric condition $X(k)$ below.

DEFINITION 3.5.3. *For $x \in M$ let $2r(x)$ be the rank of the Levi form $\mathcal{L}$ at x. Then we say that $\mathcal{L}$ satisfies the condition $X(k)$ at x when we have*

$$(3.5.3) \qquad k \notin \{r(x), r(x)+1, \ldots, d - r(x)\}.$$

For instance, the condition $X(0)$ is satisfied if, and only if, the Levi form does not vanish. Also, if M^{2n+1} is a contact manifold or a nondegenerate CR manifold then the Levi form is everywhere nondegenerate, so $r(x) = 2n$ and the $X(k)$-condition becomes $k \neq n$. In any case, we have:

PROPOSITION 3.5.4. *1) At a point $x \in M$ the Rockland condition for $\Delta_{b;k}$ is equivalent to the condition $X(k)$.*

2) The principal symbol of $\Delta_{b;k}$ is invertible if, and only if, the condition $X(k)$ is satisfied at every point. In particular, when the latter occurs $\Delta_{b;k}$ admits a parametrix in $\Psi_H^{-2}(M, \Lambda_\mathbb{C}^k H^*)$ and is hypoelliptic with gain of one derivative.

PROOF. First, thanks to Proposition 3.4.5 we only have to check that for $k = 0, \ldots, d$ at any point a the condition (3.4.22) for $\Delta_{b;k}$ is equivalent to the condition $X(k)$.

Next, let $U \subset \mathbb{R}^{d+1}$ be a Heisenberg chart around a together with an orthonormal H-frame $X_0, X_1, \ldots X_d$ of TU. Let g be the Riemannian metric of M. Then on U we can write the Levi form $\mathcal{L}$ in the form,

$$(3.5.4) \qquad \mathcal{L}(X, Y) = [X, Y] = \langle L(x)X, Y \rangle X_0 \mod H,$$

for some antisymmetric section $L(x)$ of $\text{End}_\mathbb{R} H$. In particular, if for $j, k = 1, \ldots, d$ we let $L_{jk} = \langle L X_j, X_k \rangle$ then we have

$$(3.5.5) \qquad [X_j, X_k] = L_{jk} X_0 \mod H.$$

Let $2n$ be the rank of $L(a)$. Since the condition (3.4.22) for $\Delta_{b;k}$ at a is independent of the choice of the Heisenberg chart, we may assume that U is chosen in such way that at $x = a$ we have $g(a) = 1$ and $L(a)$ is in the normal form,

$$(3.5.6) \qquad L(a) = \begin{pmatrix} 0 & D & 0 \\ -D & 0 & 0 \\ 0 & 0 & 0 \end{pmatrix}, \qquad D = \text{diag}(\lambda_1, \ldots, \lambda_n), \quad \lambda_j > 0,$$

so that $\pm i\lambda_1, \ldots, \pm i\lambda_n$ are the nonzero eigenvalues of $L(a)$ counted with multiplicity.

Let $\omega^1, \ldots, \omega^n$ be the coframe of H^* dual to $X_1, \ldots, X_d$. For a 1-form ω we let $\varepsilon(\omega)$ denote the exterior product and $\iota(\omega)$ denote the interior product with ω, that is, the contraction with the vector fields dual to ω. For an ordered subset $J = \{j_1, \ldots, j_k\} \subset \{1, \ldots, d\}$, so that $j_1 < \ldots < j_d$, we let $\omega^J = \omega^{j_1} \wedge \ldots \wedge \omega^{j_k}$ (we make the convention that $\omega^\emptyset = 1$). Then the forms ω^J's give rise to an orthonormal frame of $\Lambda_\mathbb{C}^* H^*$ over U. With respect to this frame we have

$$(3.5.7) \qquad d_b = \sum_{j=1}^d \varepsilon(\omega^j) X_j \quad \text{and} \quad d_b^* = -\sum_{l=1}^d \iota(\omega^l) X_l + \text{O}_H(1).$$

Therefore, modulo first order terms we have

$$(3.5.8) \quad \Delta_b = d_b^* d_b + d_b d_b^* = -\sum_{j,l=1}^d [\varepsilon(\omega^j)\iota(\omega^l) X_j X_l + \iota(\omega^l)\varepsilon(\omega^j) X_l X_j] =$$

$$-\frac{1}{2} \sum_{j,l=1}^d [(\varepsilon(\omega^j)\iota(\omega^l) + \iota(\omega^l)\varepsilon(\omega^j))(X_j X_l + X_l X_j) + (\varepsilon(\omega^j)\iota(\omega^l) - \iota(\omega^l)\varepsilon(\omega^j))[X_j, X_l]].$$

Combining this with (3.5.5) and the relations,

$$(3.5.9) \qquad \varepsilon(\omega^j)\iota(\omega^l) + \iota(\omega^l)\varepsilon(\omega^j) = \delta_{jl}, \qquad j, l = 1, \ldots, d,$$

3.5. INVERTIBILITY CRITERIA FOR THE MAIN DIFFERENTIAL OPERATORS

we then obtain

$$(3.5.10) \quad \Delta_b = -\sum_{j=1}^{d} X_j^2 - i\mu(x)X_0 + O_H(1), \qquad \mu(x) = \frac{1}{i}\sum_{j,l=1}^{d} \varepsilon(\omega^j)\iota(\omega^l)L_{jl}.$$

In particular, thanks to (3.5.6) at $x = a$ we have

$$(3.5.11) \quad \mu(a) = \frac{1}{i}\sum_{j=1}^{n}(\varepsilon(\omega^j)\iota(\omega^{n+j}) - \varepsilon(\omega^{n+j})\iota(\omega^j))\lambda_j.$$

For $j = 1, \ldots, n$ define $\theta^j = \frac{1}{\sqrt{2}}(\omega^j + i\omega^{n+j})$ and $\theta^{\bar{j}} = \frac{1}{\sqrt{2}}(\omega^j - i\omega^{n+j})$. Then $\frac{1}{i}(\varepsilon(\omega^j)\iota(\omega^{n+j}) - \varepsilon(\omega^{n+j})\iota(\omega^j))$ is equal to

$$(3.5.12) \quad \frac{-1}{2}[(\varepsilon(\theta^j) + \varepsilon(\theta^{\bar{j}}))(\iota(\theta^{\bar{j}}) - \iota(\theta^j)) - (\varepsilon(\theta^j) - \varepsilon(\theta^{\bar{j}}))(\iota(\theta^{\bar{j}}) + \iota(\theta^j))]$$
$$= \varepsilon(\theta^j)\iota(\theta^j) - \varepsilon(\theta^{\bar{j}})\iota(\theta^{\bar{j}}).$$

Therefore, we obtain

$$(3.5.13) \quad \mu(a) = \sum_{j=1}^{n}(\varepsilon(\theta^j)\iota(\theta^j) - \varepsilon(\theta^{\bar{j}})\iota(\theta^{\bar{j}}))\lambda_j.$$

For ordered subsets $J = \{j_1, \ldots, j_p\}$ and $\overline{K} = \{k_1, \ldots, k_q\}$ of $\{1, \ldots, n\}$ we let $\theta^J = \theta^{j_1} \wedge \ldots \wedge \theta^{j_p}$ and $\theta^{\overline{K}} = \theta^{\bar{k}_1} \wedge \ldots \wedge \theta^{\bar{k}_q}$. Then the forms $\theta^J \wedge \theta^{\overline{K}} \wedge \omega^L$ give rise to an orthonormal frame of $\Lambda_{\mathbb{C}}^*H^*$ as J and K range over all the ordered subsets of $\{1, \ldots, n\}$ and L over all the ordered subsets of $\{2n+1, \ldots, d\}$. Moreover, for $j = 1, \ldots, n$ we have

$$(3.5.14) \quad \varepsilon(\theta^j)\iota(\theta^j)(\theta^J \wedge \theta^{\overline{K}} \wedge \omega^L) = \begin{cases} \theta^J \wedge \theta^{\overline{K}} \wedge \omega^L & \text{if } j \in J, \\ 0 & \text{if } j \notin J, \end{cases}$$

$$(3.5.15) \quad \varepsilon(\theta^{\bar{j}})\iota(\theta^{\bar{j}})(\theta^J \wedge \theta^{\overline{K}} \wedge \omega^L) = \begin{cases} \theta^J \wedge \theta^{\overline{K}} \wedge \omega^L & \text{if } j \in K, \\ 0 & \text{if } j \notin K. \end{cases}$$

Combining this with (3.5.13) then gives

$$(3.5.16) \quad \mu(a)(\theta^J \wedge \theta^{\overline{K}} \wedge \omega^L) = \mu_{J,\overline{K}}(a)\theta^J \wedge \theta^{\overline{K}} \wedge \omega^L, \quad \mu_{J,\overline{K}}(a) = \sum_{j \in J}\lambda_j - \sum_{j \in K}\lambda_j.$$

This shows that $\mu(a)$ diagonalizes in the basis of $\Lambda_{\mathbb{C}}^*H_a^*$ provided by the forms of $\theta^J \wedge \theta^{\overline{K}} \wedge \omega^L$ with eigenvalues given by the numbers $\mu_{J,\overline{K}}(a)$. In particular, for $k = 0, \ldots, d$ we have

$$(3.5.17) \quad \mathrm{Sp}\,\mu(a)_{|\Lambda^k H^*} = \{\mu_{J,\overline{K}};\ |J| + |K| \leq k\}.$$

Note that we always have $|\mu_{J,K}| \leq \sum_{j=1}^{n}\lambda_j$ with equality if, and only if, one the subsets J or K is empty and the other is $\{1, \ldots, n\}$, which occurs for eigenvectors in the subspace spanned by the forms $\theta^1 \wedge \ldots \theta^n \wedge \omega^L$ and $\theta^{\bar{1}} \wedge \ldots \theta^{\bar{n}} \wedge \omega^L$ as L ranges over all the subsets of $\{2n+1, \ldots, d\}$.

Since $\lambda_1, \ldots, \lambda_n$ are the eigenvalues of $|L(a)|$, each of them counted twice, if follows that the condition (3.4.22) for $\Delta_{b;k}$ reduces to $\pm\sum_{j=1}^{n}\lambda_j \notin \mathrm{Sp}\,\mu(a)_{|\Lambda^k H^*}$. This latter condition is satisfied if, and only if, the space $\Lambda_{\mathbb{C}}^k H_a^*$ does contain any of the forms $\theta^1 \wedge \ldots \theta^n \wedge \omega^L$ and $\theta^{\bar{1}} \wedge \ldots \theta^{\bar{n}} \wedge \omega^L$ with L subset of $\{2n+1, \ldots, d\}$. Therefore, the sublaplacian $\Delta_{b;k}$ satisfies (3.4.22) at a if, and only if, the integer k

is not between n and $n + d - 2n = d - n$, that is, if, and only if, the condition $X(k)$ holds at a. The proof is thus achieved. $\square$

Suppose now that M is an orientable CR manifold of dimension $2n+1$ with Heisenberg structure $H = \Re(T_{1,0} \oplus T_{0,1})$ and let θ be a global nonvanishing section of TM/H with associated Levi form L_θ. Assume in addition that $T_\mathbb{C} M$ is endowed with a Hermitian metric compatible with its CR structure. Then it follows from (2.2.13) that $\Delta_{b;p,q}$ preserves the bidegree. In this we shall now refine the condition $X(k)$ in each degree (p,q) in terms of the following condition.

DEFINITION 3.5.5. *For $x \in M$ let $r(x)$ and $\kappa(x)$ respectively denote the rank and the number of negative eigenvalues of the Levi form L_θ at x. Then the condition $X(p,q)$ is satisfied at x when we have*

$$(3.5.18) \quad \{(p,q), (q,p)\} \cap \{(\kappa(x) + j, r(x) - \kappa(x) + k);\ j, k = 0, \ldots, n - r(x)\} = \emptyset.$$

In particular, when M is κ-strictly pseudoconvex the condition $X(p,q)$ reduces to $(p,q) \neq (\kappa, n - \kappa)$ and $(p,q) \neq (n - \kappa, \kappa)$. In any case we have:

PROPOSITION 3.5.6. *1) At any point $x \in M$ the Rockland condition for $\Delta_{b;p,q}$ is equivalent to the condition $X(p,q)$.*

2) The principal symbol of $\Delta_{b;p,q}$ is invertible if, and only if, the condition $X(p,q)$ holds at every point. In particular, when the latter occurs $\Delta_{b;p,q}$ admits a parametrix in $\Psi_H^{-2}(M, \Lambda^{p,q})$ and is hypoelliptic with gain of one derivative.

PROOF. As in the proof of Proposition 3.5.4 thanks to Proposition 3.4.5 the proof reduces to checking that at any point $a \in M$ the condition (3.4.22) for $\Delta_{b;p,q}$ is equivalent to the condition $X(p,q)$.

Near a let X_0 be a real vector field such that $\theta(X_0) = 1$ and let $Z_1, \ldots, Z_n$ be an orthonormal frame of $T_{1,0}$. Since the Hermitian form h commutes with complex conjugation $\overline{Z_1}, \ldots, \overline{Z_n}$ is an orthonormal frame of $T_{0,1}$. In addition, we write the Levi form L_θ in the form,

$$(3.5.19) \qquad L_\theta(Z, W) = h(L^c(x)Z, W), \qquad Z, W \text{ sections of } T_{1,0},$$

where $L^c(x)$ is a Hermitian section of $\operatorname{End} T_{1,0}$. In particular, if for $j, k = 1, \ldots, n$ we let $L^c_{jk}(x) = h(LZ_j, Z_k)$ then we have

$$(3.5.20) \qquad [Z_j, \overline{Z_k}] = -i L^c_{jk}(x) X_0 \mod H.$$

Let r be the rank of L_θ at a and let κ be its number of negative eigenvalues. As in the proof of Proposition 3.5.4 we may assume that $L^c(a)$ is of the form,

$$(3.5.21) \qquad L^c(a) = \operatorname{diag}(\lambda_1, \ldots, \lambda_n),$$

where $\lambda_1, \ldots, \lambda_n$ are the eigenvalues of $L^c(a)$ ordered in such way that $\lambda_j > 0$ for $j \leq r - \kappa$ and $\lambda_j > 0$ for $r - \kappa + 1 \leq j \leq r$, while $\lambda_j = 0$ for $j \geq r + 1$.

Let $\theta^1, \ldots, \theta^n$ (resp. $\theta^{\bar{1}}, \ldots, \theta^{\bar{n}}$) be the orthonormal coframe of $\Lambda^{1,0}$ (resp. $\Lambda^{0,1}$) dual to $Z_1, \ldots, Z_n$ (resp. $\overline{Z_1}, \ldots, \overline{Z_n}$). For any ordered subsets $J = \{j_1, \ldots, j_p\}$ and $\overline{K} = \{k_1, \ldots, k_q\}$ of $\{1, \ldots, n\}$ we let $\theta^{J,\overline{K}} = \theta^{j_1} \wedge \ldots \wedge \theta^{j_p} \wedge \theta^{\bar{k}_1} \wedge \ldots \wedge \theta^{\bar{k}_q}$. Then the forms $\theta^{J,\overline{K}}$ give rise to an orthonormal coframe of $\Lambda_\mathbb{C}^* H^*$.

3.5. INVERTIBILITY CRITERIA FOR THE MAIN DIFFERENTIAL OPERATORS

As shown in [**BG**] near a the operator $\Box_b$ has the form,

(3.5.22) $$\Box_b = -\frac{1}{2} \sum_{1 \leq j \leq n} (Z_j \overline{Z}_j + \overline{Z}_j Z_j) - i\nu(x)X_0 + O_H(1),$$

(3.5.23) $$\nu(x) = \sum_{1 \leq j,k \leq n} \iota(\theta^{\overline{j}})\varepsilon(\theta^{\overline{k}}) L^c_{jk}(x) - \frac{1}{2}\sum_{j=1}^{n} L^c_{jj}(x).$$

Hence $\overline{\Box}_b = -\frac{1}{2}\sum_{j=1}^n (Z_j \overline{Z}_j + \overline{Z}_j Z_j) + i\overline{\nu}(x)X_0 + O_H(1)$. Thus,

(3.5.24) $$\Delta_b = \Box_b + \overline{\Box}_b = -\frac{1}{2}\sum_{j=1}^{n}(Z_j\overline{Z}_j + \overline{Z}_j Z_j) - i\mu(x)X_0 + O_H(1),$$

where $\mu(x) = \nu(x) - \overline{\nu}(x) = \sum_{j,k=1}^{n}(\iota(\theta^{\overline{j}})\varepsilon(\theta^{\overline{k}})L^c_{jk}(x) - \iota(\theta^j)\varepsilon(\theta^k)\overline{L^c_{jk}}(x))$. In particular, for $x = a$ we obtain

(3.5.25) $$\mu(a) = \sum_{j=1}^{n}(\iota(\theta^{\overline{j}})\varepsilon(\theta^{\overline{j}}) - \iota(\theta^j)\varepsilon(\theta^j))\lambda_j = \sum_{j=1}^{n}(\varepsilon(\theta^j)\iota(\theta^j) - \varepsilon(\theta^{\overline{j}})\iota(\theta^{\overline{j}}))\lambda_j.$$

Therefore, as in (3.5.16) we have $\mu(a)(\theta^{J,\overline{K}}) = (\sum_{j \in J}\lambda_j - \sum_{k \in \overline{K}}\lambda_j)\theta^{J,\overline{K}}$. Thus,

(3.5.26) $$\mathrm{Sp}\,\mu(a)|_{\Lambda^{p,q}} = \{\sum_{j \in J}\lambda_j - \sum_{k \in \overline{K}}\lambda_j;\; |J|=p, |\overline{K}|=q\}.$$

For $j = 1,\ldots,n$ let $X_j = \frac{1}{\sqrt{2}}(Z_j + \overline{Z}_j)$ and $X_{n+j} = \frac{1}{i\sqrt{2}}(\overline{Z}_j - Z_j)$. Then $X_1,\ldots,X_{2n}$ is an orthonormal frame of H and we have

(3.5.27) $$\Delta_b = -(X_1^2 + \ldots + X_{2n}^2) - i\mu X_0 + O_H(1).$$

Let $L(x) = (L_{jk}(x))_{0 \leq jk \leq 2n}$ be such that

(3.5.28) $$[X_j, X_k] = L_{jk}X_0 \mod H.$$

Since the integrability of $T_{1,0}$ implies that $[Z_j, Z_k] = [\overline{Z}_j, \overline{Z}_k] = 0 \mod H$ one can check that $L(x)$ is of the form,

(3.5.29) $$L(x) = \begin{pmatrix} 0 & -\Re L^c(x) \\ \Re L^c(x) & 0 \end{pmatrix}.$$

This implies that $\frac{1}{2}\mathrm{Trace}\,|L(x)| = \mathrm{Trace}\,|L^c(x)|$. Thus,

(3.5.30) $$\frac{1}{2}\mathrm{Trace}\,|L(a)| = \sum_{j=1}^{n}|\lambda_j| = \sum_{j=1}^{r}|\lambda_j|.$$

Since we always have $|\sum_{j \in J}\lambda_j - \sum_{k \in \overline{K}}\lambda_j| \leq \sum_{j=1}^{r}|\lambda_j|$, in the same way as in the proof of Proposition 3.5.4 the condition (3.4.22) for $\Delta_{b;p,q}$ at a becomes

(3.5.31) $$\sum_{j=1}^{r}|\lambda_j| > |\sum_{j \in J}\lambda_j - \sum_{k \in \overline{K}}\lambda_k|,$$

for any ordered subsets J and $\overline{K}$ such that $|J| = p$ and $|\overline{K}| = q$.

In fact, we have $\sum_{j=1}^{r}|\lambda_j| = |\sum_{j \in J}\lambda_j - \sum_{k \in \overline{K}}\lambda_k|$ if, and only if, either $\overline{K}^c \subset \{1,\ldots,r-\kappa\} \subset J$ and $J^c \subset \{r-\kappa+1,\ldots,r\} \subset \overline{K}$, or $J^c \subset \{1,\ldots,r-\kappa\} \subset \overline{K}$ and $\overline{K}^c \subset \{r-\kappa+1,\ldots,r\} \subset J$. This is possible for J and $\overline{K}$ such that $|J| = p$ and $|\overline{K}| = q$ if, and only if, either p or q of the form $r - \kappa + j$ with $0 \leq j \leq n - r$ and the other is of the form $\kappa + k$ with $0 \leq k \leq n - r$, that is, if, and only if, the condition $X(p,q)$ fails at a. Therefore, the condition (3.4.22) at a for $\Delta_{b;p,q}$ is equivalent to the condition $X(p,q)$ at a. The proof is therefore complete. □

3.5.4. Gover-Graham operators. Let M^{2n+1} be a strictly pseudoconvex CR manifold equipped with a pseudohermitian strcuture θ, i.e., a contact form θ anihilating $H = \Re(T_{1,0} \oplus T_{0,1})$ and such that the associated Levi form on $T_{1,0}$ is positive definite. We endow $T_{\mathbb{C}}M$ with the associate Levi metric and for $k = 1,\ldots,n+1,n+2,n+4,\ldots$ we let $\Box_\theta^{(k)} : C^\infty(M) \to C^\infty(M)$ be the Gover-Graham of order k. In addition, we let X_0 denote the Reeb vector field of θ, so that $\imath_{X_0}\theta = 1$ and $\imath_{X_0}d\theta = 0$ and we let $\Delta_{b;0}$ denote the horizontal sublaplacian of M acting on functions.

When $k = 1$ the operator $\Box_\theta^{(1)}$ agrees with the conformal sublaplacian of Jerison-Lee [**JL1**], so its principal symbols agrees with that of $\Delta_{b;0}$ and is therefore invertible. In general, we have:

PROPOSITION 3.5.7. *1) The operator $\Box_\theta^{(k)}$ is equal to*

$$(3.5.32) \quad (\Delta_{b;0}+i(k-1)X_0)(\Delta_{b;0}+i(k-3)X_0)\cdots(\Delta_{b;0}-i(k-1)X_0)+\mathrm{O}_H(2k-1).$$

2) Unless for the value $k = n+1$ the principal symbol of $\Box_\theta^{(k)}$ is invertible.

PROOF. Let $Z_1,\ldots,Z_n$ be a local orthonormal frame for $T_{1,0}$ and for $w \in \mathbb{C}$ define $\Box_w = \overline{Z_1}Z_1 + \ldots + \overline{Z_n}Z_n + iwX_0$. The operator $\Box_\theta^{(k)}$ corresponds to the operators $P_{w,w'}$ and $\mathcal{P}_{w,w'}$ of [**GG**] with $w = w' = \frac{1}{2}(k-1-n)$ under the canonical trivializations of the density bundles $\mathcal{E}(w,w) = |\Lambda^{n,n}|^{-w/n+2}$, $w \in \mathbb{R}$. Therefore, as explained in [**GG**, pp. 15, 25], the operator $\Box_\theta^{(k)}$ is equal to

$$(3.5.33) \quad (-2\Box_{-\frac{1}{2}(n+k-1)})(-2\Box_{1-\frac{1}{2}(n+k-1)})\cdots(-2\Box_{\frac{1}{2}(k-1-n)})+\mathrm{O}_H(2k-1).$$

Next, notice that for $j = 1,\ldots,n$ we have

$$(3.5.34) \quad 2\overline{Z_j}Z_j = \overline{Z_j}Z_j + Z_j\overline{Z_j} - [Z_j,\overline{Z_j}] = \overline{Z_j}Z_j + Z_j\overline{Z_j} + iX_0 + \mathrm{O}_H(1).$$

In view of the formula (3.5.24) for $\Delta_{b;0}$ it follows that $-2\Box_w$ is equal to

$$(3.5.35) \quad -\sum_{j=1}^n(\overline{Z_j}Z_j+Z_j\overline{Z_j})-i(2w+n)X_0+\mathrm{O}_H(1) = \Delta_{b;0}-i(2w+n)X_0+\mathrm{O}_H(1).$$

Combining this with (3.5.33) then yields the formula (3.5.32).

Now, in the same way as in the proof of Proposition 3.5.6 we can show that the sublaplacian $\Delta_{b;0} + i\alpha T$, $\alpha \in \mathbb{C}$, satisfies the condition (3.4.22) iff α is not in the singular set $\pm(n+2\mathbb{N})$. Thanks to (3.5.32) we see that:

(i) For $k = 1,\ldots,n$ the principal term in (3.5.32) is the product of sublaplacians $\Delta_{b;0} + i\alpha X_0$ with $|\alpha| \leq k-1 < n$.

(ii) For $k = n+1$ the principal term in (3.5.32) contains the factors $\Delta_{b;0} \pm inX_0$, whose principal symbols are not invertible.

(iii) For $k = n+2, n+4,\ldots$ the integers $k-1$ and n have opposite parities, so the principal term in (3.5.32) is the product of sublaplacians $\Delta_{b;0} + i\alpha X_0$ with integers α which are not not in the singular set $\pm(n+2\mathbb{N})$, since their parity is the opposite to that of n.

Therefore, unless for $k = n+1$ the principal symbol of $\Box_\theta^{(k)}$ appears as the product of invertible symbols, hence is invertible. □

3.5.5. Contact Laplacian.

Let (M^{2n+1}, H) be an orientable contact manifold, let θ be a contact form on H with Reeb vector field X_0 and let J be a calibrated almost complex structure on H, so that we can endow M with the Riemannian metric $g_{\theta,J} = d\theta(., J.) + \theta^2$.

Consider the contact complex of M,

$$(3.5.36) \quad C^\infty(M) \stackrel{d_{R;0}}{\to} \ldots C^\infty(M, \Lambda_1^n) \stackrel{D_{R;n}}{\to} C^\infty(M, \Lambda_2^n) \ldots \stackrel{d_{R;2n-1}}{\to} C^\infty(M, \Lambda^{2n}).$$

Let $\Delta_{R;k} : C^\infty(M, \Lambda^k) \to C^\infty(M, \Lambda^k)$ be the contact Laplacian in degree $k \neq n$ and let $\Delta_{R;nj} : C^\infty(M, \Lambda_j^n) \to C^\infty(M, \Lambda_j^n)$, $j = 1, 2$, be the contact Laplacians in degree n as defined in (2.2.34)–(2.2.34).

The almost complex structure J of H defines a bigrading on $\Lambda_\mathbb{C}^* H^*$. More precisely, we have an orthogonal splitting $H \otimes \mathbb{C} = T_{1,0} \oplus T_{0,1}$ with $T_{1,0} = \ker(J+i)$ and $T_{0,1} = \overline{T_{1,0}} = \ker(J-i)$. Therefore, if we consider the subbundles $\Lambda^{1,0} = T_{1,0}^*$ and $\Lambda^{0,1} = T_{0,1}^*$ of $H^* \otimes \mathbb{C} \subset T_\mathbb{C}^* M$ then we have the orthogonal decomposition $\Lambda_\mathbb{C}^* H^* = \bigoplus_{0 \leq p,q \leq n} \Lambda^{p,q}$ with $\Lambda^{p,q} = (\Lambda^{1,0})^p \wedge (\Lambda^{0,1})^q$. We then get a bigrading on Λ_j^n, $j = 1, 2$, by letting

$$(3.5.37) \quad \Lambda_j^n = \bigoplus_{p+q=n} \Lambda_j^{p,q}, \qquad \Lambda_j^{p,q} = \Lambda_j^n \cap \Lambda^{p,q}.$$

As shown by Rumin [**Ru**, Prop. 7] there exist first order differential operators P_k, $k = 0, .., n-1, n+1, \ldots, 2n$ and second order differential operators $P_{p,q}^{(j)}$, $j = 1, 2$, $p + q = n$, such that

$$(3.5.38) \quad (n - k + 2)\Delta_{R;k} = (n - k)(n - k + 1)\Delta_{b;k} + P_k^* P_k, \qquad k = 0, .., n-1,$$

$$(3.5.39) \quad (k - n + 2)\Delta_{R;k} = (k - n)(k - n + 1)\Delta_{b;k} + P_k^* P_k, \qquad k = n+1, \ldots, 2n,$$

$$(3.5.40) \quad 4\Delta_{R;nj} = \Delta_{b;n}^2 + (P_{p,q}^{(j)})^* P_{p,q}^{(j)} \quad \text{on } \Lambda_\pm^{p,q} \text{ with } \sup(p, q) \geq 1,$$

$$(3.5.41) \quad \Delta_{R;nj} = (\Delta_{b;n} + iX_0)^2 \quad \text{on } \Lambda_j^{n,0},$$

$$(3.5.42) \quad \Delta_{R;nj} = (\Delta_{b;n} - iX_0)^2 \quad \text{on } \Lambda_j^{0,n}.$$

These formulas enabled him to prove:

PROPOSITION 3.5.8 ([**Ru**, p. 300]). *The operators $\Delta_{R;k}$, $k \neq n$, and $\Delta_{R;nj}$, $j = 1, 2$, satisfy the Rockland condition at every point.*

Proposition 3.5.8 allowed Rumin to apply results of Helffer-Nourrigat [**HN3**] for proving the maximal hypoellipticity of Δ_R in every degree. In particular, unlike the Kohn Laplacian and the horizontal sublaplacian, the contact Laplacian is hypoelliptic in every bidegree.

Alternatively, we may combine Proposition 3.5.8 with Theorem 3.3.18 to get:

PROPOSITION 3.5.9. *1) The contact Laplacian $\Delta_{R;k}$, $k \neq n$, has an invertible principal symbol of degree -2, hence admits a parametrix in $\Psi_H^{-2}(M, \Lambda^k)$ and is hypoelliptic with gain of one derivative.*

2) The contact Laplacian $\Delta_{R;nj}$, $j = 1, 2$, has an invertible principal symbol of degree -4, hence admits a parametrix in $\Psi_H^{-4}(M, \Lambda_j^n)$ and is hypoelliptic with gain of two derivatives.

CHAPTER 4

Holomorphic families of Ψ_HDOs

In this chapter we define holomorphic families of Ψ_HDOs and check their main properties. To this end we make use of an "almost homogeneous" approach to the Heisenberg calculus which we describe in the first section.

4.1. Almost homogeneous approach to the Heisenberg calculus

In this section we explain how the ΨDOs can be described in terms of symbols and kernels which are almost homogeneous, in the sense that there are homogeneous modulo infinite order terms.

DEFINITION 4.1.1. *A symbol* $q(x,\xi) \in C^\infty(U \times \mathbb{R}^{d+1})$ *is almost homogeneous of degree* m, $m \in \mathbb{C}$, *when we have*

(4.1.1) $\qquad q(x, \lambda.\xi) - \lambda^m q(x,\xi) \in S^{-\infty}(U \times \mathbb{R}^{d+1}) \qquad$ *for any* $\lambda > 0$.

The space of almost homogeneous symbols of degree m *is denoted* $S^m_{ah}(U \times \mathbb{R}^{d+1})$.

LEMMA 4.1.2 ([**BG**, Prop. 12.72]). *Let* $q(x,\xi) \in C^\infty(U \times \mathbb{R}^{d+1})$. *Then the following are equivalent:*

(i) $q(x,\xi)$ *is almost homogeneous of degree* m;

(ii) $q(x,\xi)$ *is in* $S^m(U \times \mathbb{R}^{d+1})$ *and we have* $q \sim p_m$ *with* $p_m \in S_m(U \times \mathbb{R}^{d+1})$, *i.e., the only nonzero term in the asymptotic expansion (3.1.13) for q is p_m.*

Granted this we shall now prove:

LEMMA 4.1.3. *Let* $p \in C^\infty(U \times \mathbb{R}^{d+1})$. *Then we have equivalence:*

(i) p *belongs to* $S^m(U \times \mathbb{R}^{d+1})$.

(ii) For $j = 0, 1, ..$ there exists $q_{m-j} \in S^{m-j}_{ah}(U \times \mathbb{R}^{d+1})$ such that $p \sim \sum_{j \geq 0} q_{m-j}$.

PROOF. Suppose that $p \sim \sum_{j \geq 0} q_{m-j}$ with $q_{m-j} \in S^{m-j}_{ah}(U \times \mathbb{R}^{d+1})$. By Lemma 4.1.2 there exists $p_{m-j} \in S_{m-j}(U \times \mathbb{R}^{d+1})$ such that $q_{m-j} \sim p_{m-j}$. Then we have $p \sim \sum_{j \geq 0} p_{m-j}$ and so p belongs to $S^m(U \times \mathbb{R}^{d+1})$. Hence (ii) implies (i).

Conversely, assume that p is in $S^m(U \times \mathbb{R}^{d+1})$. Then we have $p \sim \sum_{j \geq 0} p_{m-j}$ with $p_{m-j} \in S_{m-j}(U \times \mathbb{R}^{d+1})$. Let $\varphi \in C^\infty_c(\mathbb{R}^{d+1})$ be such that $\varphi(\xi) = 1$ near $\xi = 1$ and $\varphi(\xi) = 0$ for $\|\xi\| \leq 1$. For $j = 0, 1, ..$ set $q_{m-j}(x,\xi) = (1 - \varphi(\xi))p_{m-j}(x,\xi)$. For any $t > 0$ the symbol $q_{m-j}(x, t.\xi) - t^{m-j} q_{m-j}(x,\xi)$ is equal to

(4.1.2) $\qquad (\varphi(t.\xi) - \varphi(\xi))p_{m-j}(x,\xi) \in S^{-\infty}(U \times \mathbb{R}^{d+1})$,

so q_{m-j} belongs to $S^{m-j}_{ah}(U \times \mathbb{R}^{d+1})$. Moreover, as $q_{m-j}(x,\xi) = p_{m-j}(x,\xi)$ for $\|\xi\| > 1$ we see that $p \sim \sum_{j \geq 0} \tilde{p}_{m-j}$. Hence (i) implies (ii). $\qquad\square$

The almost homogeneous symbols have been considered in [**BG**, Sect. 12] already. In the sequel it will be important to have a "dual" notion of almost homogeneity for distributions as follows.

DEFINITION 4.1.4. *The space $\mathcal{D}'_{reg}(\mathbb{R}^{d+1})$ consists of the distributions on $\mathbb{R}^{d+1}$ that are smooth outside the origin. It is endowed with the weakest topology that makes continuous the inclusions of $\mathcal{D}'_{reg}(\mathbb{R}^{d+1})$ into $\mathcal{D}'(\mathbb{R}^{d+1})$ and $C^\infty(\mathbb{R}^{d+1}\backslash 0)$.*

DEFINITION 4.1.5. *A distribution $K(x,y) \in C^\infty(U)\hat{\otimes}\mathcal{D}'_{reg}(\mathbb{R}^{d+1})$ is said to be almost homogeneous of degree m, $m \in \mathbb{C}$, when*

$$(4.1.3) \qquad K(x,\lambda.y) - \lambda^m K(x,y) \in C^\infty(U \times \mathbb{R}^{d+1}) \quad \text{for any } \lambda > 0.$$

We let $\mathcal{K}^m_{ah}(U \times \mathbb{R}^{d+1})$ denote the space of almost homogeneous distributions of degree m.

PROPOSITION 4.1.6. *Let $K(x,y) \in C^\infty(U)\hat{\otimes}\mathcal{D}'_{reg}(\mathbb{R}^{d+1})$. Then the following are equivalent:*

(i) $K(x,y)$ belongs to $\mathcal{K}^m_{ah}(U \times \mathbb{R}^{d+1})$.

(ii) We can put $K(x,y)$ into the form,

$$(4.1.4) \qquad K(x,y) = K_m(x,y) + R(x,y),$$

for some $K_m \in \mathcal{K}_m(U \times \mathbb{R}^{d+1})$ and $R \in C^\infty(U \times \mathbb{R}^{d+1})$.

(iii) We can put $K(x,y)$ into the form,

$$(4.1.5) \qquad K(x,y) = \check{p}_{\xi \to y}(x,y) + R(x,y),$$

for some $p \in S^{\hat{m}}_{ah}(U \times \mathbb{R}^{d+1})$, $\hat{m} = -(m+d+2)$, and $R \in C^\infty(U \times \mathbb{R}^{d+1})$.

PROOF. First, if $K_m \in \mathcal{K}_m(U \times \mathbb{R}^{d+1})$ then (3.1.21) implies that, for any $\lambda > 0$, the distribution $K(x,\lambda.y) - \lambda^m K(x,y)$ is in $C^\infty(U \times \mathbb{R}^{d+1})$. Thus (ii) implies (i).

Second, let $p \in S^{\hat{m}}_{ah}(U \times \mathbb{R}^{d+1})$. By Lemma 4.1.2 there is $p_{\hat{m}} \in S_m(U \times \mathbb{R}^{d+1})$ such that $p \sim p_{\hat{m}}$. Thanks to Lemma 3.1.12 we extend $p_{\hat{m}}$ into a distribution $\tau(x,\xi)$ in $C^\infty(U)\hat{\otimes}\mathcal{S}'(\mathbb{R}^{d+1})$ such that $\check{\tau}_{\xi \to y}(x,y)$ is in $\mathcal{K}_m(U \times \mathbb{R}^{d+1})$. Let $\varphi \in C^\infty_c(\mathbb{R}^{d+1})$ be such that $\varphi = 1$ near the origin. Then we can write

$$(4.1.6) \qquad p = \tau + \varphi(p-\tau) + (1-\varphi)(p-p_{\hat{m}}).$$

Here $\varphi(\xi)(p(x,\xi) - \tau(x,\xi))$ belongs to $C^\infty(U) \otimes \mathcal{D}'(\mathbb{R}^{d+1})$ and is supported on a fixed compact set with respect to ξ, so $[\varphi(p-\tau)]^\vee_{\xi \to y}$ is smooth. Moreover, as $p \sim p_{\hat{m}}$ both $(1-\varphi)(p-p_m)$ and $[(1-\varphi)(p-p_m)]^\vee_{\xi \to y}$ are in $S^{-\infty}(U \times \mathbb{R}^{d+1})$. It then follows that $\check{p}_{\xi \to y}$ coincides with $\check{\tau}_{\xi \to y}$ up to an element of $C^\infty(U \times \mathbb{R}^{d+1})$. Since $\check{\tau}_{\xi \to y}(x,y)$ is in $\mathcal{K}_m(U \times \mathbb{R}^{d+1})$ we deduce from this that (iii) implies (ii).

To complete the proof it remains to show that (i) implies (iii). Assume that $K(x,y)$ belongs to $\mathcal{K}^m_{ah}(U \times \mathbb{R}^{d+1})$. Let $\varphi(y) \in C^\infty_c(\mathbb{R}^{d+1})$ be so that $\varphi(y) = 1$ near $y = 0$ and set $p = (\varphi K)^\wedge_{y \to \xi}(x,y)$. Then p is a smooth and has slow growth with respect to ξ. Moreover $\check{p}_{\xi \to y}(x,y)$ differs from $K(x,y)$ by the smooth function $(1-\varphi(y))K(x,y)$.

Next, using (3.1.23) we see that, for any $\lambda > 0$, the function $p(x,\lambda.\xi) - \lambda^{\hat{m}} p(x,\xi)$ is equal to

$$(4.1.7) \quad \lambda^{-(d+2)}[\varphi(\lambda^{-1}.y)K(x,\lambda^{-1}.y) - \lambda^{-m}\varphi(y)K(x,y)]^\wedge_{y \to \xi}$$
$$= \lambda^{-(d+2)}[(\varphi(\lambda^{-1}.y) - \varphi(y))K(x,\lambda^{-1}.y) + \varphi(y)(K(x,\lambda^{-1}.y) - \lambda^{-m}K(x,y))]^\wedge_{y \to \xi}.$$

Notice that $(\varphi(\lambda^{-1}.y)-\varphi(y))K(x,\lambda^{-1}.y)+\varphi(y)(K(x,\lambda^{-1}.y)-\lambda^{-m}K(x,y))$ belongs to $C^\infty(U\times\mathbb{R}^{d+1})$ and is compactly supported with respect to y, so it belongs to $S^{-\infty}(U\times\mathbb{R}^{d+1})$. Since the latter is also true for Fourier transform with respect to y we see that $p(x,\lambda.\xi)-\lambda^{\hat{m}}p(x,\xi)$ is in $S^{-\infty}(U\times\mathbb{R}^{d+1})$ for any $\lambda>0$, that is, the symbol p is almost homogeneous of degree $\hat{m}$. It then follows that the distribution K satisfies (iii). This proves that (i) implies (iii). The proof is thus achieved. $\square$

4.2. Holomorphic families of Ψ_HDOs

From now on we let Ω denote an open subset of $\mathbb{C}$.

DEFINITION 4.2.1. *A family* $(p_z)_{z\in\Omega}\subset S^*(U\times\mathbb{R}^{d+1})$ *is holomorphic when:*

(i) The order $m(z)$ of p_z depends analytically on z;

(ii) For any $(x,\xi)\in U\times\mathbb{R}^{d+1}$ the function $z\to p_z(x,\xi)$ is holomorphic on Ω;

(iii) The bounds of the asymptotic expansion (3.1.13) for p_z are locally uniform with respect to z, i.e., we have $p_z\sim\sum_{j\geq 0}p_{z,m(z)-j}$, $p_{z,m(z)-j}\in S_{m(z)-j}(U\times\mathbb{R}^{d+1})$, and for any integer N and for any compacts $K\subset U$ and $L\subset\Omega$ we have

$$(4.2.1) \qquad |\partial_x^\alpha\partial_\xi^\beta(p_z-\sum_{j<N}p_{z,m(z)-j})(x,\xi)|\leq C_{NKL\alpha\beta}\|\xi\|^{\Re m(z)-N-\langle\beta\rangle},$$

for $(x,z)\in K\times L$ and $\|\xi\|\geq 1$.

We let $\mathrm{Hol}(\Omega,S^*(U\times\mathbb{R}^{d+1}))$ denote the set of holomorphic families with values in $S^*(U\times\mathbb{R}^{d+1})$.

REMARK 4.2.2. If $(p_z)_{z\in\Omega}$ is a holomorphic family of symbols then the homogeneous symbols $p_{z,m(z)-j}$ depend analytically on z. Indeed, for $\xi\neq 0$ we have

$$(4.2.2) \qquad p_{z,m(z)}(x,\xi)=\lim_{\lambda\to\infty}\lambda^{-m(z)}p_z(x,\lambda.\xi).$$

Since the above axioms imply that the family $(\lambda^{-m(z)}p_z(x,\lambda.\xi))_{\lambda\geq 1}$ is bounded in the Fréchet-Montel space $\mathrm{Hol}(\Omega,C^\infty(U\times(\mathbb{R}^{d+1}\setminus 0))$ the convergence actually holds in $\mathrm{Hol}(\Omega,C^\infty(U\times(\mathbb{R}^{d+1}\setminus 0)))$. Hence $p_{z,m(z)}$ depends analytically on z. Moreover, as for $j=1,2,\ldots$ and for $\xi\neq 0$ we also have

$$(4.2.3) \qquad p_{j,z}(x,\xi)=\lim_{\lambda\to\infty}\lambda^{j-m(z)}(p_z(x,\lambda.\xi)-\sum_{l<j}\lambda^{m(z)-l}p_{z,m(z)-l}(x,\xi)),$$

an easy induction shows that all the symbols p_{z,m_z-j} depend analytically on z.

Recall that $\Psi^{-\infty}(U)=\mathcal{L}(\mathcal{E}'(U),C^\infty(U))$ is naturally a Fréchet space which is isomorphic to $C^\infty(U\times U)$ by the Schwartz's kernel theorem. Therefore holomorphic families of smoothing operators make sense and we may define holomorphic families of Ψ_HDOs as follows.

DEFINITION 4.2.3. *A family $(P_z)_{z\in\Omega}\subset\Psi_H^m(U)$ is holomorphic when it can be put into the form*

$$(4.2.4) \qquad P_z=p_z(x,-iX)+R_z \qquad z\in\Omega,$$

with $(p_z)_{z\in\Omega}\in\mathrm{Hol}(\Omega,S^*(U\times\mathbb{R}^{d+1}))$ and $(R_z)_{z\in\Omega}\in\mathrm{Hol}(\Omega,\Psi^{-\infty}(U))$.

We let $\mathrm{Hol}(\Omega,\Psi_H^*(U))$ denote the set of holomorphic families of Ψ_HDOs.

For technical sake it will be useful to consider the symbol class below.

DEFINITION 4.2.4. *The space $S_{\|}^k(U \times \mathbb{R}^{d+1})$, $k \in \mathbb{R}$, consists of functions $p(x,\xi)$ in $C^\infty(U \times \mathbb{R}^{d+1})$ such that, for any compact $K \subset U$, we have*

(4.2.5) $\qquad |\partial_x^\alpha \partial_\xi^\beta p(x,\xi)| \leq C_{K\alpha\beta}(1+\|\xi\|)^{k-\langle\beta\rangle}, \qquad (x,\xi) \in K \times \mathbb{R}^{d+1}.$

Its space topology is defined by the family of seminorms given by sharpest constants $C_{K\alpha\beta}$'s in the estimates (4.2.5).

Note that the estimates (3.1.13) imply that $S^m(U \times \mathbb{R}^{d+1})$, $m \in \mathbb{C}$, is contained in $S_{\|}^k(U \times \mathbb{R}^{d+1})$ for any $k \geq \Re m$.

PROPOSITION 4.2.5. *Let $(P_z)_{z \in \Omega}$ be a holomorphic family of $\Psi_H DOs$. Then:*

1) $(P_z)_{z \in \Omega}$ gives rise to holomorphic families with values in $\mathcal{L}(C_c^\infty(U), C^\infty(U))$ and $\mathcal{L}(\mathcal{E}'(U)), \mathcal{D}'(U))$.

2) Off the diagonal of $U \times U$ the distribution kernel of P_z is represented by a holomorphic family of smooth functions.

PROOF. Without any loss of generality we may suppose that $P_z = p_z(x, -iX)$, with $(p_z)_{z \in \Omega}$ in $\mathrm{Hol}(\Omega, S^*(U \times \mathbb{R}^{d+1}))$. Moreover, shrinking Ω if necessary, we may also assume that the degree m_z of p_z stays bounded, as much so $(p_z)_{z \in \Omega}$ is contained in $S_{\|}^k(U \times \mathbb{R}^{d+1})$ for some real $k \geq 0$. For $j = 1, \ldots, d$ let $\sigma_j(x,\xi)$ denote the classical symbol of $-iX_j$ and set $\sigma = (\sigma_0, \ldots, \sigma_d)$. Then the proof of [**BG**, Prop. 10.22] shows that the map $p(x,\xi) \to p^\sigma(x,\xi) := p(x, \sigma(x,\xi))$ is continuous from $S_{\|}^k(U \times \mathbb{R}^{d+1})$ to $S_{\frac{1}{2},\frac{1}{2}}^k(U \times \mathbb{R}^{d+1})$. Thus, the family $(p_z^\sigma)_{z \in \Omega}$ belongs to $\mathrm{Hol}(\Omega, S_{\frac{1}{2},\frac{1}{2}}^k(U \times \mathbb{R}^{d+1}))$.

Next, it follows from the proof of [**Hö1**, Thm. 2.2] that:

(i) The quantization map $q \to q(x,D)$ induces continuous $\mathbb{C}$-linear maps from $S_{\frac{1}{2},\frac{1}{2}}^k(U \times \mathbb{R}^{d+1})$ to $\mathcal{L}(C_c^\infty(U), C^\infty(U))$ and to $\mathcal{L}(\mathcal{E}'(U)), \mathcal{D}'(U))$;

(ii) The linear map $q(x,\xi) \to \check{q}_{\xi \to y}(x,y)$ is continuous from $S_{\frac{1}{2},\frac{1}{2}}^k(U \times \mathbb{R}^{d+1})$ to $C^\infty(U) \otimes \mathcal{D}'_{\mathrm{reg}}(\mathbb{R}^{d+1})$, in such way that for any $q \in S_{\frac{1}{2},\frac{1}{2}}^k(U \times \mathbb{R}^{d+1})$ the distribution kernel $\check{q}_{\xi \to y}(x, x-y)$ of $q(x,D)$ is represented off the diagonal by a smooth function depending continuously on q.

As a continuous $\mathbb{C}$-linear map is analytic it follows that, on the one hand, $(p_z(x, -iX))_{z \in \Omega}$ gives rise to holomorphic families with values in $\mathcal{L}(C_c^\infty(U), C^\infty(U))$ and $\mathcal{L}(\mathcal{E}'(U)), \mathcal{D}'(U))$ and, on the other hand, the distribution kernel of P_z is represented off the diagonal by a holomorphic family of smooth functions. The proposition is thus proved. $\square$

DEFINITION 4.2.6. *Let $(P_z)_{z \in \Omega} \subset \mathcal{L}(C_c^\infty(U), C^\infty(U))$ and for $z \in \Omega$ let $k_{P_z}(x,y)$ denote the distribution kernel of P_z. Then the family $(P_z)_{z \in \Omega}$ is said to be uniformly properly supported when, for any compact $K \subset U$, there exist compacts $L_1 \subset U$ and $L_2 \subset K$ such that for any $z \in \Omega$ we have*

(4.2.6) $\quad \mathrm{supp}\, k_{P_z}(x,y) \cap (U \times K) \subset L_1 \quad and \quad \mathrm{supp}\, k_{P_z}(x,y) \cap (K \times U) \subset L_2.$

Bearing in mind the above definition we have:

PROPOSITION 4.2.7. *Let $(P_z)_{z \in \Omega}$ be a holomorphic family of $\Psi_H DOs$.*

1) We can write P_z in the form $P_z = Q_z + R_z$ with $(Q_z)_{z \in \Omega} \in \mathrm{Hol}(\Omega, \Psi_H^*(U))$ uniformly properly supported and $(R_z)_{z \in \Omega} \in \mathrm{Hol}(\Omega, \Psi^{-\infty}(U))$.

2) *If the family $(P_z)_{z\in\Omega}$ is uniformly properly supported then it gives rise to holomorphic families of continuous endomorphisms of $C_c^\infty(U)$ and $C^\infty(U)$ and of $\mathcal{E}'(U)$ and $\mathcal{D}'(U)$.*

PROOF. Let $(\varphi_i)_{i\geq 0} \subset C_c^\infty(U)$ be a partition of the unity subordinated to a locally finite covering $(U_i)_{i\geq 0}$ of U by relatively compact open subsets. For each $i \geq 0$ let $\psi_i \subset C_c^\infty(U)_i$ be such that $\psi_i = 1$ near $\mathrm{supp}\,\varphi_i$ and set $\chi(x,y) = \sum \varphi_i(x)\psi_i(y)$. Then χ is a smooth function on $U \times U$ which is properly supported and such that $\chi(x,y) = 1$ near the diagonal of $U \times U$.

For $z \in \Omega$ let $k_{P_z}(x,y)$ denote the distribution kernel of P_z and let Q_z and R_z be the elements of $\mathcal{L}(C_c^\infty(U), C^\infty(U))$ with respective distribution kernels,

$$(4.2.7) \quad k_{Q_z}(x,y) = \chi(x,y)k_{P_z}(x,y) \quad \text{and} \quad k_{R_z}(x,y) = (1 - \chi(x,y))k_{P_z}(x,y).$$

Notice that since χ is properly supported the family $(Q_z)_{z\in\Omega}$ is uniformly properly supported. As by Proposition 4.2.5 the distribution $k_{P_z}(x,y)$ is represented outside the diagonal of $U \times U$ by a holomorphic family of smooth functions, we see that $(k_{R_z}(x,y))$ is a holomorphic family of smooth kernels, i.e., $(R_z)_{z\in\Omega}$ is a holomorphic family of smoothing operators. Since $Q_z = P_z - R_z$ it follows that $(Q_z)_{z\in\Omega}$ is a holomorphic family of Ψ_HDOs. Hence the first assertion.

Assume now that $(P_z)_{z\in\Omega}$ is uniformly properly supported. Thanks to Proposition 4.2.5 we already know that $(P_z)_{z\in\Omega}$ gives rise to holomorphic families with values in $\mathcal{L}(C_c^\infty(U), C^\infty(U))$ and $\mathcal{L}(\mathcal{E}'(U), \mathcal{D}'(U))$. Let K be a compact subset of U. Then (4.2.6) implies that there exists a compact $L \subset U$ such that for every $z \in \Omega$ the operator P_z maps $C_K^\infty(U)$ to $C_L^\infty(U)$ and $\mathcal{E}'_K(U)$ to $\mathcal{E}'_L(U)$, in such way that $(P_z)_{z\in\Omega}$ gives rise to holomorphic families with values in $\mathcal{L}(C_K^\infty(U), C_L^\infty(U))$ and $\mathcal{L}(\mathcal{E}'_K(U), \mathcal{E}'_L(U))$. In view of the definitions of the topologies of $C_c^\infty(U)$ and $\mathcal{E}'(U)$ as the inductive limit topologies of $C_K^\infty(U)$ and $\mathcal{E}_K(U)$ as K ranges over compacts of U, this shows that the family $(P_z)_{z\in\Omega}$ gives rise to elements of $\mathrm{Hol}(\Omega, C_c^\infty(U))$ and $\mathrm{Hol}(\Omega, \mathcal{E}'(U))$.

Next, let $(\varphi_i)_{i\geq 0} \subset C_c^\infty(U)$ be a partition of unity. For each index i let K_i be a compact neighborhood of $\mathrm{supp}\,\varphi_i$. Then (4.2.6) implies that there exists a compact $L_i \subset U$ such that $\mathrm{supp}\,k_{P_z}(x,y) \cap (K_i \times U) \subset L_i$ for every $z \in \Omega$. Let $\psi_i \in C^\infty(U)$ be such that $\psi_i = 1$ near K_i. Then we have

$$(4.2.8) \qquad P_z = \sum_{i\geq 0}\varphi_i P_z = \sum_{i\geq 0}\varphi_i P_z \psi_i.$$

Since each family $(\varphi_i P_z \psi_i)_{z\in\Omega}$ is holomorphic with values in $\mathcal{L}(C^\infty(U))$ and $\mathcal{L}(\mathcal{D}'(U))$ and the sums are locally finite this shows that $(P_z)_{z\in\Omega}$ gives rise to elements of $\mathrm{Hol}(\Omega, C^\infty(U))$ and $\mathrm{Hol}(\Omega, \mathcal{D}'(U))$. $\square$

4.3. Composition of holomorphic families of Ψ_HDOs

Let us now look at the analyticity of the composition of Ψ_HDOs. To this end we need to deal with holomorphic families of almost homogeneous symbols as follows.

DEFINITION 4.3.1. *A family $(q_z)_{z\in\Omega} \subset S^*_{ah}(U \times \mathbb{R}^{d+1})$ is holomorphic when:*

(i) The degree $m(z)$ of q_z is a holomorphic function on Ω;

(ii) The family $(q_z)_{z\in\Omega}$ belongs to $\mathrm{Hol}(\Omega, C^\infty(U \times \mathbb{R}^{d+1}))$;

(iii) $(q_z(x,t.\xi) - t^{m(z)}q_z(x,\xi))_{z\in\Omega}$ is in $\mathrm{Hol}(\Omega, S^{-\infty}(U \times \mathbb{R}^{d+1}))$ for any $t > 0$.

We let $\operatorname{Hol}(\Omega, S_{ah}^*(U \times \mathbb{R}^{d+1}))$ denote the set of holomorphic families of almost homogeneous symbols.

LEMMA 4.3.2. *Let* $(q_z)_{z \in \Omega} \in \operatorname{Hol}(\Omega, C^\infty(U \times \mathbb{R}^{d+1}))$. *Then the following are equivalent:*

(i) *The family* $(q_z)_{z \in \Omega}$ *is in* $\operatorname{Hol}(\Omega, S_{ah}^*(U \times \mathbb{R}^{d+1}))$ *and has degree degree* $m(z)$;

(ii) *The family* $(q_z)_{z \in \Omega}$ *lies in* $\operatorname{Hol}(\Omega, S^*(U \times \mathbb{R}^{d+1}))$ *and, in the sense of (4.2.1), we have* $q_z \sim p_z$ *where, for every* $z \in \Omega$, *the symbol* p_z *belongs to* $S_{m(z)}(U \times \mathbb{R}^{d+1})$.

PROOF. Assume that $(q_z)_{z \in \Omega}$ is in $\operatorname{Hol}(\Omega, S^*(U \times \mathbb{R}^{d+1}))$ and, in the sense of (4.2.1), we have $q_z \sim p_z$ where, for every $z \in \Omega$, the symbol p_z belongs to $S_{m(z)}(U \times \mathbb{R}^{d+1})$. Then the order $m(z)$ of q_z is a holomorphic function on Ω and, for any compact subset $K \subset U$, any integer N and any open $\Omega' \subset\subset \Omega$, we have

$$(4.3.1) \qquad |\partial_x^\alpha \partial_\xi^\beta (q_z - p_z)(x, \xi)| \leq C_{NK\Omega'\alpha\beta}\|\xi\|^{-N},$$

for $x \in K$, $\|\xi\| \geq 1$ and $z \in \Omega'$. Therefore, for any $t > 0$, the family $\{q_z(x, t.\xi) - t^{m(z)}q_z(x, \xi)\}_{z \in \Omega} = \{(q_z(x, t.\xi) - q_z(x, t.\xi)) - t^{m(z)}(q_z(x, \xi) - q_z(x, \xi))\}_{z \in \Omega}$ is contained in $\operatorname{Hol}(\Omega, S^{-\infty}(U \times \mathbb{R}^{d+1}))$. Hence $(q_z)_{z \in \Omega}$ belongs to $\operatorname{Hol}(\Omega, S_{ah}^*(U \times \mathbb{R}^{d+1}))$.

Conversely, suppose that $(q_z)_{z \in \Omega}$ is contained in $\operatorname{Hol}(\Omega, S_{ah}^*(U \times \mathbb{R}^{d+1}))$ and has degree $m(z)$. Then, for any $t > 0$, any compact $K \subset U$, any integer N and any open $\Omega' \subset\subset \Omega$, we have

$$(4.3.2) \qquad |\partial_x^\alpha \partial_\xi^\beta (q_z(x, t.\xi) - t^{m(z)}q_z(x, \xi))| \leq C_{tNK\Omega'\alpha\beta}(1 + \|\xi\|)^{-N},$$

for $(x, \xi) \in K \times \mathbb{R}^{d+1}$ and $z \in \Omega'$. Then replacing ξ by $s.\xi$, $s > 0$, in (4.3.2) shows that when $N \geq \sup_{z \in \Omega'} \Re \hat{m}(z)$ we have

$$(4.3.3) \qquad |\partial_x^\alpha \partial_\xi^\beta (s^{m(z)} q_z(x, st.\xi) - (st)^{m(z)} q_z(x, s.\xi))| \leq C_{tNK\Omega'\alpha\beta} s^{\Re m(z) - N}\|\xi\|^{-N}$$
$$\leq C_{tNK\Omega'\alpha\beta} s^{-1}\|\xi\|^{-N}.$$

for $(x, \xi) \in K \times \mathbb{R}^{d+1} \backslash 0$ and $z \in \Omega'$.

Next, for $k \in \mathbb{N}$ let $q_{z,k}(x, \xi) = (2^k)^{-m} q_z(x, 2^k.\xi)$. Then, for any compact $K \subset U$, any open $\Omega' \subset\subset \Omega$ and any integer $N \geq \sup_{z \in \Omega'} \Re \hat{m}(z)$, we have

$$(4.3.4) \qquad |\partial_x^\alpha \partial_\xi^\beta (q_{z,k+1}(x, \xi) - q_{z,k}(x, \xi))| \leq C_{2NK\Omega'\alpha\beta} 2^{-k}\|\xi\|^{-N},$$

for $(x, \xi) \in K \times \mathbb{R}^{d+1}\backslash 0$ and $z \in \Omega'$. This shows that the series $\sum_{k \geq 0}(q_{z,k+1} - q_{z,k})$ is convergent in $\operatorname{Hol}(\Omega, C^\infty(U \times (\mathbb{R}^{d+1}\backslash 0)))$. Hence the sequence $(q_{z,k})_{k \geq 0}$ converges in $\operatorname{Hol}(\Omega, C^\infty(U \times (\mathbb{R}^{d+1}\backslash 0)))$ to some family $(p_z)_{z \in \Omega}$. In fact, taking $s = 2^k$ in (4.3.3) and letting $k \to \infty$ with t fixed shows that q_z is homogeneous of degree $m(z)$ with respect to the ξ-variable. Moreover, for any compact $K \subset U$, any open $\Omega' \subset\subset \Omega$ and any integer $N \geq \sup_{z \in \Omega'} \Re \hat{m}(z)$, we have

$$(4.3.5) \quad |\partial_x^\alpha \partial_\xi^\beta (q_z - p_z)(x, \xi)| \leq \sum |\partial_x^\alpha \partial_\xi^\beta (q_{z,k+1} - q_{z,k})(x, \xi)| \leq C_{2NK\Omega'\alpha\beta}\|\xi\|^{-N},$$

for $(x, \xi) \in K \times (\mathbb{R}^{d+1}\backslash 0)$ and $z \in \Omega'$, i.e. we have $q_z \sim p_z$ in the sense of (4.2.1). $\square$

Using Lemma 4.3.2 and arguing as in the proof of Lemma 4.1.3 we get the following characterization of holomorphic families of symbols.

LEMMA 4.3.3. *Let* $(p_z)_{z \in \Omega} \in \operatorname{Hol}(\Omega, C^\infty(U \times \mathbb{R}^{d+1}))$. *Then we have equivalence:*

(i) *The family* $(p_z)_{z \in \Omega}$ *is in* $\operatorname{Hol}(\Omega, S^*(U \times \mathbb{R}^{d+1}))$ *and has order* $m(z)$.

(ii) For $j = 0, 1, \ldots$ there exists $(q_{j,z})_{z \in \Omega} \in \mathrm{Hol}(\Omega, S_{ah}(U \times \mathbb{R}^{d+1}))$ almost homogeneous of degree $m(z) - j$ so that we have $p_z \sim \sum_{j \geq 0} q_{j,z}$ in the sense of (4.2.1).

Next, it is shown in [**BG**, Sect. 12] that, as for homogeneous symbols in (3.1.16)–(3.1.19), there is a continuous bilinear product,

$$(4.3.6) \qquad * : S_{\|}^{k_1}(U \times \mathbb{R}^{d+1}) \times S_{\|}^{k_2}(U \times \mathbb{R}^{d+1}) \longrightarrow S_{\|}^{k_1+k_2}(U \times \mathbb{R}^{d+1}),$$

which is homogeneous in the sense that, for any $\lambda \in \mathbb{R}$, we have

$$(4.3.7) \qquad (p_1 * p_2)_\lambda = p_{1,\lambda} * p_{2,\lambda}, \qquad p_j \in S_{\|}^{k_j}(U \times \mathbb{R}^{d+1}).$$

This product is related to the product of homogeneous symbols as follows.

LEMMA 4.3.4 ([**BG**, Sect. 13]). *For $j = 1, 2$ let $p_j \in S^{m_j}(U \times \mathbb{R}^{d+1})$ have principal symbol $p_{m_j} \in S_{m_j}(U \times \mathbb{R}^{d+1})$. Then $p_1 * p_1$ lies in $S^{m_1+m_2}(U \times \mathbb{R}^{d+1})$ and has principal symbol $p_{m_1} * p_{m_2}$.*

Furthermore, this product is holomorphic, for we have:

LEMMA 4.3.5. *For $j = 1, 2$ let $(p_{j,z})_{z \in \Omega} \subset S^*(U \times \mathbb{R}^{d+1})$ be a holomorphic family of symbols. Then $(p_{1,z} * p_{2,z})_{z \in \Omega}$ is a holomorphic family of symbols as well.*

PROOF. For $j = 1, 2$ let $m_j(z)$ be the order of $p_{j,z}$. Since $m_1(z)$ and $m_2(z)$ are holomorphic functions, possibly by shrinking Ω, we may assume that we have $\sup_{z \in \Omega} m_j(z) \leq k < \infty$, so that the families $(p_{j,z})_{z \in \Omega}$ are in $\mathrm{Hol}(\Omega, S_{\|}^k(U \times \mathbb{R}^{d+1}))$. As $*$ gives rise to a continuous $\mathbb{C}$-bilinear map from $S_{\|}^k(U \times \mathbb{R}^{d+1}) \times S_{\|}^k(U \times \mathbb{R}^{d+1})$ to $S_{\|}^{2k}(U \times \mathbb{R}^{d+1})$ we see that $p_{1,z} * p_{2,z}$ is in $\mathrm{Hol}(\Omega, S_{\|}^{2k}(U \times \mathbb{R}^{d+1}))$, hence is in $\mathrm{Hol}(\Omega, C^\infty(U \times \mathbb{R}^{d+1}))$.

Now, assume that $p_{j,z}$, $j = 1, 2$, is almost homogeneous of degree $m_j(z)$. Then using (4.3.7) we see that for $\lambda > 0$ the symbol $(p_{1,z} * p_{2,z})_\lambda - \lambda^{m_1(z)+m_2(z)} p_{1,z} * p_{2,z}$ is equal to

$$(4.3.8) \quad p_{1,z,\lambda} * p_{1,z,\lambda} - \lambda^{m_1(z)+m_2(z)} p_{1,z} * p_{2,z}$$
$$= (p_{1,z,\lambda} - \lambda^{m_1(z)} p_{1,z}) * p_{2,z} + \lambda^{m_1(z)} p_{1,z,\lambda} * (p_{2,z,\lambda} - \lambda^{m_2(z)} p_{2,z}).$$

As $(p_{1,z,\lambda} - \lambda^{m_1(z)} p_{1,z})_{z \in \Omega}$ and $(p_{2,z,\lambda} - \lambda^{m_2(z)} p_{2,z})_{z \in \Omega}$ are holomorphic families with values in $S^{-\infty}(U \times \mathbb{R}^{d+1})$, combining this with the analyticity of $*$ on $S_{\|}^*(U \times \mathbb{R}^{d+1})$ shows that, for any $\lambda > 0$, the family $(p_{1,z} * p_{2,z})_\lambda - \lambda^{m_1(z)+m_2(z)} p_{1,z} * p_{2,z}$ belongs to $\mathrm{Hol}(\Omega, S^{-\infty}(U \times \mathbb{R}^{d+1}))$. Then Lemma 4.3.3 implies that $p_{1,z} * p_{2,z}$ is a holomorphic family of almost homogeneous of symbols of degree $m_1(z) + m_2(z)$.

In general, by Lemma 4.3.3 we have $p_{j,z} \sim \sum_{l \geq 0} p_{j,z,l}$, with $(p_{j,z,l})_{z \in \Omega}$ in $\mathrm{Hol}(\Omega, S_{\mathrm{ah}}^*(U \times \mathbb{R}^{d+1}))$ of degree $m_j(z) - l$ and $\sim$ taken in the sense of (4.2.1). In particular, for any integer N we have $p_{j,z}$ is equal to $\sum_{l<N} p_{j,z,l}$ modulo a family in $\mathrm{Hol}(\Omega, S_{\|}^{k-N}(U \times \mathbb{R}^{d+1}))$. Thus,

$$(4.3.9) \qquad p_{1,z} * p_{2,z} = \sum_{l+p<N} p_{1,z,l} * p_{2,z,p} \quad \mathrm{mod}\ \mathrm{Hol}(\Omega, S_{\|}^{2k-N}(U \times \mathbb{R}^{d+1})).$$

As explained above $(p_{1,z,l} * p_{2,z,p})_{z \in \Omega}$ is a holomorphic family of almost homogeneous of symbols of degree $m_1(z) + m_2(z) - l - p$. It then follows from Lemma 4.3.3 that $(p_{1,z} * p_{2,z})_{z \in \Omega}$ is a holomorphic family of symbols. $\square$

We are now ready to prove:

PROPOSITION 4.3.6. *For $j = 1, 2$ let $(P_{j,z})_{z \in \Omega}$ be in $\mathrm{Hol}(\Omega, \Psi_H^*(U))$ and suppose that at least one the families $(P_{1,z})_{z \in \Omega}$ or $(P_{2,z})_{z \in \Omega}$ is uniformly properly supported. Then the family $(P_{1,z}P_{2,z})_{z \in \Omega}$ is a holomorphic family of Ψ_HDOs.*

PROOF. By assumption $(P_{1,z})_{z \in \Omega}$ or $(P_{2,z})_{z \in \Omega}$ is uniformly properly supported, so by Proposition 4.2.7 it gives rise to holomorphic families with values in $\mathcal{L}(C^\infty(U))$ and $\mathcal{L}(\mathcal{E}'(U))$. Moreover, Proposition 4.2.7 tells us that the other family at least coincides with a uniformly properly supported holomorphic family of Ψ_HDOs up to a holomorphic family of smoothing operators. It thus follows that $(P_{1,z}P_{2,z})_{z \in \Omega}$ is the product of two uniformly properly supported holomorphic families of Ψ_HDOs up to a holomorphic families of smoothing operators.

As a consequence of this we may assume that the families $(P_{1,z})_{z \in \Omega}$ and $(P_{2,z})_{z \in \Omega}$ are both uniformly properly supported. Thanks to (4.2.8) this allows us to write

$$(4.3.10) \qquad P_{1,z}P_{2,z} = \sum_{i \geq 0} \varphi_i P_{1,z} \psi_i P_{2,z},$$

where $(\varphi_i)_{i \geq 0} \subset C_c^\infty(U)$ and $(\psi_i)_{i \geq 0} \subset C_c^\infty(U)$ are locally finite families such that (φ_i) is a partition of the unity and $\psi_i = 1$ near $\mathrm{supp}\,\varphi_i$.

Next, for $j = 1, 2$ let us write $P_{j,z} = p_{j,z}(x, -iX) + R_{j,z}$, with $(p_{j,z})_{z \in \Omega}$ in $\mathrm{Hol}(\Omega, S^*(U \times \mathbb{R}^{d+1}))$ and $(R_{j,z})_{z \in \Omega}$ in $\mathrm{Hol}(\Omega, \Psi^{-\infty}(U))$. Since by Proposition 4.2.5 each family $(p_{j,z}(x, -iX))_{z \in \Omega}$ is holomorphic with values in $\mathcal{L}(C_c^\infty(U), C^\infty(U))$ and $\mathcal{L}(\mathcal{E}'(U), \mathcal{D}'(U))$ using (4.3.10) we see that

$$(4.3.11) \quad P_{1,z}P_{2,z} = \sum \varphi_i p_{1,z}(x, -iX) \psi_i p_{2,z}(x, -iX) \quad \mathrm{mod}\,\mathrm{Hol}(\Omega, \Psi^{-\infty}(U)).$$

At this stage we make appeal to:

LEMMA 4.3.7 ([**BG**, Prop. 14.45]). *For $j = 1, 2$ let $p_j \in S_{\|}^{k_j}(U \times \mathbb{R}^{d+1})$ and let $\psi \in C_c^\infty(U)$. Then:*

$$(4.3.12) \qquad p_1(x, -iX) \psi p_2(x, -iX) = p_1 \#_\psi p_2(x, -iX),$$

where $\#_\psi$ is a continuous bilinear map from $S_{\|}^{k_1}(U \times \mathbb{R}^{d+1}) \times S_{\|}^{k_2}(U \times \mathbb{R}^{d+1})$ to $S_{\|}^{k_1+k_2}(U \times \mathbb{R}^{d+1})$. Moreover, for any integer $N \geq 1$ we have

$$(4.3.13) \qquad p_1 \#_\psi p_2 = \sum_{j < N} \sum_{\alpha\beta\gamma\delta}^{(j)} h_{\alpha\beta\gamma\delta} \psi(D_\xi^\delta p_1) * (\xi^\gamma \partial_x^\alpha \partial_\xi^\beta p_2) + R_{N,\psi}(p_1, p_2),$$

where the notation is the same as that of Proposition 3.1.9 and $R_{N,\psi}$ is a continuous bilinear map from $S_{\|}^{k_1}(U \times \mathbb{R}^{d+1}) \times S_{\|}^{k_2}(U \times \mathbb{R}^{d+1})$ to $S_{\|}^{k_1+k_2-N}(U \times \mathbb{R}^{d+1})$.

REMARK 4.3.8. The continuity content of Lemma 4.3.7 is not explicitly stated in Proposition 14.45 of [**BG**], but it follow from its proof or from a standard use of the closed graph theorem.

Now, thanks to Lemma 4.3.7 we have

$$(4.3.14) \qquad P_{1,z}P_{2,z} = p_z(x, -iX) + R_z, \qquad p_z = \sum \varphi_i p_{1,z} \#_{\psi_i} p_{2,z}.$$

Furthermore, possibly by shrinking Ω, we may assume that there is a real k such that for $j = 1, 2$ we have $\Re\mathrm{ord}\,p_{j,z} \leq k$ for any $z \in \Omega$. Then the continuity contents of Lemma 4.3.7 imply that $(p_z)_{z\in\Omega}$ belongs to $\mathrm{Hol}(\Omega, S_{\|}^{2k}(U \times \mathbb{R}^{d+1}))$ and for any integer $N \geq 1$ we can write

$$(4.3.15) \qquad p_z = \sum_{r<N} q_{r,z} + R_{N,z}, \qquad q_{j,z} = \sum_{\alpha\beta\gamma\delta}^{(r-s-t)} h_{\alpha\beta\gamma\delta}(D_\xi^\delta p_{1,z}) * (\xi^\gamma \partial_x^\alpha \partial_\xi^\beta p_{2,z}),$$

where $(R_{N,z})_{z\in\Omega} := (\sum_{i\geq 0} \varphi_i R_{N,\psi_i}(p_{1,z}, p_{2,z}))_{z\in\Omega}$ is in $\mathrm{Hol}(\Omega, S_{\|}^{2k-N}(U \times \mathbb{R}^{d+1}))$. Thanks to Lemma 4.3.5 the family $(q_{r,z})_{z\in\Omega}$ is in $\mathrm{Hol}(\Omega, S^*(U \times \mathbb{R}^{d+1}))$ and has order $m_1(z) + m_2(z) - r$. Therefore, we have $p_z \sim \sum_{j\geq 0} q_{j,z}$ in the sense of (4.2.1), which by Lemma 4.3.5 implies that $(p_z)_{z\in\Omega}$ belongs to $\mathrm{Hol}(\Omega, S^*(U \times \mathbb{R}^{d+1}))$. Combining this with (4.3.14) then shows that $(P_{1,z} P_{2,z})_{z\in\Omega}$ is a holomorphic family of Ψ_HDOs. $\square$

4.4. Kernel characterization of holomorphic families of ΨDOs

We shall now give a characterization of holomorphic families of Ψ_HDOs in terms of holomorphic families with values in $\mathcal{K}^*(U \times \mathbb{R}^{d+1})$. Since the latter is defined in terms of asymptotic expansions involving distributions in $\mathcal{K}_*(U \times \mathbb{R}^{d+1})$ a difficulty occurs, since for a family $(K_z)_{z\in\Omega} \subset \mathcal{K}_*(U \times \mathbb{R}^{d+1})$ logarithmic singularities may appear as the order of K_z crosses non-negative integer values.

This issue is resolved by making use of holomorphic families of almost homogeneous distributions as follows.

DEFINITION 4.4.1. *A family $(K_z)_{z\in\Omega} \subset \mathcal{K}_{ah}^*(U \times \mathbb{R}^{d+1})$ is holomorphic when:*

(i) The degree $m(z)$ of K_z is a holomorphic function on Ω;

(ii) The family $(K_z)_{z\in\Omega}$ belongs to $\mathrm{Hol}(\Omega, C^\infty(U) \otimes \mathcal{D}'_{reg}(\mathbb{R}^{d+1}))$;

(iii) For any $\lambda > 0$ the family $\{K_z(x, \lambda.y) - \lambda^{m(z)} K_z(x,y)\}_{z\in\Omega}$ is a holomorphic family with values in $C^\infty(U \times \mathbb{R}^{d+1})$.

We let $\mathrm{Hol}(\Omega, \mathcal{K}_{ah}^*(U \times \mathbb{R}^{d+1}))$ denote the set of holomorphic families of almost homogeneous distributions.

LEMMA 4.4.2. *Let $(K_z)_{z\in\Omega} \in \mathrm{Hol}(\Omega, C^\infty(U) \otimes \mathcal{D}'_{reg}(\mathbb{R}^{d+1}))$. Then we have equivalence:*

(i) The family $(K_z)_{z\in\Omega}$ belongs to $\mathrm{Hol}(\Omega, \mathcal{K}_{ah}^*(U \times \mathbb{R}^{d+1}))$ and has degree $m(z)$.

(ii) We can put $(K_z)_{z\in\Omega}$ into the form,

$$(4.4.1) \qquad K_z(x,y) = \check{p}_{z,\xi\to y}(x,y) + R_z(x,y), \qquad z \in \Omega,$$

with $(p_z)_{z\in\Omega}$ in $\mathrm{Hol}(\Omega, S_{ah}^*(U \times \mathbb{R}^{d+1}))$ of degree $\hat{m}(z) := -(m(z) + d + 2)$ and $(R_z)_{z\in\Omega}$ in $\mathrm{Hol}(\Omega, C^\infty(U \times \mathbb{R}^{d+1}))$.

PROOF. Assume that $(K_z)_{z\in\Omega}$ belongs to $\mathrm{Hol}(\Omega, \mathcal{K}_{ah}^*(U \times \mathbb{R}^{d+1}))$ and has degree $m(z)$. Let $\varphi \in C_c^\infty(U \times \mathbb{R}^{d+1})$ be such that $\varphi(y) = 1$ near $y = 0$ and define $p_z = (\varphi(y) K_z(x,y))_{y\to\xi}^{\wedge}$. As $(\varphi(y) K_z(x,y))_{z\in\Omega}$ is in $\mathrm{Hol}(\Omega, C^\infty(U) \otimes \mathcal{E}'_L(\mathbb{R}^{d+1}))$ with $L = \mathrm{supp}\,\varphi$, we see that $(p_z)_{z\in\Omega}$ belongs to $\mathrm{Hol}(\Omega, C^\infty(U \times \mathbb{R}^{d+1}))$. Moreover, since $((1 - \varphi(y)) K_z(x,y))_{z\in\Omega}$ is in $\mathrm{Hol}(\Omega, C^\infty(U \times \mathbb{R}^{d+1}))$ we also see that $K_z(x,y) = \check{p}_{z,\xi\to y}(x,y)$ modulo $\mathrm{Hol}(\Omega, C^\infty(U \times \mathbb{R}^{d+1}))$.

On the other hand, it follows from (3.1.23) that for any $\lambda > 0$ the symbol $\lambda^{-(d+2)}(p_z(x, \lambda.\xi) - \lambda^{\hat{m}(z)} p_z(x, \xi))$ is equal to

$$(4.4.2) \qquad [(\varphi(\lambda^{-1}.y) - \varphi(y)) K_z(x, \lambda^{-1}.y) + \varphi(y)(K_z(x, \lambda^{-1}.y) - \lambda^{-m} K_z(x,y))]_{y\to\xi}^{\wedge}.$$

This is the Fourier transform with respect to y of a holomorphic family with values in $C_c^\infty(U \times \mathbb{R}^{d+1})$, so it belongs to $\mathrm{Hol}(\Omega, S^{-\infty}(U \times \mathbb{R}^{d+1}))$.

Combining all this with Lemma 4.3.2 shows that $(p_z)_{z\in\Omega}$ is a family of almost homogeneous symbols of degree $\hat{m}(z)$.

Conversely, let $(p_z)_{z\in\Omega} \in \mathrm{Hol}(\Omega, S^*_{\mathrm{ah}}(U\times\mathbb{R}^{d+1}))$ have degree $\hat{m}(z)$. Possibly by shrinking Ω we may assume that we have $\sup_{z\in\Omega} \Re \hat{m}(z) \leq k < \infty$. Then the family $(p_z)_{z\in\Omega}$ belongs to $\mathrm{Hol}(\Omega, S^k_{\|}(U \times \mathbb{R}^{d+1}))$, hence to $\mathrm{Hol}(\Omega, S^k_{\frac{1}{2},\frac{1}{2}}(U \times \mathbb{R}^{d+1}))$. As mentioned in the proof of Proposition 4.2.5, the map $q(x,\xi) \to \check{q}_{\xi\to y}(x,y)$ is analytic from $S^k_{\frac{1}{2},\frac{1}{2}}(U\times\mathbb{R}^{d+1})$ to $C^\infty(U)\hat{\otimes}\mathcal{D}'_{\mathrm{reg}}(\mathbb{R}^{d+1})$, so $(\check{p}_{z,\xi\to y})_{z\in\Omega}$ is a holomorphic family with values in $C^\infty(U)\hat{\otimes}\mathcal{D}'_{\mathrm{reg}}(\mathbb{R}^{d+1})$.

Next, using (3.1.23) we see that, for any $\lambda > 0$, we have

(4.4.3) $\quad \check{p}_{z,\xi\to y}(x,\lambda.y) - \lambda^{m(z)}\check{p}_{z,\xi\to y}(x,y) = [p_z(x,\lambda^{-1}.\xi) - \lambda^{-\hat{m}(z)} p_z(x,\xi)]^\wedge_{\xi\to y}$.

Since by Lemma 4.3.2 the r.h.s. of (4.4.3) is the inverse Fourier transform with respect to ξ of an element of $\mathrm{Hol}(\Omega, S^{-\infty}(U \times \mathbb{R}^{d+1}))$, we see that the family $\{\check{p}_{z,\xi\to y}(x,\lambda.y) - \lambda^{m(z)}\check{p}_{z,\xi\to y}(x,y)\}_{z\in\Omega}$ is contained in $\mathrm{Hol}(\Omega, C^\infty(U\times\mathbb{R}^{d+1}))$. It then follows that $(\check{p}_{z,\xi\to y})_{z\in\Omega}$ is a holomorphic family of almost homogeneous distributions of degree $m(z)$. $\square$

DEFINITION 4.4.3. *A family* $(K_z)_{z\in\Omega} \subset \mathcal{K}^*(U\times\mathbb{R}^{d+1})$ *is holomorphic when:*

(i) *The order* m_z *of* K_z *is a holomorphic function of* z;

(ii) *For* $j=0,1,..$ *there exists* $(K_{j,z}) \in \mathrm{Hol}(\Omega,\mathcal{K}^*_{\mathrm{ah}}(U\times\mathbb{R}^{d+1}))$ *of degree* $m(z)+j$ *such that* $K_z \sim \sum_{j\geq 0} K_{j,z}$ *in the sense that, for any open* $\Omega' \subset\subset \Omega$ *and any integer* N, *as soon as* J *is large enough we have*

(4.4.4) $\quad K_z - \sum_{j\leq J} K_{z,m_z+j} \in \mathrm{Hol}(\Omega', C^N(U\times\mathbb{R}^{d+1}))$.

PROPOSITION 4.4.4. *For a family* $(K_z)_{z\in\Omega} \subset \mathcal{K}^*(U\times\mathbb{R}^{d+1})$ *the following are equivalent:*

(i) *The family* $(K_z)_{z\in\Omega}$ *is holomorphic and has order* $m(z)$.

(ii) *We can put* $(K_z)_{z\in\Omega}$ *into the form,*

(4.4.5) $\quad K_z(x,y) = (p_z)^\vee_{\xi\to y}(x,y) + R_z(x,y)$,

for some family $(p_z)_{z\in\Omega} \in \mathrm{Hol}(\Omega, S^*(U\times\mathbb{R}^{d+1}))$ *of order* $\hat{m}(z) := -(m(z)+d+2)$ *and some family* $(R_z)_{z\in\Omega} \in \mathrm{Hol}(\Omega, C^\infty(U\times\mathbb{R}^{d+1}))$.

PROOF. Assume that $(K_z)_{z\in\Omega}$ is in $\mathrm{Hol}(\Omega, \mathcal{K}^*(U\times\mathbb{R}^{d+1}))$. Let $\varphi \in C^\infty(\mathbb{R}^{d+1})$ be such that $\varphi(y) = 1$ near $y=0$ and for $z\in\Omega$ set $p_z = (\varphi(y)K_z(x,y))^\wedge_{y\to\xi}$. Since $(K_z)_{z\in\Omega}$ lies in $\mathrm{Hol}(\Omega, C^\infty(U)\hat{\otimes}\mathcal{D}'_{\mathrm{reg}}(\mathbb{R}^{d+1}))$ we see that $(p_z)_{z\in\Omega}$ belongs to $\mathrm{Hol}(\Omega, C^\infty(U\times\mathbb{R}^{d+1}))$ and we have

(4.4.6) $\quad K_z(x,y) = (p_z)^\vee_{\xi\to y}(x,y) + (1-\varphi(y))K_z(x,y)$
$\qquad\qquad\qquad = (p_z)^\vee_{\xi\to y}(x,y) \mod \mathrm{Hol}(\Omega, C^\infty(U\times U))$.

Let us write $K_z \sim \sum_{j\geq 0} K_{j,z}$ with $(K_{j,z})_{z\in\Omega} \in \mathrm{Hol}(\Omega, \mathcal{K}^*_{\mathrm{ah}}(U\times\mathbb{R}^{d+1}))$ of degree $\hat{m}(z)+j$ and $\sim$ taken in the sense of (4.4.4) and for $j=0,1,\ldots$ define $p_{j,z} = (\varphi(y)K_{j,z})_{z\in\Omega}$. Then arguing as in the proof of Lemma 4.4.2 shows that $(p_{j,z})_{z\in\Omega}$ is a family in $\mathrm{Hol}(\Omega, S^*_{\mathrm{ah}}(U\times\mathbb{R}^{d+1}))$ of degree $\hat{m}(z)-j$.

Next, in the sense of (4.4.4) we have $(p_z)^\vee_{\xi \to y}(x,y) \sim \sum_{j \geq 0}(p_{j,z})^\vee_{\xi \to y}(x,y)$. Under the Fourier transform with respect to y this shows that, for any compact $L \subset U$, any integer N and any open $\Omega' \subset\subset \Omega$, as soon as J is large enough we have estimates,

$$(4.4.7) \quad |\partial_x^\alpha \partial_\xi^\beta (p_z - \sum_{j \leq J}(\varphi(y)K_{j,z}(x,y))^\wedge_{y \to \xi})(x,\xi)| \leq C_{\Omega' N J L \alpha \beta}(1+|\xi|^2)^{-[N/2]},$$

for $(x,\xi) \in L \times \mathbb{R}^{d+1}$ and $z \in \Omega'$. Hence $p_z \sim \sum_{j \geq 0} p_{j,z}$ in the sense of (4.2.1). It thus follows that $(p_z)_{z \in \Omega}$ is in $\text{Hol}(\Omega, S^*(U \times \mathbb{R}^{d+1}))$ and has order $\hat{m}(z)$, so using (4.4.6) we see that the family $(K_z)_{z \in \Omega}$ is of the form (4.4.5).

Conversely, assume that $K_z(x,y) = (p_z)^\vee_{\xi \to y}(x,y) + R_z(x,y)$ with $(p_z)_{z \in \Omega}$ in $\text{Hol}(\Omega, S^*(U \times \mathbb{R}^{d+1}))$ of order $\hat{m}(z)$ and $(R_z)_{z \in \Omega}$ in $\text{Hol}(\Omega, C^\infty(U \times \mathbb{R}^{d+1}))$. By Lemma 4.3.3 we have $p_z \sim \sum_{j \geq 0} p_{j,z}$ with $(p_{j,z})_{z \in \Omega} \in \text{Hol}(\Omega, S^*_{\text{ah}}(U \times \mathbb{R}^{d+1}))$ of degree $m(z) - j$ and $\sim$ taken in the sense of (4.2.1). Thus, under the inverse Fourier transform with respect to ξ, we have $(p_z)^\vee_{\xi \to y} \sim \sum_{j \geq 0}(p_{j,z})^\vee_{\xi \to y}$ in the sense of (4.4.4). As Lemma 4.4.2 tells us that $((p_{j,z})^\vee_{\xi \to y})_{z \in \Omega}$ is a holomorphic family of almost homogeneous distributions of degree $\hat{m}(z) + j$, it follows that $((p_z)^\vee_{\xi \to y})_{z \in \Omega}$ belongs to $\text{Hol}(\Omega, \mathcal{K}^*(U \times \mathbb{R}^{d+1}))$ and has order $\hat{m}(z)$. Since $(K_z)_{z \in \Omega}$ agrees with $((p_z)^\vee_{\xi \to y})_{z \in \Omega}$ up to an element of $\text{Hol}(\Omega, C^\infty(U \times U))$, we see that $(K_z)_{z \in \Omega}$ is a holomorphic family with values in $\mathcal{K}^*(U \times \mathbb{R}^{d+1})$. $\square$

We are now ready to prove the kernel characterization of holomorphic families of $\Psi_H\text{DOs}$. As before for $x \in U$ we let ψ_x and ε_x respectively denote the coordinate changes to the privileged coordinates and Heisenberg coordinates at x.

PROPOSITION 4.4.5. *Let* $(P_z)_{z \in \Omega} \in \text{Hol}(\Omega, \mathcal{L}(C_c^\infty(U), C^\infty(U)))$ *have distribution kernel* $k_{P_z}(x,y)$. *Then the following are equivalent:*

(i) The family $(P_z)_{z \in \Omega}$ *is a holomorphic family of* $\Psi_H\text{DOs}$ *of order* $m(z)$.

(ii) We can put $k_{P_z}(x,y)$ *in the form,*

$$(4.4.8) \quad k_{P_z}(x,y) = |\psi'_x| P_z(x, -\varepsilon_x(y)) + R_z(x,y),$$

with $(K_z)_{z \in \Omega}$ *in* $\text{Hol}(\Omega, \mathcal{K}^*(U \times \mathbb{R}^{d+1}))$ *of order* $\hat{m}(z) := -(m(z) + d + 2)$ *and* $(R_z)_{z \in \Omega}$ *in* $\text{Hol}(\Omega, C^\infty(U \times U))$.

(iii) We can put $k_{P_z}(x,y)$ *in the form,*

$$(4.4.9) \quad k_{P_z}(x,y) = |\varepsilon'_x| P_z(x, -\varepsilon_x(y)) + R_z(x,y).$$

with $(K_{P_z})_{z \in \Omega}$ *in* $\text{Hol}(\Omega, \mathcal{K}^*(U \times \mathbb{R}^{d+1}))$ *of order* $\hat{m}(z) := -(m(z) + d + 2)$ *and* $(R_z)_{z \in \Omega}$ *in* $\text{Hol}(\Omega, C^\infty(U \times U))$.

PROOF. First, it follows from (3.1.27) and Proposition 4.4.4 that (i) and (ii) are equivalent.

Next, for $x \in U$ let ϕ_x denote the transition map from the privileged coordinates at x to the Heisenberg coordinates at x. Recall that this gives rise to an action on distributions on $U \times \mathbb{R}^{d+1}$ given by

$$(4.4.10) \quad K(x,y) \longrightarrow \phi_x^* K(x,y), \quad \phi_x^* K(x,y) = K(x, \phi_x^{-1}(y)).$$

Since ϕ_x depends smoothly on x, this action gives rise to continuous linear isomorphisms of $C^N(U \times \mathbb{R}^{d+1})$, $N \geq 0$, and $C^\infty(U \times \mathbb{R}^{d+1})$ onto themselves, hence to analytic isomorphisms. Moreover, since $\phi_x(0) = 0$ this also yields an analytic isomorphism of $C^\infty(U)\hat\otimes \mathcal{D}'_{\text{reg}}(\mathbb{R}^{d+1})$ onto itself. Combining this with the homogeneity property (3.1.32) we then deduce that (4.4.10) induces linear isomorphisms of $\text{Hol}(\Omega,\mathcal{K}^*_{\text{ah}}(U\times\mathbb{R}^{d+1}))$ and $\text{Hol}(\Omega,\mathcal{K}^*(U\times\mathbb{R}^{d+1}))$ onto themselves. Together with (3.1.29) this shows that the statements (ii) and (iii) are equivalent. □

4.5. Holomorphic families of ΨDOs on a general Heisenberg manifold

Let us now define holomorphic families of Ψ_HDOs a general Heisenberg manifold. To this end we will need the following lemma.

LEMMA 4.5.1. *Let $(K_z)_{z\in\Omega} \in \text{Hol}(\Omega,\mathcal{K}*(U\times\mathbb{R}^{d+1}))$ and assume there exists an integer N such that $\inf_{z\in\Omega}\Re\text{ord}K_z \geq 2N$. Then the family $(K_z)_{z\in\Omega}$ is contained in $\text{Hol}(\Omega,C^N(U\times\mathbb{R}^{d+1}))$.*

PROOF. Thanks to Proposition 4.4.5 we may assume that K_z is of the form $K_z(x,y) = \check{p}_{z,\xi\to y}(x,y)$ with $(p_z)_{z\in\Omega}$ in $\text{Hol}(\Omega,S^*(U\times\mathbb{R}^{d+1}))$. Since by assumption we have $-(\Re\text{ord}p_z + d + 2) = \Re\text{ord}K_z \geq 2N$ we see that $(p_z)_{z\in\Omega}$ is contained in $\text{Hol}(\Omega,S_{\|}^{-(2N+d+2)}(U\times\mathbb{R}^{d+1}))$. Since the map $p\to \check{p}_{\xi\to y}$ is continuous from $S_{\|}^{-(2N+d+2)}(U\times\mathbb{R}^{d+1})$ to $C^\infty(U)\hat\otimes C^N(U\times\mathbb{R}^{d+1})$ (see Lemma A.1 in Appendix A), it follows that $(K_z)_{z\in\Omega}$ lies in $\text{Hol}(\Omega,C^N(U\times\mathbb{R}^{d+1}))$. □

PROPOSITION 4.5.2. *Let $\tilde U$ be an open subset of $\mathbb{R}^{d+1}$ together with a hyperplane bundle $\tilde H \subset T\tilde U$ and a $\tilde H$-frame of $T\tilde U$ and let $\phi : (U,H) \to (\tilde U,\tilde H)$ be a Heisenberg diffeomorphism. Then for any family $(\tilde P_z)_{z\in\Omega} \in \text{Hol}(\Omega,\Psi^*_{\tilde H}(\tilde U))$ the family $(P_z)_{z\in\Omega} := (\phi^*\tilde P_z)_{z\in\Omega}$ is contained in $\text{Hol}(\Omega,\Psi^*_H(U))$.*

PROOF. For $x\in U$ and $\tilde x \in \tilde U$ let ε_x and $\tilde\varepsilon_{\tilde x}$ denote the coordinate changes to the Heisenberg coordinates at x and $\tilde x$ respectively. Then by Proposition 4.4.5 the distribution kernel $k_{\tilde P_z}(\tilde x,\tilde y)$ of $\tilde P_z$ is of the form

$$(4.5.1) \qquad k_{\tilde P_z}(\tilde x,\tilde y) = |\tilde\varepsilon'_{\tilde x}|K_{\tilde P_z}(\tilde x,-\tilde\varepsilon_{\tilde x}(\tilde y)) + \tilde R_z(\tilde x,\tilde y),$$

with $(K_{\tilde P_z})_{z\in\Omega}$ in $\text{Hol}(\Omega,\mathcal{K}^*(\tilde U\times \tilde U))$ and $(\tilde R_z)_{z\in\Omega}$ in $\text{Hol}(\Omega,C^\infty(\tilde U\times\mathbb{R}^{d+1}))$. Then the proof of Proposition 3.1.18 in Appendix A shows that the distribution kernel $k_{P_z}(x,y)$ of P_z takes the form,

$$(4.5.2) \quad k_{P_z}(x,y) = |\varepsilon'_x|K_{P_z}(x,-\varepsilon_x(y))$$
$$+ (1-\chi(x,-\varepsilon_x(y)))|\tilde\varepsilon'_{\phi(x)}|K_{\tilde P_z}(\phi(x),-\tilde\varepsilon_{\phi(x)}(\phi(y))) + \tilde R_z(\phi(x),\phi(y)),$$

where we have let

$$(4.5.3) \qquad K_{P_z}(x,y) = \chi(x,y)|\partial_y\Phi(x,y)|K_{\tilde P_z}(\phi(x),\Phi(x,y)),$$
$$(4.5.4) \qquad \Phi(x,y) = -\tilde\varepsilon_{\phi(x)}\circ\phi\circ\varepsilon_x^{-1}(-y),$$

and the function $\chi(x,y) \in C^\infty(U\times\mathbb{R}^{d+1})$ has a supported contained in the open $\mathcal{U} = \{(x,y)\in U\times\mathbb{R}^{d+1};\ \varepsilon_x^{-1}(-y)\in U\}$, is properly supported with respect to x and is such that $\chi(x,y) = 1$ near $U\times\{0\}$. In particular, we have

$$(4.5.5) \qquad k_{P_z}(x,y) = |\varepsilon'_x|K_{P_z}(x,-\varepsilon_x(y)) \quad \text{mod } \text{Hol}(\Omega,C^\infty(U\times U)).$$

Let us now prove that $(K_{P_z})_{z\in\Omega}$ belongs to $\text{Hol}(\Omega,\mathcal{K}^*(U\times\mathbb{R}^{d+1}))$. To this end, possibly by shrinking Ω, we may assume that $\inf_{z\in\Omega}\Re\hat m(z) \geq \mu > -\infty$. Moreover, the proof of Proposition 3.1.18 in Appendix A also shows that for any integer N we have

$$K_{P_z}(x,y) = \sum_{\langle\alpha\rangle < N} \sum_{\frac{3}{2}\langle\alpha\rangle \leq \langle\beta\rangle < \frac{3}{2}N} K_{\alpha\beta,z}(x,y) + \sum_{j=1}^{3} R_{N,z}^{(j)}(x,y), \tag{4.5.6}$$

$$K_{\alpha\beta,z}(x,y) = a_{\alpha\beta}(x) y^\beta (\partial_{\tilde{y}}^\alpha K_{\tilde{P}_z})(\phi(x), \phi'_H(x)y), \tag{4.5.7}$$

where the smooth functions $a_{\alpha\beta}(x)$ are as in Proposition 3.1.18 and the remainder terms $R_{N,z}^{(j)}(x,y)$, $j = 1,2,3$, take the forms:

- $R_{N,z}^{(2)}(x,y) = \sum_{\langle\alpha\rangle < N} \sum_{\langle\beta\rangle = \frac{3}{2}N} r_{M\alpha}(x,y) y^\beta (\partial_{\tilde{y}}^\alpha K_{\tilde{P}_z})(\phi(x), \phi'_H(x)y)$ for some functions $r_{\alpha\beta}(x,y)$ in $C^\infty(U \times \mathbb{R}^{d+1})$;

- $R_{N,z}^{(3)}(x,y) = \sum_{\langle\alpha\rangle = N} \sum_{\langle\beta\rangle = \frac{3}{2}N} \int_0^1 r_{\alpha\beta}(t,x,y) (\tilde{y}^\beta \partial_{\tilde{y}}^\alpha K_{\tilde{P}_z})(\phi(x), t\Phi(x,y) + (1-t)\phi'_H(x)y) dt$, for some functions $r_{\alpha\beta}(t,x,y)$ in $C^\infty([0,1] \times U \times \mathbb{R}^{d+1})$;

- $R_{N,z}^{(3)}(x,y) = \sum_{\langle\alpha\rangle < N} \sum_{\frac{3}{2}\langle\alpha\rangle \leq \langle\beta\rangle < \frac{3}{2}N} (1 - \chi(x,y)) K_{\alpha\beta,z}(x,y)$.

Observe that the map $\Phi(x,y) = (\phi(x), \phi'_H(x)y)$ is a smooth diffeomorphism from $U \times \mathbb{R}^{d+1}$ onto $\tilde{U} \times \mathbb{R}^{d+1}$ such that $\Phi(x,0) = (\phi(x), 0)$ and for any $\lambda \in \mathbb{R}$ we have $\Phi(x, \lambda.y) = (\phi(x), \lambda.\phi'_H(x)y)$, so along similar lines as that of the proof of Proposition 4.4.5 we can prove that the map

$$\mathcal{D}'(U \times \mathbb{R}^{d+1}) \ni K(x,y) \longrightarrow K(\phi(x), \phi'_H(x)y) \in \mathcal{D}'(\tilde{U} \times \mathbb{R}^{d+1}) \tag{4.5.8}$$

gives rise to a linear map from $\mathrm{Hol}(\Omega, \mathcal{K}^*(U \times \mathbb{R}^{d+1}))$ to $\mathrm{Hol}(\Omega, \mathcal{K}^*(\tilde{U} \times \mathbb{R}^{d+1}))$ preserving the order. Therefore $(K_{\alpha\beta,z}(x,y))_{z\in\Omega}$ is in $\mathrm{Hol}(\Omega, \mathcal{K}^*(\tilde{U} \times \mathbb{R}^{d+1}))$ and has order $\hat{m}(z) + \langle\beta\rangle - \langle\alpha\rangle$. Incidentally $(R_{N,z}^{(3)})_{z\in\Omega}$ belongs to $\mathrm{Hol}(\Omega, C^\infty(U \times \mathbb{R}^{d+1}))$.

On the other hand, if $\frac{3}{2}\langle\alpha\rangle \leq \langle\beta\rangle = \frac{3}{2}N$ then the order $\hat{m}_{\alpha\beta}(z) = \hat{m}(z) + \langle\beta\rangle - \langle\alpha\rangle$ of $\tilde{y}^\beta \partial_{\tilde{y}}^\alpha K_{\tilde{P}_z}$ is such that $\Re \hat{m}_{\alpha\beta}(z) \geq \Re \hat{m}(z) + \frac{1}{3}\langle\beta\rangle \geq \mu + \frac{N}{2}$. Therefore, Lemma 4.5.1 implies that, for any integer J, as soon as N is large enough $(\tilde{y}^\beta \partial_{\tilde{y}}^\alpha K_{\tilde{P}_z})_{z\in\Omega}$ is in $\mathrm{Hol}(\Omega, C^J(\tilde{U} \times \mathbb{R}^{d+1}))$, so that remainder terms $(R_{N,z}^{(2)})_{z\in\Omega}$ and $(R_{N,z}^{(3)})_{z\in\Omega}$ are in $\mathrm{Hol}(\Omega, C^J(U \times \mathbb{R}^{d+1}))$.

All this shows that we have $K_{P_z}(x,y) \sim \sum_{\frac{3}{2}\langle\alpha\rangle \leq \langle\beta\rangle} K_{\alpha\beta,z}(x,y)$ in the sense of (4.4.4), which implies that $(K_{P_z})_{z\in\Omega}$ belongs to $\mathrm{Hol}(\Omega, \mathcal{K}^*(U \times \mathbb{R}^{d+1}))$. Combining this with (4.5.5) and Proposition 4.4.5 then shows that $(P_z)_{z\in\Omega}$ is a holomorphic family of Ψ_HDOs. □

Now, let (M^{d+1}, H) be a Heisenberg manifold and let $\mathcal{E}$ be a smooth vector bundle over M. Then Proposition 4.5.2 allows us to define holomorphic families with values in $\Psi_H^*(M, \mathcal{E})$ as follows.

DEFINITION 4.5.3. *A family $(P_z)_{z\in\Omega} \subset \Psi_H^*(M, \mathcal{E})$ is holomorphic when:*

(i) The order $m(z)$ of P_z is a holomorphic function of z;

(ii) For φ and ψ in $C_c^\infty(M)$ with disjoint supports $(\varphi P_z \psi)_{z\in\Omega}$ is a holomorphic family of smoothing operators, i.e., is given by a holomorphic family of smooth distribution kernels);

(iii) For any trivialization $\tau : \mathcal{E}_{|U} \to U \times \mathbb{C}^r$ over a local Heisenberg chart $\kappa : U \to V \subset \mathbb{R}^{d+1}$ the family $(\kappa_* \tau_*(P_{z|U}))_{z\in\Omega}$ belongs to $\mathrm{Hol}(\Omega, \Psi_H^*(V, \mathbb{C}^r)) := \mathrm{Hol}(\Omega, \Psi_H^*(V)) \otimes \mathrm{End}\,\mathbb{C}^r$.

All the preceding properties of holomorphic families of Ψ_HDOs on an open subset of $\mathbb{R}^{d+1}$ hold *verbatim* for holomorphic families with values in $\Psi_H^*(M, \mathcal{E})$. Moreover, we have:

PROPOSITION 4.5.4. *The principal symbol map* $\sigma_* : \Psi_H^*(M, \mathcal{E}) \to S_*(\mathfrak{g}^*M, \mathcal{E})$ *is analytic, in the sense that for any holomorphic family* $(P_z)_{z \in \Omega} \subset \Psi_H^*(M, \mathcal{E})$ *the family of symbols* $(\sigma_*(P_z))_{z \in \Omega}$ *is in* $\mathrm{Hol}(\Omega, C^\infty(\mathfrak{g}^*M \setminus 0, \mathrm{End}\,\mathcal{E}))$.

PROOF. Let $(P_z)_{z \in \Omega} \subset \Psi_H^*(M, \mathcal{E})$ be a holomorphic family of Ψ_HDOs of order $m(z)$ and let us show that the family of symbols $(\sigma_*(P_z))_{z \in \Omega}$ belongs to $\mathrm{Hol}(\Omega, C^\infty(\mathfrak{g}^*M \setminus 0, \mathrm{End}\,\mathcal{E}))$. Since this a purely local issue we may as well assume that $(P_z)_{z \in \Omega}$ is a holomorphic family of scalar Ψ_HDOs on a local trivializing Heisenberg chart $U \subset \mathbb{R}^{d+1}$.

By Proposition 4.4.5 we can put the distribution kernel of P_z into the form,

$$(4.5.9) \qquad k_{P_z}(x,y) = |\varepsilon_x'| K_{P_z}(x, -\varepsilon_x(y)) + R_z(x,y),$$

with $(K_{P_z})_{z \in \Omega} \in \mathrm{Hol}(\Omega, \mathcal{K}^*(U \times \mathbb{R}^{d+1}))$ of order $\hat{m}(z) = -(m(z) + d + 2)$ and $(R_z)_{z \in \Omega} \in \mathrm{Hol}(\Omega, C^\infty(U \times U))$. Let $\varphi \in C^\infty(\mathbb{R}^{d+1})$ be such that $\varphi(y) = 1$ near $y = 0$ and let $p_z = (\varphi(y) K_{P_z}(x,y))^{\wedge}_{\xi \to y}$. Then the proof of Proposition 4.4.4 shows that $(p_z)_{z \in \Omega}$ is a holomorphic family of symbols. Moreover, we have

$$(4.5.10) \quad K_{P_z}(x,y) = (p_z)^{\vee}_{\xi \to y}(x,y) + (1 - \varphi(y)) K_{P_z}(x,y)$$
$$= (p_z)^{\vee}_{\xi \to y}(x,y) \mod \mathrm{Hol}(\Omega, C^\infty(U \times U)).$$

Let $z \in \Omega$ and let $K_{\hat{m}(z)} \in \mathcal{K}_{\hat{m}(z)}(U \times \mathbb{R}^{d+1})$ be the principal kernel of K_{P_z}. Then (4.5.10) and Proposition 3.1.14 show that the leading symbol of p_z is the restriction to $U \times (\mathbb{R}^{d+1} \setminus 0)$ of $(K_{\hat{m}(z)})^{\wedge}_{y \to \xi}$. Since the latter is equal to $\sigma_{m(z)}(P_z)$, we see that the leading symbol of p_z is just $\sigma_{m(z)}(P_z)$. Since $(p_z)_{z \in \Omega}$ is a holomorphic family of symbols it then follows from Remark 4.2.2 that the family $(\sigma_{m(z)}(P_z))_{z \in \Omega}$ belongs to $\mathrm{Hol}(\Omega, C^\infty(U \times (\mathbb{R}^{d+1} \setminus 0)))$. The proof is thus achieved. $\square$

4.6. Transposes and adjoints of holomorphic families of Ψ_HDOs

Let us now look at the analyticity and anti-analyticity of transposes and adjoints of holomorphic family of Ψ_HDOs.

PROPOSITION 4.6.1. *Let* $(P_z)_{z \in \Omega} \in \mathrm{Hol}(\Omega, \Psi_H^*(M, \mathcal{E}))$. *Then the transpose family* $(P_z^t) \subset \Psi_H^*(M, \mathcal{E}^*)$ *is a holomorphic family of* Ψ_HDOs.

PROOF. For $z \in \Omega$ let $k_{P_z}(x,y)$ denote the distribution kernel of P_z. The distribution kernel of P_z^t is $k_{P_z^t}(x,y) = k_{P_z}(y,x)^t$, hence is represented outside the diagonal by a holomorphic family of smooth kernels. Therefore, we need only to prove the statement for a holomorphic family of scalar Ψ_HDOs on a Heisenberg chart $U \subset \mathbb{R}^{d+1}$, as we shall now suppose that the family $(P_z)_{z \in \Omega}$ is. In addition, there is no loss of generality in assuming that the order $m(z)$ of P_z is such that there exists $\mu \in \mathbb{R}$ so that $\Re \hat{m}(z) \geq \mu$ for any $z \in \Omega$.

Next, thanks to Proposition 4.4.5 the kernel of P_z is of the form

$$(4.6.1) \qquad k_{P_z} = |\varepsilon_x'| K_{P_z}(x, -\varepsilon_x(y)) + R_z(x,y),$$

with $(K_{P_z})_{z \in \Omega}$ in $\mathrm{Hol}(\Omega, \mathcal{K}^*(U \times \mathbb{R}^{d+1}))$ and $(R_z)_{z \in \Omega}$ in $\mathrm{Hol}(\Omega, C^\infty(U \times U))$. Then the proof of Proposition 3.1.21 in Appendix B shows that $k_{P_z}(x,y)$ is equal to

$$(4.6.2) \quad |\varepsilon_x'| K_{P_z^t}(x, -\varepsilon_x(y)) + (1 - \chi(x, -\varepsilon_x(y)))|\varepsilon_y'| K_{P_z}(y, -\varepsilon_y(x)) + R_z(y,x),$$

4.6. TRANSPOSES AND ADJOINTS OF HOLOMORPHIC FAMILIES OF Ψ_HDOs

where we have let

(4.6.3) $$K_{P_z^t}(x,y) = \chi(x,y)|\varepsilon'_x|^{-1}|\varepsilon'_y|K_{P_z}(\varepsilon_x^{-1}(-y), -\varepsilon_{\varepsilon_x^{-1}(-y)}(x)),$$

and the function $\chi(x,y) \in C^\infty(U \times \mathbb{R}^{d+1})$ has a support contained in the open subset $\mathcal{U} = \{(x,y) \in U \times \mathbb{R}^{d+1}; \; \varepsilon_x^{-1}(-y) \in U\}$, is properly supported with respect x and is equal to 1 near $U \times \{0\}$. In particular, we have

(4.6.4) $$k_{P_z^t}(x,y) = |\varepsilon'_x|K_{P_z^t}(x, -\varepsilon_x(y)) \qquad \mod \text{Hol}(\Omega, C^\infty(U \times U)).$$

Moreover, it follows from the proof of Proposition 3.1.21 in Appendix B that for any integer N we can write

(4.6.5) $$K_{P_z^t}(x,y) = \sum_{\alpha,\beta,\gamma,\delta}^{(N)} K_{\alpha\beta\gamma\delta,z} + \sum_{j=1}^{4} R_{N,z}(x,y),$$

(4.6.6) $$K_{\alpha\beta\gamma\delta,z} = a_{\alpha\beta\gamma\delta}(x) y^{\beta+\delta} (\partial_x^\gamma \partial_y^\alpha K_{P_z})(x,-y),$$

where the smooth functions $a_{\alpha\beta\gamma\delta}(x)$ are as in Proposition 3.1.21, the summation goes over all the multi-orders α, β, γ and δ such that $\langle \alpha \rangle < N$, $\frac{3}{2}\langle \alpha \rangle \leq \langle \beta \rangle < \frac{3}{2}N$ and $|\gamma| \leq |\delta| \leq 2|\gamma| < 2N$ and the remainder terms $R_{N,z}^{(j)}(x,y)$ take the forms:

- $R_{N,z}^{(1)} = \sum_{\langle \alpha \rangle = N} \sum_{\langle \beta \rangle \doteq \frac{3}{2}\langle \alpha \rangle} \frac{|\varepsilon'_y|}{|\varepsilon'_x|} \int_0^1 r_{N\alpha\beta}(t,x,y)(y^\beta \partial_y^\alpha K_{P_z})(\varepsilon_x^{-1}(-y), \Phi_t(x,y)),$

where the functions $r_{N\alpha\beta}(t,x,y)$ are in $C^\infty([0,1] \times U \times \mathbb{R}^{d+1})$, the equality $k \doteq \frac{3}{2}l$ means that k is equal to $\frac{3}{2}l$ if $\frac{3}{2}l$ is integer and to $\frac{3}{2}l + \frac{1}{2}$ otherwise, and we have let $\Phi_t(x,y) = -y + t(y - \varepsilon_{\varepsilon_x^{-1}(-y)}(x));$

- $R_{N,z}^{(2)}(x,y) = \sum_{\langle \beta \rangle \doteq \frac{3}{2}N} r_{N\alpha}(x,y) y^\beta (\partial_y^\alpha K_{P_z})(\varepsilon_x^{-1}(-y), -y),$ with $r_{N\alpha}(x,y)$ in $C^\infty(U \times \mathbb{R}^{d+1});$

- $R_{N,z}^{(3)}(x,y) = \sum_{|\gamma|=N} \sum_{N \leq |\delta| \leq 2N} a_{\alpha\beta\gamma\delta}(x) y^{\beta+\delta} \int_0^1 (1-t)^{N-1} (\partial_x^\gamma \partial_y^\alpha K_{P_z})(x + t(\varepsilon_x^{-1}(-y) - x), -y) dt,$ with $a_{\alpha\beta\gamma\delta}(x) \in C^\infty(U);$

- $R_{N,z}^{(4)}(x,y) = \sum_{\alpha,\beta,\gamma,\delta}^{(N)} (1 - \chi(x,y)) K_{\alpha\beta\gamma\delta,z}(x,y).$

Each family $(K_{\alpha\beta\gamma\delta,z})_{z \in \Omega}$ belongs to $\text{Hol}(\Omega, \mathcal{K}^*(U \times \mathbb{R}^{d+1}))$. Moreover, the remainder term $R_{N,z}^{(4)}$ belongs to $\text{Hol}(\Omega, C^\infty(U \times \mathbb{R}^{d+1}))$ and, along similar lines as that of the proof of Proposition 4.5.2, we can show that for any integer J the other remainder terms $(R_{N,z}^{(j)})_{z \in \Omega}$ are in $\text{Hol}(\Omega, C^J(U \times \mathbb{R}^{d+1}))$ as soon N is large enough. Therefore, we have $K_{P_z^t} \sim \sum_{\frac{3}{2}\langle \alpha \rangle \leq \langle \beta \rangle} \sum_{|\gamma| \leq |\delta| \leq 2|\gamma|} K_{\alpha\beta\gamma\delta,z}$ in the sense of (4.4.4), which means that $K_{P_z^t}$ belongs to $\text{Hol}(\Omega, \mathcal{K}^*(U \times \mathbb{R}^{d+1}))$. Combining this with (4.6.4) and Proposition 4.4.5 then shows that $(P_z^t)_{z \in \Omega}$ is a holomorphic family of Ψ_HDOs. $\square$

Assume now that M is endowed with a density > 0 and $\mathcal{E}$ with a Hermitian metric. Then Proposition 4.6.1 allows us to carried out the proof of Proposition 3.2.12 in the setting of holomorphic families, so that we get:

PROPOSITION 4.6.2. *Let $(P_z)_{z \in \Omega} \subset \Psi_H^*(M, \mathcal{E})$ be a holomorphic family of Ψ_HDOs. Then the family $(P_z^*)_{z \in \Omega} \subset \Psi_H^*(M, \mathcal{E}^*)$ is an anti-holomorphic family of Ψ_HDOs, in the sense that $(P_{\bar z}^*)_{z \in \Omega}$ is a holomorphic family of Ψ_HDOs.*

CHAPTER 5

Heat Equation and Complex Powers of Hypoelliptic Operators

In this chapter we deal with complex powers of hypoelliptic differential operators in connection with the heat equation. Due to the lack of microlocality of the Heisenberg we cannot carry out in the Heisenberg setting the standard approach of Seeley [**Se**] to the complex powers of elliptic operators. Instead we rely on the pseudodifferential representation of the heat kernel of [**BGS**], which is especially suitable for dealing with positive differential operators.

In Section 5.1 we recall the pseudodifferential representation of the heat kernel of an hypoelliptic operator in terms of the Volterra-Heisenberg calculus of [**BGS**] and we extend to this setting the intrinsic approach of Chapter 3. We then specialize the results to sublaplacians and integer powers of sublaplacians in Section 5.2.

In Section 5.3 we make use of the framework of Section 5.1 to prove that the complex powers of a positive differential operators form a holomorphic family of Ψ_HDOs, provided that the principal symbol of the corresponding heat operator is invertible in the Volterra-Heisenberg calculus.

The other two sections are devoted to applications of the above result. First, in Section 5.4 we make use of it to extend Theorem 3.3.18 to non-integer Ψ_HDOs and to show that the invertibility of the principal symbol of the heat operator is implied by the Rockland condition when the bracket condition $H + [H, H] = TM$ holds.

Second, in Section 5.5 we construct the weighted Sobolev spaces $W_H^s(M, \mathcal{E})$ and check their main properties. In particular, we prove that they yield sharp regularity results for Ψ_HDOs.

Throughout this chapter we let (M^{d+1}, H) be a compact Heisenberg manifold equipped with a (smooth) density > 0 and let $\mathcal{E}$ be a Hermitian vector bundle over M of rank r.

5.1. Pseudodifferential representation of the heat kernel

In this section we recall the pseudodifferential representation of the heat kernel of a hypoelliptic operator of [**BGS**], which extends to the Heisenberg setting the approach to the heat kernel asymptotics of [**Gre**].

Let $P : C^\infty(M, \mathcal{E}) \to C^\infty(M, \mathcal{E})$ be a differential operator of even (Heisenberg) order v which is selfadjoint and bounded from below. We also assume that the principal symbol of P is an invertible principal symbol, which by Theorem 3.3.18 is equivalent to say that P satisfies the Rockland condition at every point. In particular, P is hypoelliptic with gain of $\frac{v}{2}$-derivatives by Theorem 3.3.1.

Since P is bounded from below it generates on $L^2(M, \mathcal{E})$ a heat semigroup e^{-tP}, $t \geq 0$. In fact, the hypoellipticity implies for $t > 0$ the operator e^{-tP} is smoothing, i.e., is given a smooth kernel $k_t(x,y)$ in $C^\infty(M \times M, \mathcal{E} \boxtimes (\mathcal{E} \otimes |\Lambda|(M)))$, where $|\Lambda|(M)$ denotes the bundle of densities on M.

On the other hand, the heat semi-group allows us to invert the heat equation. Indeed, the operator given by

$$(5.1.1) \qquad Q_0 f(x,t) = \int_0^\infty e^{-sP} f(x, t-s) ds, \qquad f \in C_c^\infty(M \times \mathbb{R}, \mathcal{E}),$$

maps continuously $C_c^\infty(M \times \mathbb{R}, \mathcal{E})$ into $C^0(\mathbb{R}, L^2(M, \mathcal{E})) \subset \mathcal{D}(M \times \mathbb{R}, \mathcal{E})$ and satisfies

$$(5.1.2) \qquad (P + \partial_t) Q_0 f = Q_0 (P + \partial_t) f = f \qquad \forall f \in C_c^\infty(M \times \mathbb{R}, \mathcal{E}).$$

Notice that the operator Q_0 has the *Volterra property* of [**Pi**], i.e., it is translation invariant and satisfies the causality principle with respect to the time variable, or equivalently, has a distribution kernel of the form $K_{Q_0}(x, y, t-s)$ with $K_{Q_0}(x, y, t)$ supported outside the region $\{t < 0\}$. Indeed, at the level of distribution kernels the formula (5.1.1) implies that we have

$$(5.1.3) \qquad K_{Q_0}(x,y,t) = \begin{cases} k_t(x,y) & \text{if } t > 0, \\ 0 & \text{if } t < 0. \end{cases}$$

The above equalities are the main motivation for using pseudodifferential operators to study the heat kernel $k_t(x,y)$. The idea is to consider a class of Ψ_HDOs, the Volterra Ψ_HDOs, taking into account:

(i) The aforementioned Volterra property;

(ii) The parabolic homogeneity of the heat operator $P + \partial_t$, i.e., the homogeneity with respect to the dilations of $\mathbb{R}^{d+2} = \mathbb{R}^{d+1} \times \mathbb{R}$ defined by

$$(5.1.4) \qquad \lambda.(\xi, \tau) = (\lambda.\xi, \lambda^v \tau), \qquad \lambda \neq 0.$$

In the sequel for $g \in \mathcal{S}'(\mathbb{R}^{d+2})$ and $\lambda \neq 0$ we let g_λ denote the element of $\mathcal{S}'(\mathbb{R}^{d+2})$ such that

$$(5.1.5) \quad \langle g_\lambda(\xi, \tau), f(\xi, \tau) \rangle = |\lambda|^{-(d+2+v)} \langle g(\xi, \tau), f(\lambda^{-1} \xi, \lambda^{-v} \tau) \rangle, \quad f \in \mathcal{S}(\mathbb{R}^{d+2}).$$

DEFINITION 5.1.1. *A distribution $g \in \mathcal{S}'(\mathbb{R}^{d+2})$ is parabolic homogeneous of degree m, $m \in \mathbb{Z}$, when we have $g_\lambda = \lambda^m g$ for any $\lambda \neq 0$.*

Let $\mathbb{C}_-$ denote the complex halfplane $\{\Im \tau < 0\}$ with closure $\overline{\mathbb{C}_-} \subset \mathbb{C}$. Then we define Volterra symbols and Volterra Ψ_HDOs as follows.

DEFINITION 5.1.2. *The space $S_{v,m}(\mathbb{R}_{(v)}^{d+2})$, $m \in \mathbb{Z}$, consists of functions $q(\xi, \tau)$ in $C^\infty(\mathbb{R}^{d+2} \setminus 0)$ such that:*

(i) $q(\lambda.\xi, \lambda^v \tau) = \lambda^m q(x, \xi, \tau)$ for any $\lambda \neq 0$;

(ii) $q(\xi, \tau)$ extends to a function in $C^0((\mathbb{R}^{d+1} \times \overline{\mathbb{C}_-}) \setminus 0)$ whose restriction to $\mathbb{R}^{d+1} \times \mathbb{C}_-$ belongs to $C^\infty(\mathbb{R}^{d+1}) \hat{\otimes} \operatorname{Hol}(\mathbb{C}_-)$.

We also endow $S_{v,m}(\mathbb{R}_{(v)}^{d+2})$ with the Fréchet space topology inherited from that of $C^\infty(\mathbb{R}^{d+2} \setminus 0) \cap C^0((\mathbb{R}^{d+1} \times \overline{\mathbb{C}_-}) \setminus 0) \cap [C^\infty(\mathbb{R}^{d+1}) \hat{\otimes} \operatorname{Hol}(\mathbb{C}_-)]$.

The interest of the above definition stems from:

LEMMA 5.1.3 ([**BGS**, Prop. 1.9]). *Let $q(\xi, \tau) \in S_{v,m}(\mathbb{R}_{(v)}^{d+2})$. Then there exists a unique distribution $g \in \mathcal{S}'(\mathbb{R}^{d+2})$ agreeing with q on $\mathbb{R}^{d+2} \setminus 0$ such that:*

(ii) g is parabolic homogeneous of degree m;

(iii) The inverse Fourier transform $\check{g}(x,t)$ vanishes for $t < 0$.

5.1. PSEUDODIFFERENTIAL REPRESENTATION OF THE HEAT KERNEL

Let U be an open subset of $\mathbb{R}^{d+1}$ together with a hyperplane bundle $H \subset TU$ and H-frame $X_0, \ldots, X_d$ of TU.

DEFINITION 5.1.4. $S_{v,m}(U \times \mathbb{R}^{d+2}_{(v)})$, $m \in \mathbb{Z}$, consists of functions $q(x, \xi, \tau)$ in $C^\infty(U \times (\mathbb{R}^{d+2} \setminus 0))$ such that:

(i) $q(x, \lambda.\xi, \lambda^v \tau) = \lambda^m q(x, \xi, \tau)$ for any $\lambda \neq 0$;

(ii) $q(x, \xi, \tau)$ extends to an element of $C^\infty(U) \hat\otimes C^0((\mathbb{R}^{d+1} \times \overline{\mathbb{C}_-}) \setminus 0)$ whose restriction to $U \times \mathbb{R}^{d+1} \times \mathbb{C}_-$ belongs to $C^\infty(U \times \mathbb{R}^{d+1}) \hat\otimes \operatorname{Hol}(\mathbb{C}_-)$.

DEFINITION 5.1.5. $S_v^m(U \times \mathbb{R}^{d+2}_{(v)})$, $m \in \mathbb{Z}$, consists of functions $q(x, \xi, \tau)$ in $C^\infty(U \times \mathbb{R}^{d+2})$ with an asymptotic expansion $q \sim \sum_{j \geq 0} q_{m-j}$ with q_{m-j} in $S_{v,m-j}(U \times \mathbb{R}^{d+2}_{(v)})$ and $\sim$ taken in the sense that, for any integer N and any compact $K \subset U$, we have

$$(5.1.6) \quad |\partial_x^\alpha \partial_\xi^\beta \partial_\tau^k (q - \sum_{j < N} q_{m-j})(x, \xi, \tau)| \leq C_{NK\alpha\beta k}(\|\xi\| + |\tau|^{1/v})^{m-N-\langle\beta\rangle-vk},$$

for $x \in K$ and $|\xi| + |\tau|^{\frac{1}{v}} > 1$.

DEFINITION 5.1.6. Let $q(x, \xi, \tau) \in S_{v,m}(U \times \mathbb{R}^{d+2}_{(v)})$ and let g be the distribution in $C^\infty(U) \hat\otimes \mathcal{S}'(\mathbb{R}^{d+1})$ be the unique homogeneous extension of q provided by Lemma 5.1.3. Then we let $\check{q}(x, y, t)$ denote the inverse Fourier transform of $g(x, \xi, \tau)$ with respect to the variables (ξ, τ).

REMARK 5.1.7. The above definition makes sense since it follows from the proof of Lemma 5.1.3 in [**BGS**] that the extension process of Lemma 5.1.3 applied to every symbol $q(x,.,.)$, $x \in U$, is smooth with respect to x, so really gives rise to an element of $C^\infty(U) \hat\otimes \mathcal{S}'(\mathbb{R}^{d+2})$.

DEFINITION 5.1.8. $\Psi_{H,v}^m(U \times \mathbb{R}_{(v)})$, $m \in \mathbb{Z}$, consists of continuous operators $Q: C_c^\infty(U_x \times \mathbb{R}_t) \to C^\infty(U_x \times \mathbb{R}_t)$ such that Q has the Volterra property and can be put into the form

$$(5.1.7) \qquad Q = q(x, -iX, D_t) + R,$$

with q in $S_v^m(U \times \mathbb{R}^{d+2}_{(v)})$ and R in $\Psi^{-\infty}(U \times \mathbb{R})$.

REMARK 5.1.9. It is immediate to extend the properties of Ψ_HDOs on U alluded to in Section 3.1 to Volterra Ψ_HDOs on $U \times \mathbb{R}$ except for the asymptotic completeness as in Lemma 3.1.7, which is crucial for carrying out the standard parametrix construction. The issue is that the cut-off arguments of the classical proof the asymptotic completeness of standard ΨDOs cannot be carried through in Volterra setting because we require analyticity with respect to the time covariable. A proof of the asymptotic completeness of Volterra ΨDOs is given in [**Pi**], but simpler proofs which can be carried out *verbatim* for Volterra Ψ_HDOs can be found in [**Po5**].

Let $a \in U$. Then, as for Heisenberg symbols, the convolution on the groups $G^{(a)} \times \mathbb{R}$, gives rise to a continuous bilinear product,

$$(5.1.8) \qquad *^{(a)} : S_{v,m_1}(\mathbb{R}^{d+2}_{(v)}) \times S_{v,m_2}(\mathbb{R}^{d+2}_{(v)}) \to S_{v,m_1+m_2}(\mathbb{R}^{d+2}_{(v)}).$$

Here again $*^a$ depends smoothly on a and so we get a bilinear product,

(5.1.9) $\quad S_{v,m_1}(U \times \mathbb{R}^{d+2}_{(v)}) \times S_{v,m_2}(U \times \mathbb{R}^{d+2}_{(v)}) \to S_{v,m_1+m_2}(U \times \mathbb{R}^{d+2}_{(v)}),$

(5.1.10) $\quad q_1 * q_2(x,\xi,\tau) = (q_1(x,.,.) *^x q_2(x,.,.))(\xi,\tau), \quad q_j \in S_{v,m_j}(U \times \mathbb{R}^{d+2}_{(v)}).$

PROPOSITION 5.1.10. *For $j = 1, 2$ let $Q_j \in \Psi^{m_j}_{H,v}(U \times \mathbb{R}_{(v)})$ and assume that one of these operators is properly supported with respect to the space variable x. Then $Q_1 Q_2$ belongs to $\Psi^{m_j}_{H,v}(U \times \mathbb{R}_{(v)})$ and if $q_j \sim \sum_{k \geq 0} q_{j,m-k}$ denote the symbol of Q_j then $Q_1 Q_2$ has symbol $q \sim \sum_{k \geq 0} q_{m_1+m_2-k}$ where, using the notation of Proposition 3.1.9, we have*

$$(5.1.11) \quad q_{m_1+m_2-k} = \sum_{k_1+k_2 \leq k} \sum_{\alpha,\beta,\gamma,\delta}^{(k-k_1-k_2)} h_{\alpha\beta\gamma\delta}(D^\delta_\xi q_{1,m_1-k_1}) * (\xi^\gamma \partial^\alpha_x \partial^\beta_\xi q_{2,m_2-k_2}).$$

REMARK 5.1.11. Since $G^{(a)} \times \mathbb{R}$ is Abelian with respect to the time variable, the product $*^a$ is merely the pointwise product with respect to τ, e.g., we have $\tau * q = \tau \dot{*} q = \tau q$ for any $q \in S_{v,m}(U \times \mathbb{R}^{d+2}_{(v)})$. In particular, the Volterra-Heisenberg calculus, while not microlocal with respect to the space variable, is to a large extent microlocal with respect to the time variable.

On the other hand, thanks to the Volterra property the kernels of Ψ_HDOs can be characterized as follows.

DEFINITION 5.1.12. $\mathcal{K}_{v,m}(U \times \mathbb{R}^{d+2}_{(v)})$, $m \in \mathbb{Z}$, consists of distributions $K(x,y,t)$ in $C^\infty(U) \hat{\otimes} \mathcal{S}'_{reg}(\mathbb{R}^{d+2})$ such that:

(i) *The support of $K(x,y,t)$ is contained in $U \times \mathbb{R}^{d+1} \times \mathbb{R}_+$;*

(ii) $K(x, \lambda.y, \lambda^v t) = (\text{sign } \lambda)^d \lambda^m K(x,y,t)$ *for any $\lambda \in \mathbb{R} \setminus 0$.*

DEFINITION 5.1.13. $\mathcal{K}^m_v(U \times \mathbb{R}^{d+2}_{(v)})$, $m \in \mathbb{Z}$, is the space of distributions $K(x,y,t)$ in $\mathcal{D}'(U \times \mathbb{R}^{d+2})$ which admit an asymptotic expansion $K \sim \sum_{j \geq 0} K_{m+j}$ with K_{m+j} in $\mathcal{K}_{v,m+j}(U \times \mathbb{R}^{d+2})$ and $\sim$ taken in the sense of (3.1.24).

In the sequel, for $x \in U$ we let ψ_x and ε_x respectively denote the changes of variable to the privileged coordinates and to the Heisenberg coordinates at x. Then, along the same lines as that of the proofs of Proposition 3.1.15 and Proposition 3.1.16, we obtain the following characterization of Volterra Ψ_HDOs.

PROPOSITION 5.1.14. *Let $Q : C^\infty_c(U_x \times \mathbb{R}_t) \to C^\infty(U_x \times \mathbb{R}_t)$ be a continuous operator with distribution kernel $k_Q(x,t;y,s)$. Then the following are equivalent:*

(i) *The operator Q belongs to $\Psi^m_{H,v}(U \times \mathbb{R})$;*

(ii) *The kernel of Q can be put into the form,*

(5.1.12) $\quad k_Q(x,t;y,s) = |\psi'_x| K(x, -\psi_x(y), t-s) + R(x,y,t-s),$

with K in $\mathcal{K}^{\hat{m}}_v(U \times \mathbb{R}^{d+2}_{(v)})$, $\hat{m} = -(m+d+2+v)$, and R in $C^\infty(U \times \mathbb{R}^{d+2})$.

(iii) *The kernel of Q can be put into the form,*

(5.1.13) $\quad k_Q(x,t;y,s) = |\varepsilon'_x| K_Q(x, -\varepsilon_x(y), t-s) + R(x,y,t-s),$

with K_Q in $\mathcal{K}^{\hat{m}}_v(U \times \mathbb{R}^{d+2}_{(v)})$, $\hat{m} = -(m+d+2+v)$, and R in $C^\infty(U \times \mathbb{R}^{d+2})$.

An interesting consequence of Proposition 5.1.14 is the following small time asymptotics for the kernel of a Volterra Ψ_HDO.

5.1. PSEUDODIFFERENTIAL REPRESENTATION OF THE HEAT KERNEL

PROPOSITION 5.1.15 ([**BGS**, Thm. 4.5]). *Let* $Q \in \Psi_{H,\mathrm{v}}^m(U \times \mathbb{R}_{(v)})$ *have symbol* $q \sim \sum_{j \geq 0} q_{m-j}$ *and kernel* $k_Q(x, y, t-s)$. *Then as* $t \to 0^+$ *the following asymptotics holds in* $C^\infty(U)$,

$$k_Q(x, x, t) \sim t^{-\frac{2[\frac{m}{2}]+d+4}{v}} \sum_{j \geq 0} t^j |\varepsilon_x'| \check{q}_{2[\frac{m}{2}]-2j}(x, 0, 1). \tag{5.1.14}$$

Let $\tilde{U}$ be an open subset of $\mathbb{R}^{d+1}$ together with a hyperplane bundle $\tilde{H} \subset T\tilde{U}$ and a $\tilde{H}$-frame of $T\tilde{U}$ and let $\phi : (U, H) \to (\tilde{U}, \tilde{H})$ be a Heisenberg diffeomorphism. Then using Proposition 5.1.14 and arguing along similar lines that of the proof of Proposition 3.1.18 allows us to prove:

PROPOSITION 5.1.16. *Let* $\tilde{Q} \in \Psi_{\tilde{H},\mathrm{v}}^m(\tilde{U} \times \mathbb{R}_{(v)})$ *and set* $Q = (\phi \oplus 1_\mathbb{R})^* \tilde{Q}$.

1) The operator Q *belongs to* $\Psi_{H,\mathrm{v}}^m(U \times \mathbb{R}_{(v)})$.

2) If the distribution kernel of $\tilde{P}$ *is of the form (5.1.13) with* $K_{\tilde{Q}}(\tilde{x}, \tilde{y}, t)$ *in* $\mathcal{K}_{\mathrm{v}}^{\hat{m}}(\tilde{U} \times \mathbb{R}_{(v)}^{d+2})$ *then the distribution kernel of* P *can be written in the form (5.1.13) with* $K_Q(x, y, t)$ *in* $\mathcal{K}_{\mathrm{v}}^{\hat{m}}(U \times \mathbb{R}_{(v)}^{d+2})$ *such that*

$$K_Q(x, y, t) \sim \sum_{\langle \beta \rangle \geq \frac{3}{2}\langle \alpha \rangle} \frac{1}{\alpha! \beta!} a_{\alpha\beta}(x) y^\beta (\partial_{\tilde{y}}^\beta K_{\tilde{Q}})(\phi(x), \phi_H'(x)y, t), \tag{5.1.15}$$

where the functions $a_{\alpha\beta}(x)$ *are as in (3.1.36).*

This allows us to define Volterra Ψ_HDOs on the manifold $M \times \mathbb{R}$ and acting on the sections of the bundle $\mathcal{E}$ (or rather on the sections of the pullback of $\mathcal{E}$ by the projection $M \times \mathbb{R} \to M$, again denoted $\mathcal{E}$).

DEFINITION 5.1.17. $\Psi_{H,\mathrm{v}}^m(M \times \mathbb{R}_{(v)}, \mathcal{E})$, $m \in \mathbb{Z}$, *consists of continuous operators* $Q : C_c^\infty(M \times \mathbb{R}, \mathcal{E}) \to C^\infty(M \times \mathbb{R}, \mathcal{E})$ *such that:*

(i) Q *has the Volterra property;*

(ii) The distribution kernel of Q *is smooth off the diagonal of* $(M \times \mathbb{R}) \times (M \times \mathbb{R})$;

(iii) For any trivialization $\tau : \mathcal{E}_{|U} \to U \times \mathbb{C}^r$ *of* $\mathcal{E}$ *over a local Heisenberg chart* $\kappa : U \to V \subset \mathbb{R}^{d+1}$ *the operator* $(\kappa \otimes \mathrm{id})_* \tau_*(Q_{|U \times \mathbb{R}})$ *belongs to* $\Psi_{H,\mathrm{v}}^m(V \times \mathbb{R}_{(v)}, \mathbb{C}^r) := \Psi_{H,\mathrm{v}}^m(V \times \mathbb{R}_{(v)}) \otimes \mathrm{End}\,\mathbb{C}^r$.

Using Proposition 5.1.16 we can define the global principal symbol of a Volterra Ψ_HDO as follows. Let $\mathfrak{g}^*M$ denote the dual bundle of the Lie algebra bundle $\mathfrak{g}M$ of M and consider the canonical projection $\pi : \mathfrak{g}^*M \times \mathbb{R} \to M$.

In the sequel, depending on the context, 0 denotes either the zero section of $\mathfrak{g}^*M$ or the zero section of $\mathfrak{g}^*M$ crossed with $\{0\} \subset \bar{\mathbb{C}}_-$.

DEFINITION 5.1.18. $S_{\mathrm{v},m}(\mathfrak{g}^*M \times \mathbb{R}_{(v)}, \mathcal{E})$, $m \in \mathbb{Z}$, *consists of sections* $q(x, \xi, \tau)$ *in* $C^\infty((\mathfrak{g}^*M \times \mathbb{R}) \setminus 0, \pi^* \mathrm{End}\,\mathcal{E})$ *such that:*

(i) $q(x, \lambda.\xi, \lambda^v \tau) = \lambda^m q(x, \xi, \tau)$ *for any* $\lambda \in \mathbb{R} \setminus 0$;

(ii) $q(x, \xi, \tau)$ *extends to a section of* $\pi^* \mathrm{End}\,\mathcal{E}$ *over* $(\mathfrak{g}^*M \times \bar{\mathbb{C}}_-) \setminus 0$ *which is smooth with respect to the base space variable and continous with respect to the others and which restricts on* $\mathfrak{g}^*M \times \mathbb{C}_-$ *to an element of* $C^\infty(\mathfrak{g}^*M, \pi^* \mathrm{End}\,\mathcal{E}) \hat{\otimes} \mathrm{Hol}(\mathbb{C}_-)$.

Using (5.1.15) and arguing as in the proof of Proposition 3.2.2 we get:

PROPOSITION 5.1.19. *For any $Q \in \Psi_{H,v}^m(M \times \mathbb{R}_{(v)}, \mathcal{E})$ there is a unique symbol $\sigma_m(Q) \in S_{v,m}(\mathfrak{g}^*M \times \mathbb{R}_{(v)}, \mathcal{E})$ such that, if in a local trivializing Heisenberg chart $U \subset \mathbb{R}^{d+1}$ we let $K_{Q,\hat{m}} \in \mathcal{K}_{\hat{m}}(U \times \mathbb{R}^{d+1})$ be the leading kernel for the kernel K_Q in the form (5.1.13) for Q, then for $(x, \xi, \tau) \in U \times [(\mathbb{R}^{d+1} \times \mathbb{R}) \setminus 0]$ we have*

(5.1.16) $$\sigma_m(Q)(x, \xi, \tau) = [K_{Q,\hat{m}}]^{\wedge}_{(y,t) \to (\xi,\tau)}(x, \xi, \tau).$$

Equivalently, on any trivializing Heisenberg coordinates centered at $a \in M$ the symbol $\sigma_m(Q)(a, ., .)$ coincides with the (local) principal symbol of Q at $x = 0$.

DEFINITION 5.1.20. *For $Q \in \Psi_{H,v}^m(M \times \mathbb{R}_{(v)}, \mathcal{E})$ the symbol $\sigma_m(Q)(x, \xi, \tau)$ provided by Proposition 5.1.19 is called the (global) principal symbol of Q.*

Extending Definition 5.1.6 to Volterra-Heisenberg symbols on $M \times \mathbb{R}$ we can define the model operator of a Volterra Ψ_HDO as follows.

DEFINITION 5.1.21. *Let $Q \in \Psi_{H,v}^m(M \times \mathbb{R}_{(v)}, \mathcal{E})$ have principal symbol $\sigma_m(Q)$ and let $a \in M$. Then the model operator Q^a of Q at a is the left-convolution operator by $\sigma_m(Q)^{\vee}(a, ., .)$, that is, Q^a is the continuous endomorphism of $\mathcal{S}(G_aM \times \mathbb{R}, \mathcal{E}_a)$ such that, for any $f \in \mathcal{S}(G_aM, \mathcal{E}_a)$, we have*

(5.1.17) $$Q^a f(x, t) = \langle \sigma_m(Q)^{\vee}(a, y, t), f(x.y^{-1}, t - s) \rangle.$$

REMARK 5.1.22. *The model operator Q^a can be defined as an endomorphism of $\mathcal{S}(G_aM, \mathcal{E}_a)$, not just as an endomorphism of $\mathcal{S}_0(G_aM, \mathcal{E}_a)$ as in Definition 3.2.7, because $\sigma_m(Q)^{\vee}_{(\xi,\tau) \to (y,t)}(a, ., .)$ makes sense as an element of $\mathcal{S}'(G_aM, \mathcal{E}_a)$.*

PROPOSITION 5.1.23. *The group law on the fibers of $GM \times \mathbb{R}$ gives rise to a convolution product $*$ from $S_{v,m_1}(\mathfrak{g}^*M \times \mathbb{R}_{(v)}, \mathcal{E}) \times S_{v,m_2}(\mathfrak{g}^*M \times \mathbb{R}_{(v)}, \mathcal{E})$ to $S_{v,m_1+m_2}(\mathfrak{g}^*M \times \mathbb{R}_{(v)}, \mathcal{E})$ such that, for $q_{m_j} \in S_{v,m_j}(\mathfrak{g}^*M \times \mathbb{R}_{(v)}, \mathcal{E})$, we have*

(5.1.18) $$q_1 * q_2(x, \xi, \tau) = [q_1(x, ., .) *^x q_2(x, ., .)](\xi, \tau),$$

where $*^x$ denote the convolution product for symbols on $G_xM \times \mathbb{R}$.

In a local trivializing Heisenberg chart the symbolic calculus for Volterra Ψ_HDOs reduces the existence of a Volterra Ψ_HDO parametrix to the invertibility of the local and global principal symbols. Therefore, we obtain:

PROPOSITION 5.1.24. *Let $Q \in \Psi_{H,v}^m(M \times \mathbb{R}_{(v)}, \mathcal{E})$, $m \in \mathbb{Z}$. Then we have equivalence:*

(i) The principal symbol of Q is invertible with respect to the product (5.1.18) of Volterra-Heisenberg symbols;

(ii) The operator Q admits a parametrix in $\Psi_{H,v}^{-m}(M \times \mathbb{R}_{(v)}, \mathcal{E})$.

In the case of the heat operator $P + \partial_t$, comparing a parametrix with the inverse (5.1.1) and using (5.1.3) allows us to obtain the pseudodifferential representation of the heat kernel of P below.

THEOREM 5.1.25 ([**BGS**, pp. 362–363]). *Suppose that the principal symbol of $P + \partial_t$ is an invertible Volterra-Heisenberg symbols. Then:*

1) The heat operator $P + \partial_t$ has an inverse $(P + \partial_t)^{-1}$ in $\Psi_{H,v}^{-v}(M \times \mathbb{R}_{(v)}, \mathcal{E})$.

2) Let $K_{(P+\partial_t)^{-1}}(x, y, t - s)$ denote the kernel of $(P + \partial_t)^{-1}$. Then the heat kernel $k_t(x, y)$ of P satisfies

(5.1.19) $$k_t(x, y) = K_{(P+\partial_t)^{-1}}(x, y, t) \quad \text{for } t > 0.$$

Combining this with Proposition 5.1.15 then gives the heat kernel asymptotics for P in the form below.

THEOREM 5.1.26 ([**BGS**, Thm. 5.6]). *If the principal symbol of $P + \partial_t$ is an invertible Volterra-Heisenberg symbol, then as $t \to 0^+$ the following asymptotics holds in $C^\infty(M, (\operatorname{End}\mathcal{E}) \otimes |\Lambda|(M))$,*

$$(5.1.20) \quad k_t(x,x) \sim t^{-\frac{d+2}{v}} \sum t^{\frac{2j}{v}} a_j(P)(x), \quad a_j(P)(x) = |\varepsilon'_x|(q_{-v-2j})^\vee(x,0,1),$$

where the equality on the right shows how to compute $a_j(P)(x)$ in a local trivializing Heisenberg chart by means of the symbol $q_{-v-2j}(x,\xi,\tau)$ of degree $-v - 2j$ of any parametrix of $P + \partial_t$ in $\Psi^{-v}_{H,\mathrm{v}}(M \times \mathbb{R}_{(v)}, \mathcal{E})$.

5.2. Heat equation and sublaplacians

In this section we specialize the results of the previous sections to sublaplacians and their integer powers. In particular, we complete the treatment in [**BGS**] of the heat kernel of a sublaplacian.

Throughout this section we let k be an integer ≥ 1 and we set $v = 2k$. In order to deal with sublaplacians it will be convenient to enlarge the definition of homogeneous Volterra-Heisenberg symbols as follows.

Let $U \subset \mathbb{R}^{d+1}$ be a Heisenberg chart with H-frame $X_0, \ldots, X_d$ and let Θ be an open angular sector whose closure contains $\mathbb{R}$ and which is contained in $\mathbb{C} \setminus i[0, \infty)$.

DEFINITION 5.2.1. *For $m \in \mathbb{Z}$ and $N \in \mathbb{N} \cup \{\infty\}$ we let $S_{\mathrm{v},m}(U \times \mathbb{R}^{d+1} \times \Theta^N_{(v)})$ be the space of functions $q(x, \xi, \tau)$ on $U \times (\mathbb{R}^{d+2} \setminus 0)$ such that:*

(i) $q(x, \xi, \tau)$ is C^∞ on $U \times (\mathbb{R}^{d+2} \setminus 0)$ and near the region $\{\tau = 0\}$ is C^∞ with respect to x and ξ and C^N with respect to τ;

(ii) We have $q(\lambda.\xi, \lambda^v \tau) = \lambda^m q(\xi, \tau)$ for any $\lambda \in \mathbb{R} \setminus 0$;

(iii) $q(x, \xi, \tau)$ extends to a continuous function on $U \times [(\mathbb{R}^{d+1} \times \overline{\Theta}) \setminus 0]$ whose restriction to $U \times \mathbb{R}^{d+1} \times \Theta$ belongs to $C^\infty(\mathbb{R}^{d+1}) \hat\otimes \operatorname{Hol}(\Theta)$.

In particular, when $\Theta = \mathbb{C}_-$ and $N = \infty$ we recover the class $S_{\mathrm{v},m}(U \times \mathbb{R}^{d+2}_{(v)})$ defined in the previous section.

As alluded to in Remark 5.1.11 the Volterra-Heisenberg calculus is microlocal with the respect to the time variable. As we shall now see this allow us to extend the product for homogeneous Volterra-Heisenberg symbols to the symbols in the class $S^N_{\mathrm{v},m}(U \times \mathbb{R}^{d+1} \times \Theta^N_{(v)})$.

DEFINITION 5.2.2. $S^{-\infty}(U \times \mathbb{R}^{d+1} \times \Theta^N)$ *consists of functions $q(x, \xi, \tau)$ on $U \times \mathbb{R}^{d+2}$ such that:*

(i) $q(x, \xi, \tau)$ is smooth on $U \times \mathbb{R}^{d+1} \times (\mathbb{R} \setminus 0)$ and for any integer N', any compact $K \subset U$ and any $c > 0$ we have estimates,

$$(5.2.1) \quad |\partial_x^\alpha \partial_\xi^\beta \partial_\tau^k q(x, \xi, \tau)| \leq C_{N'K c \alpha \beta k}(1 + |\xi| + |\tau|)^{-N'}, \quad (x, \xi) \in K \times \mathbb{R}^{d+1}, |\tau| > c,$$

with α, β and k arbitrary.

(ii) $q(x, \xi, \tau)$ is smooth with respect to x and ξ and is C^N with respect to τ near $U \times \mathbb{R}^{d+1} \times 0$ and for any integer N', any compact $K \subset U$ we have estimates,

$$(5.2.2) \quad |\partial_x^\alpha \partial_\xi^\beta \partial_\tau^k q(x, \xi, \tau)| \leq C_{N'K\alpha\beta k}(1 + |\xi| + |\tau|)^{-N'}, \quad (x, \xi, \tau) \in K \times \mathbb{R}^{d+2},$$

with α and β arbitrary and with $k \leq N$.

(iii) $q(x,\xi,\tau)$ extends to a continuous function on $U \times \mathbb{R}^{d+1} \times \overline{\Theta}$ whose restriction to $U \times \mathbb{R}^{d+1} \times \Theta$ belongs to $C^{\infty}(U \times \mathbb{R}^{d+1}) \hat{\otimes} \mathrm{Hol}(\Theta)$ and such that, for any integer N' and any compact $K \subset U$, we have estimates,

(5.2.3) $\quad |\partial_x^{\alpha} \partial_{\xi}^{\beta} q(x,\xi,\tau)| \leq C_{N'K\alpha\beta}(1 + |\xi| + |\tau|)^{-N'}, \quad (x,\xi,\tau) \in K \times \mathbb{R}^{d+1} \times \overline{\Theta},$

with α and β arbitrary.

DEFINITION 5.2.3. $S_{ah,v}^m(U \times \mathbb{R}^{d+1} \times \Theta^N)$, $m \in \mathbb{Z}$, consists of functions $q(x,\xi,\tau)$ on $U \times \mathbb{R}^{d+2}$ such that:

(i) $q(x,\xi,\tau)$ is smooth on $U \times (\mathbb{R}^{d+2} \setminus 0)$ and near the region $U \times (\mathbb{R}^{d+1} \setminus 0) \times 0$ it is smooth with respect to x and ξ and is C^N with respect to τ;

(ii) $q(x,\xi,\tau)$ extends to a continuous function on $U \times [(\mathbb{R}^{d+1} \times \overline{\Theta}) \setminus 0]$ whose restriction to $U \times \mathbb{R}^{d+1} \times \Theta$ belongs to $C^{\infty}(U \times \mathbb{R}^{d+1}) \hat{\otimes} \mathrm{Hol}(\Theta)$.

(iii) For any $\lambda \in \mathbb{R} \setminus 0$ the function $q(x, \lambda.\xi, \lambda^v \tau) - \lambda^m q(x,\xi,\tau)$ belongs to $S^{-\infty}(U \times \mathbb{R}^{d+1} \times \Theta^N)$.

Along similar lines as that of the proof of Lemma 4.3.2 and of the proof of [**Po5**, Prop. 3.3] we obtain:

LEMMA 5.2.4. *1) Let $q \in S_{ah,v}^m(U \times \mathbb{R}^{d+1} \times \Theta^N)$. Then q admits a unique homogeneous part, i.e., there is a unique symbol $q(x,\xi,\tau)$ in $S_{v,m}(U \times \mathbb{R}^{d+1} \times \Theta_{(v)}^N)$ such that on $U \times (\mathbb{R}^{d+2} \setminus 0)$ we have*

(5.2.4) $\quad q_m(x,\xi,\tau) = \lim_{\lambda \to 0} \lambda^{-m} q(x, \lambda.\xi, \lambda^v \tau).$

2) Let $q_m \in S_{v,m}(U \times \mathbb{R}^{d+1} \times \Theta_{(v)}^N)$. Then there exists $q \in S_{ah,v}^m(U \times \mathbb{R}^{d+1} \times \Theta^N)$ with homogeneous part q_m. Moreover q is unique up to the addition of a symbol in $S^{-\infty}(U \times \mathbb{R}^{d+1} \times \Theta^N)$.

In particular, this lemma implies that (5.2.4) gives rise to a linear isomorphism,

(5.2.5) $\quad S_{v,m}(U \times \mathbb{R}^{d+1} \times \Theta_{(v)}^N) \simeq S_{ah,v}^m(U \times \mathbb{R}^{d+1} \times \Theta^N)/S^{-\infty}(U \times \mathbb{R}^{d+1} \times \Theta^N).$

Let $q \in S_{ah,v}^m(U \times \mathbb{R}^{d+1} \times \Theta^N)$. For any compact $K \subset U$ we have estimates,

(5.2.6) $\quad |\partial_x^{\alpha} \partial_{\xi}^{\beta} q(x,\xi,\tau)| \leq C_{K\alpha\beta}(1 + \|\xi\| + |\tau|^v), \quad (x,\xi,\tau) K \times \mathbb{R}^{d+1} \times \overline{\Theta},$

Combining this with the inequalities,

(5.2.7) $\quad (1+|\tau|^v)^{\frac{1}{2}}(1+\|\xi\|)^{\frac{1}{2}} \leq 1 + \|\xi\| + |\tau|^v \leq (1+\|\xi\|)(1+|\tau|^v),$

(5.2.8) $\quad (1+\|\xi\|+|\tau|^v)^m \leq (1+\|\xi\|)^m(1+|\tau|^v)^{|m|},$

the latter being the Peetre's inequality, we see that we can regard $(q(.,.,\tau))_{\tau \in \overline{\Theta}}$ as a continuous family with values in $S_{\|}^m(U \times \mathbb{R}^{d+1})$ which is a $O(|\tau|^{|m|})$ in this Fréchet space as $|\tau| \to \infty$.

In fact, this family is also holomorphic on Θ, smooth on $\mathbb{R} \setminus 0$ and C^N near $\tau = 0$ and its τ-derivatives too are $O(|\tau|^{|m|})$ in $S_{\|}^m(U \times \mathbb{R}^{d+1})$ as τ becomes large.

Recall that the convolution products on the groups $G^{(x)}$, $x \in U$, give rise to a smooth family of bilinear products,

(5.2.9) $\quad *^{(x)} : S_{\|}^{m_1}(\mathbb{R}^{d+1}) \times S_{\|}^{m_2}(\mathbb{R}^{d+1}) \longrightarrow S_{\|}^{m_1+m_2}(\mathbb{R}^{d+1}).$

Therefore, if $q_j \in S_{\text{ah,v}}^{m_j}(U \times \mathbb{R}^{d+1} \times \Theta^N)$ then we define a family with values in $S_{\text{ah,v}}^{m_1+m_2}(U \times \mathbb{R}^{d+1} \times \Theta^N)$ by letting

$$(5.2.10) \quad q_1 * q_2(x, \xi, \tau) = (q_1(x, ., \tau) *^{(x)} q_2(x, ., \tau))(\xi), \quad (x, \xi, \tau) \in U \times \mathbb{R}^{d+1} \times \overline{\Theta}.$$

This family is continuous on $\overline{\Theta}$, is holomorphic on Θ, is smooth on $\mathbb{R} \setminus 0$ and is C^N near $\tau = 0$. Moreover, along with its derivatives it is a $O(|\tau|^{|m|})$ in $S_\parallel^m(U \times \mathbb{R}^{d+1})$ as τ becomes large.

On the other hand, it also follows from the inequalities (5.2.7) that if q is a symbol in $S^{-\infty}(U \times \mathbb{R}^{d+1} \times \Theta_{(v)}^N)$ then we can regard $(q(., ., \tau))_{\tau \in \overline{\Theta}}$ as a continuous family with values in $S^{-\infty}(U \times \mathbb{R}^{d+1})$ which is holomorphic on Θ, is smooth on $\mathbb{R} \setminus 0$, is C^N near $\tau = 0$ and which, for any integer N', together with its derivatives is a $O(|\tau|^{-N'})$ in $S^{-\infty}(U \times \mathbb{R}^{d+1})$ as τ becomes large. Therefore, if in (5.2.10) we replace q_1 or q_2 by an element in $S^{-\infty}(U \times \mathbb{R}^{d+1} \times \Theta_{(v)}^N)$ then the resulting symbol belongs to $S^{-\infty}(U \times \mathbb{R}^{d+1} \times \Theta_{(v)}^N)$.

Now, for $j = 1, 2$ let $q_j \in S_{\text{ah,v}}^{m_j}(U \times \mathbb{R}^{d+1} \times \Theta^N)$. Then thanks to (4.3.7) for any $\lambda \in \mathbb{R} \setminus 0$ the symbol $q_1 * q_2(x, \lambda.\xi, \lambda^v \tau) - \lambda^{m_1+m_2} q_1 * q_2(x, \xi, \tau)$ is equal to

$$(5.2.11) \quad [q_1(x, \lambda.\xi, \lambda^v \tau) - \lambda^{m_1} q_1(x, \xi, \tau)] * q_2(x, \lambda.\xi, \lambda^v \tau)$$
$$+ \lambda^{m_1} q_1(x, \xi, \tau) * [q_2(x, \lambda.\xi, \lambda^v \tau) - \lambda^{m_2} q_2(x, \xi, \tau)],$$

and so belongs to $S^{-\infty}(U \times \mathbb{R}^{d+1} \times \Theta_{(v)}^N)$ by the observations above. This shows that $q_1 * q_2$ belongs to $S_{\text{ah,v}}^{m_1+m_2}(U \times \mathbb{R}^{d+1} \times \Theta^N)$. Therefore, the formula (5.2.10) defines a bilinear product $*$ from $S_{\text{ah,v}}^{m_1}(U \times \mathbb{R}^{d+1} \times \Theta^N) \times S_{\text{ah,v}}^{m_2}(U \times \mathbb{R}^{d+1} \times \Theta^N)$ to $S_{\text{ah,v}}^{m_1+m_2}(U \times \mathbb{R}^{d+1} \times \Theta^N)$.

Moreover, if $\tilde{q}_j \in S_{\text{ah,v}}^{m_j}(U \times \mathbb{R}^{d+1} \times \Theta^N)$ has the same homogeneous part as that of q_j then thanks to the equality $q_1 * q_2 - \tilde{q}_1 * \tilde{q}_2 = (q_1 - \tilde{q}_1) * q_1 + q_1 * (q_2 - \tilde{q}_2)$ we see that $q_1 * q_2$ and $\tilde{q}_1 * \tilde{q}_2$ agree up to an element in $S^{-\infty}(U \times \mathbb{R}^{d+1} \times \Theta_{(v)}^N)$ and so have same homogeneous part. Therefore, using the isomorphism (5.2.5) we get:

LEMMA 5.2.5. *The convolution products on the groups $G^{(a)}$, $a \in U$, give rise to a bilinear product $*$ from $S_{v,m_1}(U \times \mathbb{R}^{d+1} \times \Theta_{(v)}^N) \times S_{v,m_2}(U \times \mathbb{R}^{d+1} \times \Theta_{(v)}^N)$ to $S_{v,m_1+m_2}(U \times \mathbb{R}^{d+1} \times \Theta_{(v)}^N)$.*

DEFINITION 5.2.6. *For $m \in \mathbb{Z}$ and $N \in \mathbb{N} \cup \{\infty\}$ the space $S_{v,m}(\mathfrak{g}^*M \times \Theta_{(v)}^N, \mathcal{E})$ consists of sections $q(x, \xi, \tau)$ of $\mathcal{E}$ over $(\mathfrak{g}^*M \times \mathbb{R}) \setminus 0$ such that:*

(i) $q(x, \xi, \tau)$ is C^∞ on $\mathfrak{g}^*M \times (\mathbb{R} \setminus 0)$ and near the region $\mathfrak{g}^*M \times 0$ it is C^∞ with respect to x and ξ and C^N with respect to τ;

(ii) We have $q(\lambda.\xi, \lambda^v \tau) = \lambda^m q(\xi, \tau)$ for any $\lambda \in \mathbb{R} \setminus 0$;

(iii) $q(x, \xi, \tau)$ extends to a section of $\pi^* \operatorname{End} \mathcal{E}$ over $(\mathfrak{g}^*M \times \overline{\Theta}) \setminus 0$ which is smooth with respect to the base space variable and continous with respect to the others and which restricts on $\mathfrak{g}^*M \times \mathbb{C}_-$ to an element of $C^\infty(\mathfrak{g}^*M, \pi^* \operatorname{End} \mathcal{E}) \hat{\otimes} \operatorname{Hol}(\Theta)$.

For instance, if $P : C^\infty(M, \mathcal{E}) \to C^\infty(M, \mathcal{E})$ is a differential operator of Heisenberg order v then the symbol $\sigma_v(P)(x, \xi) + i\tau$ belongs to $S_{v,v}(\mathfrak{g}^* \times \Theta_{(v)}^\infty, \mathcal{E})$ for any open angular sector Θ whose closure containing $\mathbb{R}$.

Along the same lines as that of the proof of Proposition 3.2.8, using (3.2.20) and Lemma 5.2.5 we obtain:

LEMMA 5.2.7. *The convolution products on the tangent groups G_aM, $a \in M$, give rise to a bilinear product $*$ from $S_{v,m_1}(\mathfrak{g}^*M \times \Theta^N_{(v)}, \mathcal{E}) \times S_{v,m_2}(\mathfrak{g}^*M \times \Theta^N_{(v)}, \mathcal{E})$ to $S_{v,m_1+m_2}(\mathfrak{g}^*M \times \Theta^N_{(v)}, \mathcal{E})$.*

In particular, for a symbol $q \in S_{v,m}(\mathfrak{g}^*M \times \Theta^N_{(v)}, \mathcal{E})$ it makes sense to speak about its inverse in $S_{v,-m}(\mathfrak{g}^*M \times \Theta^N_{(v)}, \mathcal{E})$.

Next, let $U \subset \mathbb{R}^{d+1}$ be a trivializing Heisenberg chart together with a H-frame $X_0, \ldots, X_d$ and let $L(x) = (L_{jk}(x))$ be the matrix of the Levi form $\mathcal{L}$ with respect to this H-frame, so that for $j, k = 1, \ldots, d$ we have

(5.2.12) $$\mathcal{L}(X_j, X_k) = [X_j, X_k] = L_{jk}X_0 \mod H.$$

Then we have the following extension of Theorem 5.22 of [**BGS**].

PROPOSITION 5.2.8. *For $\omega \in (-\frac{\pi}{2}, \frac{\pi}{2})$ define*

(5.2.13) $$\Omega_\omega = \{(\mu, x) \in M_r(\mathbb{C}) \setminus U; \; [(\cos \omega)^{-1} \Re(e^{i\omega} \operatorname{Sp} \mu)] \cap \Lambda_x = \emptyset\},$$

where the singular set Λ_x is defined as in (3.4.15)–(3.4.16). Then Ω_ω is an open set and there exists $q_\mu^{(\omega)}(x, \xi, \tau) \in C^\infty(\Omega_\omega, S_{v,-2}(\mathbb{R}^{d+2}_{(2)}, \mathbb{C}^r))$ such that:

(i) $q_\mu^{(\omega)}(x, \xi, \mu)$ depends analytically on μ;

(ii) For any $(\mu, x) \in \Omega$ the symbol $q_\mu^{(\omega)}(x, ., .)$ inverts $|\xi'|^2 + i\mu\xi_0 + ie^{-i\omega}\tau$, i.e.,

(5.2.14) $$q_\mu^{(\omega)}(x, .) *^x (|\xi'|^2 + i\mu\xi_0 + ie^{-i\omega}\tau) = (|\xi'|^2 + i\mu\xi_0 + ie^{-i\omega}\tau) *^x q_\mu^{(\omega)}(x, .) = 1.$$

Moreover, if $(\mu, x) \in \Omega_{\omega_1} \cap \Omega_{\omega_2}$ then we have

(5.2.15) $$q_\mu^{(\omega_1)}(x, \xi, e^{i\omega_1}\tau) = q_\mu^{(\omega_2)}(x, \xi, e^{i\omega_2}\tau),$$

for any ξ in $\mathbb{R}^{d+1}$ and any τ in $e^{-i\omega_1}\mathbb{C}_- \cap e^{-i\omega_2}\mathbb{C}_-$

PROOF. First, in the same way as in the proof of Proposition 3.4.4 we can show that Ω_ω is an open subset of $M_r(\mathbb{C}) \times U$.

Second, let us assume that $r = 1$. For $x \in U$ define

(5.2.16) $$\Lambda_x^0 = (-\infty, -\frac{1}{2}\operatorname{Trace}|L(a)|] \cup [\frac{1}{2}\operatorname{Trace}|L(a)|, \infty)$$

For $k = 1, 2, \ldots$ let $\Lambda_x^k = \Lambda_x^0$ if $\operatorname{rk} \mathcal{L}_x < d$ and otherwise let

(5.2.17) $$\Lambda_x^k = \{\pm(\frac{1}{2}\operatorname{Trace}|L(x)| + 2 \sum_{1 \leq j \leq n} \alpha_j|\lambda_j|); \alpha_j \in \mathbb{N}, \; \alpha_j \leq k\}.$$

Then for $k = 0, 1, \ldots$ we let Ω_ω^k denote the subset of $\mathbb{C} \times U$ defined as in (5.2.13) with $r = 1$ and with Ω_ω replaced by Ω_ω^k. Again Ω_ω^k is open and when $\operatorname{rk} \mathcal{L}_x < d$ for every $x \in U$ we have $\Omega_\omega = \Omega_\omega^0 = \Omega_\omega^k$.

We are now going to prove:

CLAIM. *For $k = 0, 1, \ldots$ there exists $q_\mu^{(\omega)}(x, \xi, \tau)$ in $C^\infty(\Omega_\omega^k, S_{v,-2}(\mathbb{R}^{d+2}_{(2)}))$ satisfying (i) and (ii) on Ω_ω^k.*

For $\omega = 0$ and $k = 0$ the claim follows from [**BGS**, Thm. 5.22] and in this case the symbol $q_\mu^{(0)} \in C^\infty(\Omega_0^0, S_{v,-2}(\mathbb{R}^{d+2}_{(2)}))$ is given by the formula,

(5.2.18) $$q_\mu^{(0)}(x, \xi, \tau) = \int_0^\infty e^{-t\mu\xi_0 - it\tau} G(x, \xi, t) dt,$$

where $G(x,\xi,t)$ is as in (3.4.2).

Similarly, for $\omega \neq 0$ we define a symbol $q_\mu^{(\omega)}(x,\xi,\tau) \in C^\infty(\Omega_\omega^0, S_{\mathrm{v},-2}(\mathbb{R}_{(2)}^{d+2}))$ satisfying (i) and (ii) by letting

$$(5.2.19) \qquad q_\mu^{(0)}(x,\xi,\tau) = \int_{e^{i\omega}(0,\infty)} e^{-t\mu\xi_0 - ite^{-i\omega}\tau} G(x,\xi,t)dt.$$

Notice also that thanks to the analyticity of $G(\xi,t)$ with respect to t, if (μ,x) is in $\Omega_{\omega_1}^0 \cap \Omega_{\omega_2}^0$ then, for any $\xi \in \mathbb{R}^{d+1}$ and any $\tau \in e^{-i\omega_1}\mathbb{C}_- \cap e^{-i\omega_2}\mathbb{C}_-$, we have

$$(5.2.20) \qquad q_\mu^{(\omega_1)}(x,\xi,e^{i\omega_1}\tau) = q_\mu^{(\omega_2)}(x,\xi,e^{i\omega_2}\tau).$$

For $k \geq 1$ the claim can be proved by making integration by parts in the integral (5.2.19) as in [**BG**]. Moreover, as in the case $k=0$ if $(\mu,x) \in \Omega_{\omega_1}^k \cap \Omega_{\omega_2}^k$ then the equality (5.2.20) holds for any $\xi \in \mathbb{R}^{d+1}$ and any $\tau \in e^{-i\omega_1}\mathbb{C}_- \cap e^{-i\omega_2}\mathbb{C}_-$.

All this shows that the lemma is true for $r=1$. The case $r \geq 2$ is then deduced from the case $r=1$ by arguing as in the proof of Proposition 3.4.4. $\square$

In the sequel given open angular sectors Θ_1 and Θ_2 we write $\Theta_1 \subset\subset \Theta_2$ to mean that $\overline{\Theta_1} \setminus 0$ is contained in Θ_2.

PROPOSITION 5.2.9. *Let $\Delta : C^\infty(M,\mathcal{E}) \to C^\infty(M,\mathcal{E})$ be a selfadjoint sublaplacian which is bounded from below and satisfies the condition (3.4.22) at every point. Then the principal symbol of $\Delta + \partial_t$ admits an inverse in $S_{\mathrm{v},-2}(\mathfrak{g}^*M \times \Theta_{(2)}^\infty, \mathcal{E})$ for any open angular sector $\Theta \subset\subset \mathbb{C} \setminus i[0,\infty)$ containing $\mathbb{R}$.*

PROOF. It is enough to prove the proposition in a local trivializing Heisenberg chart $U \subset \mathbb{R}^{d+1}$ with a H-frame $X_0, \ldots, X_d$ with respect to which Δ is of the form,

$$(5.2.21) \qquad \Delta = -\sum_{j=1}^d X_j^2 - i\mu(x)X_0 + O_H(1).$$

Since Δ is selfadjoint the matrix $\mu(x)$ is selfadjoint for every $x \in U$ and so the condition (3.4.22) implies that, with the notation of Proposition 5.2.8, the pair $(x,\mu(x))$ is in Ω_ω for any $x \in U$ and any $\omega \in (-\frac{\pi}{2},\frac{\pi}{2})$. Therefore, thanks to Proposition 5.2.8 we define an inverse $q_{-2} \in S_{\mathrm{v},-2}(U \times \mathbb{R}_{(2)}^{d+2}, \mathbb{C}^r)$ for the principal symbol $|\xi'|^2 + i\mu(x)\xi_0 + i\tau$ of Δ by letting

$$(5.2.22) \qquad q_{-2}(x,\xi,\tau) = q_{\mu(x)}^{(0)}(x,\xi,\tau), \quad (x,\xi,\tau) \in U \times [(\mathbb{R}^{d+1} \times \overline{\mathbb{C}_-}) \setminus 0].$$

For $\omega \in (-\frac{\pi}{2},\frac{\pi}{2})$ let $\Theta_\omega = \mathbb{C}_- \cup (e^{-i\omega}\mathbb{C}_-)$. Since (5.2.15) shows that $q_{-2}(x,\xi,\tau) = q_{\mu(x)}^{(\omega)}(x,\xi,e^{i\omega}\tau)$ when τ is in $\mathbb{C}_- \cap (e^{-i\omega}\mathbb{C}_-)$, we see that we can extend the definition of q_{-2} to $U \times \mathbb{R}^{d+1} \times \Theta_\omega$ by letting

$$(5.2.23) \qquad q_{-2}(x,\xi,\tau) = q_{\mu(x)}^{(\omega)}(x,\xi,e^{i\omega}\tau), \quad (x,\xi,\tau) \in U \times \mathbb{R}^{d+1} \times (e^{-i\omega}\mathbb{C}_-).$$

This defines an inverse for $|\xi'|^2 + i\mu(x)\xi_0 + i\tau$ in $S_{\mathrm{v},-2}(U \times \mathbb{R}^{d+1} \times \Theta_{\omega(2)}^\infty, \mathbb{C}^r)$. The proof is now completed by noticing that any angular sector $\Theta \subset\subset \mathbb{C} \setminus i[0,\infty)$ containing $\mathbb{R}$ is of the form $\Theta = \Theta_{\omega_1} \cup \Theta_{\omega_2}$ with $-\frac{\pi}{2} < \omega_1 < 0 < \omega_2 < \frac{\pi}{2}$. $\square$

REMARK 5.2.10. It can be shown that any selfadjoint sublaplacian with an invertible principal symbol is bounded from below (see [**Po1**], [**Po12**]). Therefore, the assumption on the boundedness from below of Δ in Proposition 5.2.9 in not necessary.

EXAMPLE 5.2.11. Proposition 5.2.9 is true for the following sublaplacians:

(a) A selfadjoint sum of squares $\Delta_{\nabla,X} = \nabla_{X_1}\nabla_{X_1}^* + \ldots + \nabla_{X_1}\nabla_{X_1}^*$, where ∇ is a connection on $\mathcal{E}$ and the vector fields $X_1, \ldots, X_m$ span H, under the bracket condition $H + [H,H] = TM$;

(b) The Kohn Laplacian on a CR manifold acting on (p,q)-forms under the condition $Y(q)$;

(c) The horizontal sublaplacian on a Heisenberg manifold acting on horizontal forms of degree k under the condition $X(k)$;

(d) The horizontal sublaplacian on a CR manifold acting on (p,q)-forms under the condition $X(p,q)$.

In particular, Proposition 5.2.9 allows us to complete the treatment of the heat kernel of the Kohn Laplacian in [**BGS**] because, as with the invertibility of its principal symbol in Section 3.4, we really need the version for systems provided by Proposition 5.2.8, but not established in [**BGS**].

Next, assume that the bracket condition $H + [H,H] = TM$ holds, that is, the Levi form of (M,H) has positive rank everywhere, and consider the sum of squares,

$$(5.2.24) \qquad \Delta_{\nabla,X} = -(\nabla_{X_1}^*\nabla_{X_1} + \ldots + \nabla_{X_m}^*\nabla_{X_m}),$$

where ∇ is a connection on $\mathcal{E}$ and the vector fields $X_1, \ldots, X_m$ span H.

PROPOSITION 5.2.12. *The principal symbol of $\Delta_{\nabla,X}^k + \partial_t$ admits an inverse in $S_{v,-2k}(\mathfrak{g}^*M \times \mathbb{R}_{(2k)}, \mathcal{E})$ and so Theorems 5.1.25 and 5.1.26 hold for $\Delta_{\nabla,X}^k$.*

PROOF. Since proving the above statement is a purely local issue and $\Delta_{\nabla,X}$ is a scalar operator modulo lower order terms, we may assume that $\mathcal{E}$ is the trivial line bundle and proceed on $U \times \mathbb{R}$ where U is a Heisenberg chart with H-frame $Y_0, \ldots, Y_d$ with respect to which $\Delta_{\nabla,X}$ takes the form,

$$(5.2.25) \qquad \Delta_{\nabla,X} = -(Y_1^2 + \ldots + Y_d^2) + \mathrm{O}_H(1).$$

In particular the principal symbol of Δ is just $|\xi'|^2$. In addition, let $p_{2k}(x,\xi)$ be the local symbol of $\Delta_{\nabla,X}^k$ on U.

Let $\arg z$ be the continuous determination of the argument on $\mathbb{C}\setminus 0$ with values in $[0, 2\pi)$ and for $z \in \mathbb{C} \setminus 0$ let $z^{\frac{1}{k}} = |z|^{\frac{1}{k}} e^{\frac{i}{k}\arg z}$ and set $0^{\frac{1}{k}} = 0$, so that the function $z \to z^{\frac{1}{k}}$ is analytic on $\mathbb{C} \setminus [0, \infty)$ and is continuous at the origin. Set $\omega = e^{\frac{2i\pi}{k}}$. Then for any $z \in \mathbb{C}$ we have the polynomial identity,

$$(5.2.26) \qquad T^k - z = (T - z^{\frac{1}{k}})(T - \omega z^{\frac{1}{k}}) \ldots (T - \omega^{k-1} z^{\frac{1}{k}}).$$

Therefore, for $T = |\xi'|^2$ and $z = -i\tau$ in $S_{v,-2k}(U \times \mathbb{R}^{d+2}_{(2k)})$ we get

$$(5.2.27) \qquad p_{2k} + i\tau = (|\xi'|^2 + i(i(-i\tau)^{\frac{1}{k}}) * \ldots * (|\xi'|^2 + i(i\omega^{k-1}(-i\tau)^{\frac{1}{k}})).$$

Let Θ be the angular sector $|\arg(-i\tau)| > \frac{\pi}{2k}$. As $\Theta \subset\subset \mathbb{C}\setminus i[0,\infty)$ and we have $H + [H,H] = TM$, so that the condition (3.4.22) for $\Delta_{\nabla,X}$ is satisfied at every point, Proposition 5.2.9 tells us that $|\xi'|^2 + i\tau$ admits an inverse $q_{(-2)}(x,\xi,\tau)$ in $S_{v,-2}(U \times \mathbb{R}^{d+1} \times \Theta_{(2)})$.

Notice that for $j = 0, 1, \ldots, k-1$ if $\Im\tau < 0$ then $i\omega^j(-i\tau)^{\frac{1}{k}}$ is in Θ, so for $(x,\xi,\tau) \in U \times [(\mathbb{R}^{d+1} \times \mathbb{C}_-) \setminus 0]$ we can let

$$(5.2.28) \qquad \tilde{q}_{(-2),j}(x,\xi,\tau) = q_{(-2)}(x,\xi,i\omega^j(-i\tau)^{\frac{1}{k}})).$$

In the sense of Definition 5.2.2 this gives rise to a symbol in $S_{v,-2}(U\times\mathbb{R}^{d+1}\times\mathbb{C}^0_{-(2k)})$. Therefore, it follows from (5.2.27) and Lemma 5.2.5 that we get an inverse for $p_{2k}+i\tau$ in $S_{v,-v}(U\times\mathbb{R}^{d+1}\times\mathbb{C}^0_{-(2k)})$ by letting

$$(5.2.29) \qquad q_{(-2k)} = \tilde{q}_{(-2),0} * \cdots * \tilde{q}_{(-2),k-1}.$$

Let us now show that $q_{(-2k)}$ is in $S_{v,-2k}(U\times\mathbb{R}^{d+1}\times\mathbb{C}^\infty_{-(2k)}) = S_{v,-2k}(U\times\mathbb{R}^{d+2}_{(2k)})$.

CLAIM. *For $j=1,2$ let $q_j \in S_{v,m_j}(U\times\mathbb{R}^{d+1}\times\mathbb{C}^0_{-(2k)})$. Then, regarding q_1, q_2 and $q_1 * q_2$ as smooth families with values in $C^\infty(U\times\mathbb{R}^{d+1})$ over $\mathbb{R}\setminus 0$, we have*

$$(5.2.30) \qquad \partial_\tau(q_1 * q_2) = (\partial_\tau q_1) * q_2 + q_1 * (\partial_\tau q_2).$$

PROOF OF THE CLAIM. It follows from (5.2.10) that the Leibniz formula (5.2.30) holds for almost homogeneous symbols in $S^*_{\mathrm{ah},v}(U\times\mathbb{R}^{d+1}\times\mathbb{C}^1_{-(2k)})$. Since ∂_τ and $*$ are homogeneous maps the formula remains true for homogeneous symbols in $S_{v,*}(U\times\mathbb{R}^{d+1}\times\mathbb{C}^1_{-(2k)})$. $\square$

Now, differentiating the equality $1 = (p_{2k}+i\tau) * q_{(-2k)}$ using (5.2.30) we get $0 = (p_{2k}+i\tau) * (\partial_\tau q_{(-2k)}) + iq_{(-2k)}$, which after multiplication by $q_{(-2k)}$ gives

$$(5.2.31) \qquad \partial_\tau q_{(-2k)} = -iq_{(-2k)} * q_{(-2k)}.$$

This equality holds in $C^\infty(\mathbb{R}\setminus 0, C^\infty(U\times\mathbb{R}^{d+1}))$ but, since $q_{(-2k)} * q_{(-2k)}$ belongs to $S_{v,-2v}(U\times\mathbb{R}^{d+1}\times\mathbb{C}^0_{-(2k)})$, this actually shows that the symbol $q_{(-2k)}$ belongs to $S_{v,-2k}(U\times\mathbb{R}^{d+1}\times\mathbb{C}^1_{-(2k)})$.

Finally, an induction shows that $q_{(-2k)}$ is contained in $S_{v,-2k}(U\times\mathbb{R}^{d+1}\times\mathbb{C}^N_{-(2k)})$ for any integer N, hence belongs to $S_{v,-2k}(U\times\mathbb{R}^{d+2}_{(2k)})$. This proves that the principal symbol $p_{(2k)}$ of $\Delta^k+\partial_t$ has an inverse in $S_{v,-2k}(U\times\mathbb{R}^{d+1}\times\mathbb{C}^1_{-(2k)})$. The proof is thus achieved. $\square$

5.3. Complex powers of hypoelliptic differential operators

In this section we show that the complex powers of a positive hypoelliptic differential operator, *a priori* defined as unbounded operators on $L^2(M,\mathcal{E})$, give rise to a holomorphic family of Ψ_HDOs.

Let $P: C^\infty(M,\mathcal{E}) \to C^\infty(M,\mathcal{E})$ be a selfadjoint differential operator of even (Heisenberg) order v such that P has an invertible principal symbol and is positive, i.e., we have $\langle Pu, u\rangle \geq 0$ for any $u \in C^\infty(M,\mathcal{E})$.

Let $\Pi_0(P)$ be the orthogonal projection onto $\ker P$ and set $P_0 = (1-\Pi_0(P))P + \Pi_0(P)$. Then P_0 is selfadjoint with spectrum contained in $[c,\infty)$ for some $c > 0$. Thus by standard functional calculus, for any $s \in \mathbb{C}$, the power P_0^s is a well defined unbounded operator on $L^2(M,\mathcal{E})$. We then define the power P^s, $s \in \mathbb{C}$, by letting

$$(5.3.1) \qquad P^s = (1-\Pi_0(P))P_0^s = P_0^s - \Pi_0(P),$$

so that P^s coincides with P_0^s on $(\ker P)^\perp$ and is zero on $\ker P$. In particular, we have $P^0 = 1 - \Pi_0(P)$ and P^{-1} is the partial inverse of P.

The main result of this section is the following.

THEOREM 5.3.1. *Suppose that the principal symbol of $P+\partial_t$ admits an inverse in $S_{v,-v}(\mathfrak{g}^*M\times\mathbb{R}_{(v)},\mathcal{E})$. Then:*

(i) For any $s \in \mathbb{C}$ the operator P^s defined by (5.3.1) is a Ψ_HDO of order vs;

(ii) The family $(P^s)_{s\in\mathbb{C}}$ forms a holomorphic 1-parameter group of Ψ_HDOs.

PROOF. Let us first assume that $\mathcal{E}$ is the trivial line bundle over M, so that P is a scalar operator. For $\Re s > 0$ the function $x \to x^{-s}$ is bounded on $[0,\infty)$, so the operators P_0^{-s} and P^{-s} are bounded. Moreover, by the Mellin formula we have

$$(5.3.2) \qquad P^{-s} = (1 - \Pi_0(P))P_0^s = \frac{1}{\Gamma(s)} \int_0^\infty t^s (1 - \Pi_0(P)) e^{-tP} \frac{dt}{t}.$$

This leads us to define

$$(5.3.3) \qquad A_s = \int_0^1 t^{s-1} e^{-tP} dt, \qquad \Re s > 0.$$

Then we have

$$(5.3.4) \qquad \begin{aligned} \Gamma(s) P^{-s} - A_s &= \int_0^1 t^{s-1} \Pi_0(P) e^{-tP} dt + \int_1^\infty t^{s-1}(1 - \Pi_0(P)) e^{-tP} dt, \\ &= \frac{1}{2}\Pi_0(P) + e^{-P/2}\left(\int_0^\infty (1+t)^{s-1} e^{-tP} dt\right) e^{-P/2}. \end{aligned}$$

Since $\Pi_0(P)$ and $e^{-P/2}$ are smoothing operators and $(\int_0^\infty (1+t)^{s-1} e^{-tP} dt)_{\Re s > 0}$ is a holomorphic family of bounded operators on $L^2(M)$, we get

$$(5.3.5) \qquad (\Gamma(s) P^{-s} - A_s)_{\Re s > 0} \in \mathrm{Hol}(\Re s > 0, \Psi^{-\infty}(M)).$$

Let us now show that $(A_s)_{\Re s > 0}$ defined by (5.3.3) is a holomorphic family of Ψ_HDOs such that $\mathrm{ord}\, A_s = -vs$. To this end observe that, in terms of distribution kernels, the formula (5.3.3) means that A_s has distribution kernel

$$(5.3.6) \qquad k_{A_s}(x,y) = \int_0^1 t^{s-1} k_t(x,y) dt.$$

where $k_t(x,y)$ denotes the heat kernel of P.

On the other hand, since P is bounded from below and the principal symbol of $P + \partial_t$ is an invertible Volterra-Heisenberg symbol, Theorem 5.1.25 tells us that $P + \partial_t$ has an inverse $Q_0 := (P + \partial_t)^{-1}$ in $\Psi_{H,v}^{-v}(M \times \mathbb{R}_{(v)}, \mathcal{E})$ and that the distribution kernel $K_{Q_0}(x, y, t-s)$ of Q_0 is related to the heat kernel of P by

$$(5.3.7) \qquad K_{Q_0}(x, y, t) = k_t(x, y) \qquad \text{for } t > 0.$$

Therefore, for $\Re s > 0$ we have

$$(5.3.8) \qquad k_{A_s}(x,y) = \int_0^1 t^{s-1} K_{Q_0}(x, y, t) dt..$$

Let φ and ψ be smooth functions on M with disjoint supports. Then using (5.3.8) we see that $\varphi A_s \psi$ has distribution kernel

$$(5.3.9) \qquad k_{\varphi A_s \psi}(x,y) = \int_0^1 t^{s-1} \varphi(x) K_{Q_0}(x, y, t) \psi(y) dt.$$

Since the distribution kernel of a Volterra-Ψ_HDO is smooth off the diagonal of $(M \times \mathbb{R}) \times (M \times \mathbb{R})$ the distribution $K_{Q_0}(x, y, t)$ is smooth on the region $\{x \neq y\} \times \mathbb{R}$, so (5.3.9) defines a holomorphic family of smooth kernels. Thus,

$$(5.3.10) \qquad (\varphi A_s \psi)_{\Re s > 0} \in \mathrm{Hol}(\Re s > 0, \Psi^{-\infty}(M)).$$

Next, the following holds.

5.3. COMPLEX POWERS OF HYPOELLIPTIC DIFFERENTIAL OPERATORS 95

LEMMA 5.3.2. *Let $V \subset \mathbb{R}^{d+1}$ be a Heisenberg chart, let $Q \in \Psi_{H,\mathrm{v}}^{-v}(V \times \mathbb{R}_{(v)})$ have distribution kernel $K_Q(x, y, t-s)$ and for $\Re s > 0$ let $B_s : C_c^\infty(V) \to C^\infty C(V)$ be given by the distribution kernel,*

$$(5.3.11) \qquad k_{B_s}(x,y) = \int_0^1 t^{s-1} K_Q(x,y,t) dt, \quad \Re s > 0.$$

Then $(B_s)_{\Re s > 0}$ is a holomorphic family of $\Psi_H DOs$ such that $\mathrm{ord} B_s = -vs$.

PROOF OF THE LEMMA. Let ε_x denote the change to the Heisenberg coordinates at x. By Proposition 5.1.14 on $V \times V \times \mathbb{R}$ the distribution $K_Q(x,y,t)$ is of the form

$$(5.3.12) \qquad K_Q(x,y,t) = |\epsilon_x'| K(x, -\varepsilon_x(y), t) + R(x,y,t),$$

for some $K \in \mathcal{K}_\mathrm{v}^{-(d+2)}(V \times \mathbb{R}_{(v)}^{d+2})$ and some $R \in C^\infty(U \times U \times \mathbb{R})$. Let us write $K \sim \sum_{j \geq 0} K_{j-(d+2)}$ with $K_l \in \mathcal{K}_{\mathrm{v},l}(V \times \mathbb{R}_{(v)}^{d+2})$. Thus, for any integer N, as soon as J large enough we have

$$(5.3.13) \quad K(x,y,t) = \sum_{j \leq J} K_{j-(d+2)}(x,y,t) + R_{NJ}(x,y,t), \quad R_{NJ} \in C^N(U \times \mathbb{R}^{d+2}).$$

In particular, on $V \times V$ we can write

$$(5.3.14) \quad k_{B_s}(x,y) = |\epsilon_x'| K_s(x, \varepsilon_x(y)) + R_s(x,y), \quad K_s(x,y) = \int_0^1 t^{s-1} K(x,y,t) dt,$$

with $(R_s)_{\Re s > 0}$ in $\mathrm{Hol}(\Re s > 0, C^\infty(V \times V))$. Moreover, $K_s(x,y)$ is of the form

$$(5.3.15) \qquad K_s = \sum_{j \leq J} K_{j,s} + R_{NJ,s}, \quad K_{j,s}(x,y) = \int_0^1 t^{s-1} K_{j-(d+2)}(x,y,t) dt,$$

with $(R_{NJ,s})_{\Re s > 0}$ contained in $\mathrm{Hol}(\Re s > 0, C^N(V \times V))$.

Notice that $K_{j-(d+2)}(x,y,t)$ is in $C^\infty(V) \hat{\otimes} \mathcal{D}_{\mathrm{reg}}'(\mathbb{R}^{d+1} \times \mathbb{R})$ and is parabolic homogeneous of degree $j - (d+2) \geq -(d+2)$. Thus the family $(K_{j,s})_{\Re s > 0}$ belongs to $\mathrm{Hol}(\Re s > 0, C^\infty(U) \hat{\otimes} \mathcal{D}_{\mathrm{reg}}'(\mathbb{R}^{d+1}))$. Moreover, for any $\lambda > 0$, the difference $K_{j,s}(x, \lambda.y) - \lambda^{vs+j-(d+2)} K_{j,s}(x,y)$ is equal to

$$(5.3.16) \qquad \int_1^{\lambda^2} t^{s-1} K(x,y,t) dt \in \mathrm{Hol}(\Re s > 0, C^\infty(V \times \mathbb{R}^{d+2})).$$

Hence $(K_{j,s})_{\Re s > 0}$ is a holomorphic family of almost homogeneous distributions of degree $vs - (d+2) + j$. Combining this with (5.3.15) then shows that $(K_s)_{\Re s > 0}$ belongs to $\mathrm{Hol}(\Re s > 0, \mathcal{K}^*(V \times \mathbb{R}^{d+1}))$ and has order $vs - (d+2)$. Therefore, using (5.3.14) and Proposition 4.4.5 we see that $(B_s)_{\Re s > 0}$ is a holomorphic family of $\Psi_H DOs$ such that $\mathrm{ord} B_s = -(\mathrm{ord} K_s + d + 2) = -vs$. □

It follows from Lemma 5.3.2 that for any local Heisenberg chart $\kappa : U \to V$ the family $(\kappa_* A_{s|U})_{\Re s > 0}$ is a holomorphic family of $\Psi_H DOs$ on V of order $-vs$. Combining this with (5.3.10) and (5.3.5) then shows that $(A_s)_{\Re s > 0}$ and $(P^s)_{\Re s < 0}$ are holomorphic families of $\Psi_H DOs$ of orders $-vs$ and vs respectively.

Now, let $s \in \mathbb{C}$ and let k be a positive integer such that $k > \Re s$. Then on $C^\infty(M, \mathcal{E})$ we have $P^s = P^{s-k} P^k$. Since P^k is a differential operator and P^{s-k} is a $\Psi_H DO$ of order $v(s-k)$ this shows that P^s is a $\Psi_H DO$ of order vs. In fact, as by Proposition 4.3.6 the product of $\Psi_H DOs$ is analytic this proves that $(P^s)_{s \in \mathbb{C}}$ is a holomorphic family of $\Psi_H DOs$ such that $\mathrm{ord} P^s = vs$ for every $s \in \mathbb{C}$.

Now, let $s \in \mathbb{C}$ and let k be a positive integer such that $k > \Re s$. Then on $C^\infty(M, \mathcal{E})$ we have

(5.3.17) $\qquad P^s u = P^{s-k} P^k u \quad \text{for any } u \in C^\infty(M, \mathcal{E}).$

As P^k is a differential operator and P^{s-k} is a Ψ_HDO of order $m(s-k)$ this proves that P^s is a Ψ_HDO of order ms. In fact, as by Proposition 4.3.6 the product of Ψ_HDOs is analytic this actually shows that $(P^s)_{s \in \mathbb{C}}$ is a holomorphic family of Ψ_HDOs such that $\mathrm{ord} P^s = vs$ for every $s \in \mathbb{C}$.

Finally, when $\mathcal{E}$ is a general vector bundle we can similarly prove that the complex powers P^s, $s \in \mathbb{C}$, forms a holomorphic family of Ψ_HDOs such that $\mathrm{ord} P^s = vs$ for any $s \in \mathbb{C}$. The proof is thus complete. $\square$

EXAMPLE 5.3.3. Thanks to Proposition 5.2.9 Theorem 5.3.1 holds for the following sublaplacians:

(a) A selfadjoint sums of squares $\Delta_{\nabla, X} = \nabla_{X_1}^* \nabla_{X_1} + \ldots + \nabla_{X_m}^* \nabla_{X_m}$, where $X_1, \ldots, X_m$ span H and ∇ is a connection on $\mathcal{E}$, under the condition that the Levi form is nonvanishing.

(b) The Kohn Laplacian $\square_{b;p,q}$ on a CR manifold acting on (p,q)-forms under the condition $Y(q)$.

(c) The horizontal sublaplacian $\Delta_{b;k}$ on a Heisenberg manifold (M^{d+1}, H) acting on sections of $\Lambda_\mathbb{C}^k H^*$ when the condition $X(k)$ holds everywhere.

(d) The horizontal sublaplacian $\Delta_{b;p,q}$ on a CR manifold acting on (p,q)-forms under the condition $X(p,q)$.

In the next section we will make use of Theorem 5.3.1 to show that when the bracket condition $H + [H, H] = TM$ holds, the Rockland condition insures us that the principal symbol of $P + \partial_t$ admits an inverse in $S_{v, -v}(\mathfrak{g}^* M \times \mathbb{R}_{(v)}, \mathcal{E})$ (see Theorem 5.4.10). Therefore, we obtain:

THEOREM 5.3.4. *Assume that the bracket condition $H + [H, H] = TM$ holds and that P satisfies the Rockland condition at every point of M. Then the complex powers P^s, $s \in \mathbb{C}$, of P form a holomorphic 1-parameter group of $\Psi_H DOs$ such that $\mathrm{ord} P^s = vs \; \forall s \in \mathbb{C}$.*

EXAMPLE 5.3.5. Assume that (M, H) is an orientable contact manifold. Then:

- In degree $k \neq n$ the complex powers of the contact Laplacian $\Delta_{R,k}$ form a holomorphic 1-parameter group of Ψ_HDOs such that $\mathrm{ord} \Delta_{R,k}^s = 2s \; \forall s \in \mathbb{C}$.

- In degree n the complex powers of the contact Laplacians $\Delta_{R,nj}$, $j = 1, 2$, form holomorphic 1-parameter groups of Ψ_HDOs such that $\mathrm{ord} \Delta_{R,nj}^s = 4s \; \forall s \in \mathbb{C}$.

REMARK 5.3.6. The above example allows us to fill a technical gap in [**JK**] in the proof that the complex powers of the contact Laplacian are Ψ_HDOs (see [**Po10**]). The latter result is one ingredient in [**JK**], among others, in a proof of the Baum-Connes conjecture for $SU(n,1)$.

REMARK 5.3.7. Given a (stratified) graded nilpotent group G it was shown by Folland that some complex powers of left invariant sum of squares are left invariant ΨDOs. This was made via the use of the fundamental solution of the associated heat operators. These results have later been extended to more general operators in [**CGGP**]. Using a different approach Geller [**Ge2**] dealt with general positive left invariant ΨDOs on the Heisenberg group $\mathbb{H}^{2n+1}$ provided that the Rockland condition is satisfied. In fact, on each tangent group $G_a M$ these results can be recovered from Theorem 5.3.4 by looking at the principal symbols of the complex powers.

5.4. Rockland condition and the heat equation

In this section we make use of Theorem 5.3.1 in connection with the Rockland condition.

First, we extend Theorem 3.3.18 and Proposition 3.3.20 to Ψ_HDOs with non-integer order as follows.

THEOREM 5.4.1. *Assume that the bracket condition $H + [H, H] = TM$ holds and let $P : C_c^\infty(M, \mathcal{E}) \to C^\infty(M, \mathcal{E})$ be a $\Psi_H DO$ of order $m \in \mathbb{C}$. Then the following are equivalent:*

(i) The principal symbol of P is invertible;

(ii) P and P^t satisfy the Rockland condition at every point $a \in M$.

(iii) P and P^* satisfy the Rockland condition at every point $a \in M$. Furthermore, when (i) or (ii) holds the operator P admits a parametrix in $\Psi_H^{-m}(M, \mathcal{E})$.

PROOF. Consider a sum of squares $\Delta_{\nabla, X} = \nabla_{X_1}^* \nabla_{X_1} + \ldots + \nabla_{X_m}^* \nabla_{X_m}$, where ∇ is a connection on $\mathcal{E}$ and the vector fields $X_1, \ldots, X_m$ span H. Since by assumption the bracket condition $H + [H, H] = TM$ holds, we know from Proposition 5.2.9 that the principal symbol of $\Delta_{\nabla, X} + \partial_t$ admits an inverse in $S_{\text{v},-2}(\mathfrak{g}^* M \times \mathbb{R}_{(2)}, \mathcal{E})$, so by Theorem 5.3.1 the operator $\Delta_{\nabla, X}^{-\frac{m}{2}}$ is a Ψ_HDO of order $-m$ with an invertible principal symbol. Therefore, along the same lines as that of the proof of Theorem 3.3.18 we can show that the conditions (i), (ii) and (iii) are equivalent. □

PROPOSITION 5.4.2. *Assume that the bracket condition $H + [H, H] = TM$ holds and let $P : C_c^\infty(M, \mathcal{E}) \to C^\infty(M, \mathcal{E})$ be a $\Psi_H DO$ of order $m \in \mathbb{C}$ with $\Re m \geq 0$ and such that P satisfies the Rockland condition at every point. Then P is hypoelliptic with gain of $\frac{1}{2} \Re m$ derivatives.*

PROOF. Since P satisfies the Rockland condition at every point, we can argue as in the proof of Proposition 3.3.20, using Theorem 5.4.1 instead of Theorem 3.3.18, to show that the principal symbol of P is left-invertible. Therefore, P admits a left-parametrix in $\Psi_H^{-m}(M, \mathcal{E})$ and it then follows from Remark 3.3.5 that P is is hypoelliptic with gain of $\frac{1}{2} \Re m$ derivatives. □

Let us now show that the Rockland condition for a (positive) differential operator P implies the invertibility of the principal symbol of $P + \partial_t$. This will show that the frameworks of [**BGS**] and of Theorem 5.3.1 apply to a large class of operators.

To reach aim it will be convenient to enlarge the class of Volterra Ψ_HDOs as follows.

DEFINITION 5.4.3. *Let U be an open subset of $\mathbb{R}^{d+1}$. Then:*

1) $S_m(U \times \mathbb{R}^{d+2}_{(v)})$, $m \in \mathbb{Z}$, consists of functions $q(x, \xi, \tau)$ in $C^\infty(U \times (\mathbb{R}^{d+2} \setminus 0))$ such that $q(x, \lambda.\xi, \lambda^v \tau) = \lambda^m q(x, \xi, \tau)$ for any $\lambda > 0$.

2) $S^m(U \times \mathbb{R}^{d+2}_{(v)})$, $m \in \mathbb{Z}$, consists of functions $q(x, \xi, \tau)$ in $C^\infty(U \times \mathbb{R}^{d+2})$ admitting an asymptotic expansion $q \sim \sum_{j \geq 0} q_{m-j}$, $q_{m-j} \in S_{m-j}(U \times \mathbb{R}^{d+1} \times \mathbb{R}_{(v)})$, with $\sim$ taken in the sense of (5.1.6).

DEFINITION 5.4.4. $\Psi_H^m(M \times \mathbb{R}_{(v)}, \mathcal{E})$, $m \in \mathbb{Z}$, consists of continuous operators Q from $C_c^\infty(M \times \mathbb{R}, \mathcal{E})$ to $C_c^\infty(M \times \mathbb{R}, \mathcal{E})$ such that:

(i) Q is translation invariant with respect to the time variable;

(ii) The kernel of Q is smooth outside the diagonal;

(iii) In any local trivializing chart $U \subset \mathbb{R}^{d+1}$ with H-frame $X_0, \ldots, X_d$ we can write Q in the form,

$$(5.4.1) \qquad Q = q(x, -iX, D_t) + R,$$

with q in $S^m(U \times \mathbb{R}^{d+2}_{(v)}, \mathbb{C}^r)$ and R smoothing operator.

The main properties of Ψ_HDOs and of Volterra Ψ_HDOs hold *mutatis mutandis* for these operators. In particular, it makes sense to define then on $M \times \mathbb{R}$ and the substitute to the class $\mathcal{K}_m(U \times \mathbb{R}^{d+1})$ and $\mathcal{K}_{v,m}(U \times \mathbb{R}^{d+2}_{(v)})$ is given by the distributions below.

DEFINITION 5.4.5. $\mathcal{K}_m(U \times \mathbb{R}^{d+2}_{(v)})$, $m \in \mathbb{Z}$, consists of distributions $K(x,y,t)$ in $C^\infty \hat{\otimes} \mathcal{D}_{reg}(\mathbb{R}^{d+2} \setminus 0)$ for which there are $c_{\alpha,l} \in C^\infty(U)$, $\langle \alpha \rangle + vl = m$, such that

$$(5.4.2) \quad K(x, \lambda.y, \lambda^v t) = \lambda^m K(x,y,t) + \lambda^m \log \lambda \sum_{\langle \alpha \rangle + vl = m} c_\alpha(x) y^\alpha t^l \qquad \forall \lambda > 0.$$

Moreover, in this context the principal symbol of a Ψ_HDO in $\Psi_H^m(M \times R_{(v)}, \mathcal{E})$ makes sense as an element of the following symbol class.

DEFINITION 5.4.6. $S_m(\mathfrak{g}^* M \times \mathbb{R}_{(v)}, \mathcal{E})$, $m \in \mathbb{Z}$, consists of sections $q(x, \xi, \tau)$ in $C^\infty((\mathfrak{g}^* M \times \mathbb{R}) \setminus 0, \operatorname{End} \pi^* \mathcal{E})$ such that $q(x, \lambda.\xi, \lambda^v \tau) = \lambda^m q(x, \xi, \tau)$.

This allows us to define the model operator and the Rockland condition at a point $a \in M$ in the same way as for Ψ_HDOs on M.

Moreover, the class $\Psi_H^m(M \times \mathbb{R}_{(v)}, \mathcal{E})$ is closed under taking adjoints, so if Q is a Volterra Ψ_HDO in $\Psi_{H,v}^m(M \times \mathbb{R}_{(v)}, \mathcal{E})$ then its adjoint is not a Volterra Ψ_HDOs but at least it belongs to $\Psi_H^m(M \times \mathbb{R}_{(v)}, \mathcal{E})$.

We can now establish the analogue of Theorem 3.3.10 in the Volterra setting.

PROPOSITION 5.4.7. *Let* $Q \in \Psi_H^0(M \times \mathbb{R}_{(v)}, \mathcal{E})$. *The following are equivalent:*

(i) *For any* $a \in M$ *the model operator* Q^a *is invertible on* $L^2(G_a M \times \mathbb{R}_{(v)}, \mathcal{E}_a)$.

(ii) Q *and* Q^* *satisfy the Rockland condition at every point of* M.

(iii) *The principal symbol of* Q *is invertible in* $S_0(\mathfrak{g}^* M \times \mathbb{R}_{(v)}, \mathcal{E})$.

Moreover, if the principal symbol of Q *is invertible and belongs to* $S_0(\mathfrak{g}^* M \times \mathbb{R}_{(v)}, \mathcal{E})$ *then its inverse is in* $S_0(\mathfrak{g}^* M \times \mathbb{R}_{(v)}, \mathcal{E})$ *too.*

PROOF. The proof of the equivalence between the conditions (i), (ii) and (iii) follows along the same lines as that of Theorem 3.3.10, since the results of [**CGGP**], [**Ch1**], [**Ch2**], [**FS2**] and [**KS**] used in the proof of Theorem 3.3.10 hold in this setting.

It remains to show that if the principal symbol of Q is invertible and belongs to $S_0(\mathfrak{g}^* M \times \mathbb{R}_{(v)}, \mathcal{E})$ then its inverse is in $S_0(\mathfrak{g}^* M \times \mathbb{R}_{(v)}, \mathcal{E})$. In view of the proof of Lemma 3.3.17 in order to reach this aim it is enough to show that, for any Heisenberg chart $U \subset \mathbb{R}^{d+1}$, the Volterra class $\mathcal{K}_{v,-(d+2+v)}(U \times \mathbb{R}^{d+2}_{(v)})$ is a closed subalgebra of $\mathcal{K}_{-(d+2+v)}(U \times \mathbb{R}^{d+2}_{(v)})$.

5.4. ROCKLAND CONDITION AND THE HEAT EQUATION

It is clear that if K_1 and K_2 are in $\mathcal{K}_{v,-(d+2+v)}(U \times \mathbb{R}_{(v)}^{d+2})$ then $K_1 * K_2$ is also in $\mathcal{K}_{v,-(d+2+v)}(U \times \mathbb{R}_{(v)}^{d+2})$. Moreover, if a sequence $(K_j)_{j \geq 0} \subset \mathcal{K}_{v,-(d+2+v)}(U \times \mathbb{R}_{(v)}^{d+2})$ converges to K in $\mathcal{K}_{-(d+2+v)}(U \times \mathbb{R}_{(v)}^{d+2})$ then K_j converges to K in $\mathcal{D}'(U \times \mathbb{R}^{d+2})$ and so we have $\operatorname{supp} K \subset \cup \operatorname{supp} K_j \subset U \times \mathbb{R}^{d+1} \times [0, \infty)$, that is, K belongs to $\mathcal{K}_{v,-(d+2+v)}(U \times \mathbb{R}_{(v)}^{d+2})$. Thus $\mathcal{K}_{v,-(d+2+v)}(U \times \mathbb{R}_{(v)}^{d+2})$ is a closed subalgebra of $\mathcal{K}_{-(d+2+v)}(U \times \mathbb{R}_{(v)}^{d+2})$ and henceforth the proof is achieved. $\square$

In the sequel we will need the notion of positive symbols below.

DEFINITION 5.4.8. *A symbol $p \in S_m(\mathfrak{g}^* M, \mathcal{E})$ is positive when it can be put into the form $p = \bar{q} * q$ for some symbol $q \in S_{\frac{m}{2}}(\mathfrak{g}^* M, \mathcal{E})$.*

The interest of this definition stems from:

LEMMA 5.4.9. *1) The principal symbol of P is positive if, and only if, there exist $Q \in \Psi_H^{\frac{v}{2}}(M, \mathcal{E})$ and $R \in \Psi_H^{v-1}(M, \mathcal{E})$ so that $P = Q^*Q + R$.*

2) If P satisfies the Rockland condition at every point and has a positive principal symbol, then the operators $P \pm \partial_t$ satisfy the Rockland condition at every point.

PROOF. Assume that the principal symbol $p = \sigma_v(P)$ of P is positive, so that there exists $q_{\frac{v}{2}} \in S_{\frac{v}{2}}(\mathfrak{g}^* M, \mathcal{E})$ such that $p = \overline{q_{\frac{v}{2}}} * q_{\frac{v}{2}}$. By Proposition 3.2.6 the principal symbol map $\sigma_{\frac{v}{2}} : \Psi_H^{\frac{v}{2}}(M, \mathcal{E}) \to S^{\frac{v}{2}}(\mathfrak{g}^* M, \mathcal{E})$ is surjective, so there exists $Q \in \Psi_H^{\frac{v}{2}}(M, \mathcal{E})$ such that $\sigma_{\frac{v}{2}}(Q) = q_{\frac{v}{2}}$. Then by Proposition 3.2.9 and Proposition 3.2.12 the operator Q^*Q has principal symbol $\overline{q_{\frac{v}{2}}} * q_{\frac{v}{2}} = \sigma_v(P)$, hence coincides with P modulo $\Psi_H^{v-1}(M, \mathcal{E})$.

Conversely, if P is of the form $P = Q^*Q + R$ with Q in $\Psi_H^{\frac{v}{2}}(M, \mathcal{E})$ and R in $\Psi_H^{v-1}(M, \mathcal{E})$ then P and Q^*Q have same principal symbol. Therefore, we have $\sigma_v(P) = \overline{\sigma_{\frac{v}{2}}(Q)} * \sigma_{\frac{v}{2}}(Q)$, that is, $\sigma_v(P)$ is a positive symbol.

Finally, suppose that P satisfies the Rockland condition at every point and has a positive principal symbol, i.e., there exists $\tilde{p}_{\frac{v}{2}} \in S_{\frac{v}{2}}(U \times G)$ so that $\sigma_v(P) = \overline{\tilde{p}_{\frac{v}{2}}} * \tilde{p}_{\frac{v}{2}}$. Therefore, if we let $\tilde{P}^a$ be the left-convolution operator with symbol $\tilde{p}_{\frac{v}{2}}(a, .)$ then by Proposition 3.2.9 and Proposition 3.2.12 we have $P^a = (\tilde{P}^a)^* \tilde{P}^a$. Thus, by Proposition 3.3.6 for every non-trivial irreducible representation π of G we have $\overline{\pi_{P^a}} = (\overline{\pi_{\tilde{p}^a}})^* \overline{\pi_{\tilde{p}^a}}$, which shows that $\overline{\pi_P^a}$ is positive. We then can argue as in the proof of [**FS2**, Lem. 4.21] to show that the operators $P^a \pm \partial_t$ satisfy the Rockland condition on $G_a M \times \mathbb{R}_{(v)}$. Hence $P + \partial_t$ and $P - \partial_t$ satisfy the Rockland condition at every point of M. $\square$

We are now ready to prove the main result of this section.

THEOREM 5.4.10. *Assume that the bracket condition $H + [H, H] = TM$ holds and that P satisfies the Rockland condition at every point.*

1) P is bounded from below if, and only if, it has a positive principal symbol.

2) If P has a positive principal symbol, then the principal symbol $P + \partial_t$ has an inverse in $S_{v,-v}(\mathfrak{g}^* M \times \mathbb{R}_{(v)}, \mathcal{E})$. Hence Theorems 5.1.25 and 5.1.26 hold for P.

PROOF. Let us first assume that P has a positive principal symbol. Since P satisfies the Rockland condition at every point, Lemma 5.4.9 tells us that the heat operator $P + \partial_t$ and its adjoint satisfy the Rockland condition at every point.

On the other hand, let $k = \frac{v}{2}$ and let $\Delta_{\nabla,X} : C^\infty(M,\mathcal{E}) \to C^\infty(M,\mathcal{E})$ be a sum of squares of the form,

$$\Delta_{\nabla,X} = \nabla_{X_1}^* \nabla_{X_1} + \ldots + \nabla_{X_m}^* \nabla_{X_m}, \quad (5.4.3)$$

where ∇ is a connection on $\mathcal{E}$ and the vector fields $X_1, \ldots, X_m$. Since by assumption the bracket condition $H + [H,H] = TM$ holds it follows from Proposition 5.2.12 that the principal symbol of $\Delta_{\nabla,X}^k + \partial_t$ admits an inverse in $S_{v,-v}(\mathfrak{g}^*M \times \mathbb{R}_{(v)}, \mathcal{E})$ and that $(\Delta^k + \partial_t)^{-1}$ belongs to $\Psi_{H,v}^{-v}(M \times \mathbb{R}_{(v)}, \mathcal{E})$. In particular, since $(\Delta^{\frac{v}{2}} + \partial_t)^{-1}$ has an invertible principal symbol, together with its adjoint it satisfies the Rockland condition at every point.

Let $Q_1 = (\Delta^k + \partial_t)^{-1}(P + \partial_t)$ and $Q_2 = (P + \partial_t)(\Delta^k + \partial_t)^{-1}$. They are elements of $\Psi_{H,v}^0(M \times \mathbb{R}_{(v)}, \mathcal{E})$ which together with their adjoints satisfy the Rockland condition at every point, since they are products of such operators. We may therefore apply Proposition 5.4.7 to deduce that the principal symbols of Q_1 and Q_2 are invertible in $S_{0,v}(\mathfrak{g}^*M \times \mathbb{R}_{(v)}, \mathcal{E})$. In the same way as in the proof of Theorem 5.4.7 this implies that the principal symbol of $P + \partial_t$ is an invertible Volterra-Heisenberg symbol.

It remains to show that P is bounded from below if, and only if, its principal symbol is positive.

Suppose that P is bounded from below. Possibly replacing P by $P + c$ with $c > 0$ large enough we may assume that P is positive. Notice that P^2 is selfadjoint and satisfies the Rockland condition at every point, since so does P. Therefore, we may apply the first part of the proof to deduce that the principal symbol of $P^2 + \partial_t$ is an invertible Volterra-Heisenberg symbol. As P^2 is positive we then may use Theorem 5.3.1 to see that $(P^2)^{\frac{1}{4}} = P^{\frac{1}{2}}$ is a Ψ_HDO of order $k = \frac{v}{2}$. Since $P = (P^{\frac{1}{2}})^2$ it then follows from Lemma 5.4.9 that the principal symbol of P is a positive symbol.

Conversely, suppose that P has a positive principal symbol. Then thanks to Lemma 5.4.9 the operator P can be written as $P = Q^*Q + R$ with $Q \in \Psi_H^k(M, \mathcal{E})$ and $R \in \Psi_H^{v-1}(M, \mathcal{E})$. Let $P_1 = Q^*Q$. Since P and P_1 have same principal symbol, the operator P_1 also satisfies the Rockland condition at every point.

Next, since P_1 is positive for $\lambda < -1$ we have

$$(P_1 - \lambda)^{-1} = \int_0^\infty e^{-tP_1} e^{t\lambda} dt, \quad (5.4.4)$$

where the integral converges in $\mathcal{L}(L^2(M, \mathcal{E}))$. Let $\alpha = \frac{v-1}{v}$. Then we have

$$R(P_1 - \lambda)^{-1} = RP_1^{-\alpha} R_{(\lambda)}, \quad R_{(\lambda)} = \int_0^\infty P_1^\alpha e^{-tP_1} e^{t\lambda} dt. \quad (5.4.5)$$

Since $P_1 = Q^*Q$ has a positive principal symbol and satisfies the Rockland condition the first part of the proof tells us that the principal symbol of $P_1 + \partial_t$ is an invertible Volterra-Heisenberg symbol. This allows us to apply Theorem 5.3.4 to deduce that $P_1^{-\alpha}$ is a Ψ_HDO of order $-v\alpha = -(v-1)$. Therefore, the operator $RP_1^{-\alpha}$ is a Ψ_HDO of order 0, hence is bounded on $L^2(M, \mathcal{E})$.

On the other hand, we have

$$\|R_{(\lambda)}\|_{L^2} \leq \int_0^\infty t^{-\alpha} \|(tP_1)^\alpha e^{-tP_1}\| e^{t\lambda} dt. \quad (5.4.6)$$

As $\alpha \in (0,1)$ the function $x \to x^\alpha e^{-x}$ maps $[0,\infty)$ to $[0,1]$. Therefore, we have $\|(tP_1)^\alpha e^{-tP_1}\| \leq 1$, from which we get

$$\|R_{(\lambda)}\|_{L^2} \leq \int_0^\infty t^{-\alpha} e^{t\lambda} dt = |\lambda|^{\alpha-1} \int_0^\infty u^{-\alpha} e^{-u} du. \tag{5.4.7}$$

Now, by combining (5.4.5), (5.4.6) and (5.4.7) together we obtain

$$\|R(P_1 - \lambda)^{-1}\|_{L^2} \leq \|RP_1^{-\alpha}\|_{L^2} \|R_{(\lambda)}\|_{L^2} \leq C_\alpha |\lambda|^{\alpha-1}, \tag{5.4.8}$$

where the constant C_α does not depend on λ. Since $\alpha < 1$ it follows that for λ negatively large enough we have $\|R(P_1 - \lambda)^{-1}\|_{L^2} \leq \frac{1}{2}$, so that $1 + R(P_1 - \lambda)^{-1}$ is invertible. Since we have

$$P - \lambda = P_1 + R - \lambda = (P_1 + R - \lambda)(P_1 - \lambda)^{-1}, \tag{5.4.9}$$

it follows that $P - \lambda$ has a right inverse for λ negatively large enough. Since we can similarly show that for λ negatively large enough $P - \lambda$ is left-invertible, we deduce that as soon as λ is negatively large enough $P - \lambda$ admits a bounded two-sided inverse. This means that the spectrum of P is contained in some interval $[c, \infty)$, that is, P is bounded from below. The proof is now complete. □

EXAMPLE 5.4.11. Assume that (M, H) is an orientable contact manifold. Then the 2nd part of Theorem 5.4.10 applies to the contact Laplacian in every degree. Thus, in degree $k \neq n$ the principal symbol of $\Delta_{R,k} + \partial_t$ admits an inverse in $S_{v,-2}(\mathfrak{g}^*M \times \mathbb{R}_{(2)}, \Lambda^k)$ and in degree n the principal symbol of $\Delta_{R,nj} + \partial_t$, $j = 1, 2$, admits an inverse in $S_{v,-4}(\mathfrak{g}^*M \times \mathbb{R}_{(4)}, \Lambda_j^n)$.

Finally, let M^{2n+1} be a compact orientable strictly pseudoconvex CR manifold endowed with a pseudohermitian contact form θ and the associated Levi metric. For $k = 1, \ldots, n+1$ and for $k = n+2, n+4, \ldots$ let $\square_\theta^{(k)} : C^\infty(M) \to C^\infty(M)$ be the Gover-Graham operator of order k. We know that $\square_\theta^{(k)}$ is selfadjoint and satisfies the Rockland condition at every point. Furthemore, we have:

PROPOSITION 5.4.12. *For $k \neq n+1$ the operator $\square_\theta^{(k)}$ is bounded from below and the principal symbol of $\square_\theta^{(k)} + \partial_t$ is invertible in $S_{v,-2k}(\mathfrak{g}^*M \times \mathbb{R}_{(2k)})$. Hence Theorems 5.1.25 and 5.1.26 hold for $\square_\theta^{(k)}$ with $v = 2k$.*

PROOF. Assume $k \neq n+1$. Since $\square_\theta^{(k)}$ is selfadjoint and by Proposition 3.5.7 it satisfies the Rockland condition at every point, thanks to Theorem 5.4.10 we are reduced to show that its principal symbol is positive. Let $\Delta_{b;0}$ be the horizontal sublaplacian on functions and let X_0 be the Reeb vector field of θ. Then by Proposition 3.5.7 the principal symbol of $\square_\theta^{(k)}$ agrees with that of

$$\square = (\Delta_{b;0} + i(k-1)X_0)(\Delta_{b;0} + i(k-3)X_0) \cdots (\Delta_{b;0} - i(k-1)X_0). \tag{5.4.10}$$

Since $k \neq n+1$ it follows from the proof of Proposition 3.5.7 that each factor $\Delta_{b;0} + i\alpha X_0$ in (5.4.10) is a selfadjoint sublaplacian satisfying the condition (3.4.22) at every point. It can be shown that any such sublaplacian is bounded from below (see [**Po1**], [**Po12**]). Since the product of selfadjoint operators which are bounded from below is bounded from below provided that the product is itself selfadjoint, we see that the operator $\frac{1}{2}(\square + \square^*)$ is bounded from below.

Notice that $\frac{1}{2}(\square + \square^*)$ has same principal as $\square$ and $\square_\theta^{(k)}$, hence satisfies the Rockland condition at every point. Therefore, Theorem 5.4.10 tells us that the principal symbol of $\frac{1}{2}(\Box + \Box^*)$ is positive. Incidentally, the operator $\Box_\theta^{(k)}$ has a positive principal. The proof is thus complete. $\square$

5.5. Weighted Sobolev Spaces

In this section we construct weighted Sobolev spaces $W_H^s(M)$, $s \in \mathbb{R}$, which provide us with sharp regularity estimates for Ψ_HDOs. To this end we assume throughout the section that the bracket condition $H + [H, H] = TM$ holds.

Let $X_1, \ldots, X_m$ be real vector fields spanning H and consider the positive sum of squares,

$$(5.5.1) \qquad \Delta_X = X_1^* X_1 + \ldots + X_m^* X_m.$$

As mentioned in Example 5.3.3 (a), since the the bracket condition $H + [H, H] = TM$ holds Theorem 5.3.1 is valid for $1 + \Delta_X$. Thus, the complex powers $(1 + \Delta_X)^s$, $s \in \mathbb{C}$, gives rise to a holomorphic 1-parameter group of Ψ_HDOs such that we have $\mathrm{ord}(1 + \Delta_X)^s = 2s$ for any $s \in \mathbb{C}$.

DEFINITION 5.5.1. $W_H^s(M)$, $s \in \mathbb{R}$, consists of all distributions $u \in \mathcal{D}'(M)$ such that $(1 + \Delta_X)^{\frac{s}{2}} u$ is in $L^2(M)$. It is endowed with the Hilbert norm given by

$$(5.5.2) \qquad \|u\|_{W_H^s} = \|(1 + \Delta_X)^{\frac{s}{2}} u\|_{L^2}, \qquad u \in W_H^s(M).$$

PROPOSITION 5.5.2. *1) Neither $W_H^s(M)$, nor its topology, depend on the choice of the vector fields $X_1, \ldots, X_m$.*

2) We have the following continuous embeddings:

$$(5.5.3) \qquad \begin{array}{llll} L_s^2(M) & \hookrightarrow & W_H^s(M) & \hookrightarrow & L_{s/2}^2(M) & \text{if } s \geq 0, \\ L_{s/2}^2(M) & \hookrightarrow & W_H^s(M) & \hookrightarrow & L_s^2(M) & \text{if } s < 0. \end{array}$$

PROOF. 1) Let $Y_1, \ldots, Y_p$ be other vector fields spanning H. The operator $(1 + \Delta_Y)^s (1 + \Delta_X)^{-s}$ is a Ψ_HDO of order 0, hence is bounded on $L^2(M)$ by Proposition 3.1.8. Therefore, we get the estimates,

$$(5.5.4) \quad \|(1 + \Delta_Y)^s u\|_{L^2} = \|(1 + \Delta_Y)^s (1 + \Delta_X)^{-s} (1 + \Delta_X)^s u\|_{L^2}$$
$$\leq C_{XYs} \|(1 + \Delta_X)^s u\|_{L^2},$$

which hold for any $u \in C^\infty(M)$. Interchanging the roles of the X_j's and of the Y_k's also gives the estimates

$$(5.5.5) \qquad \|(1 + \Delta_X)^s u\|_{L^2} \leq C_{YXs} \|(1 + \Delta_Y)^s u\|_{L^2}, \qquad u \in C^\infty(M).$$

Therefore, whether we use the X_j's or the Y_k's to define the $W_H^s(M)$ changes neither the space, nor its topology.

2) Let $s \in [0, \infty)$. Since $(1 + \Delta_X)^{\frac{s}{2}}$ is a Ψ_HDO of order s, Proposition 3.1.8 tells us that it is bounded from $L_s^2(M)$ to $L^2(M)$. Thus,

$$(5.5.6) \qquad \|u\|_{W_H^s} = \|(1 + \Delta_X)^{\frac{s}{2}} u\|_{L^2} \leq C_s \|u\|_{L_s^2}, \qquad u \in W_H^s(M),$$

which shows that $L_s^2(M)$ embeds continuously into $W_H^s(M)$.

On the other hand, as $(1 + \Delta_X)^{-\frac{s}{2}}$ has order $-s$ Proposition 3.1.8 also tells us that $(1 + \Delta_X)^{-\frac{s}{2}}$ is bounded from $L^2(M)$ to $L_{s/2}^2(M)$. Therefore, on $W_H^s(M)$ we have the estimates

$$(5.5.7) \quad \|u\|_{L_{s/2}^2} = \|(1 + \Delta_X)^{-\frac{s}{2}} (1 + \Delta_X)^{\frac{s}{2}} u\|_{L_{s/2}^2}$$
$$\leq C_s \|(1 + \Delta_X)^{-\frac{s}{2}}\|_{L^2} = C_s \|u\|_{W_H^s},$$

which shows that $W_H^s(M)$ embeds continuously into $L^2_{s/2}(M)$.

Finally, when $s < 0$ we can similarly show that we have continuous embeddings $L^2_{s/2}(M) \hookrightarrow W_H^s(M)$ and $W_H^s(M) \hookrightarrow L^2_s(M)$. $\square$

As an immediate consequence of Proposition 5.5.2 we obtain:

PROPOSITION 5.5.3. *The following equalities between topological spaces hold:*
$$(5.5.8) \quad C^\infty(M) = \cap_{s \in \mathbb{R}} W_H^s(M) \quad \text{and} \quad \mathcal{D}'(M) = \cup_{s \in \mathbb{R}} W_H^s(M).$$

Let us now compare the weighted Sobolev spaces $W_H^s(M)$ to the weighted Sobolev spaces $S_k^2(M)$, $k = 1, 2, \ldots$, of Folland-Stein [**FS1**].

In we sequel we let $\mathbb{N}_m = \{1, \ldots, m\}$ and for any $I = (i_1, \ldots, i_k)$ in $\mathbb{N}_m^k$ we set
$$(5.5.9) \quad X_I = X_{i_1} \ldots X_{i_l}.$$

DEFINITION 5.5.4 ([**FS1**]). *The Hilbert space $S_k^2(M)$, $k \in \mathbb{N}$, consists of functions $u \in L^2(M)$ such that $(X_I)u \in L^2(M)$ for any $I \in \cup_{j=1}^k \mathbb{N}_m^j$. It is endowed with the Hilbertian norm given by*
$$(5.5.10) \quad \|u\|_{S_k^2}^2 = \|u\|_{L^2}^2 + \sum_{1 \leq j \leq k} \sum_{I \in \mathbb{N}_m^j} \|X_I u\|_{L^2}^2, \quad u \in S_k^2(M).$$

PROPOSITION 5.5.5. *For $k = 1, 2, \ldots$ the weighted Sobolev spaces $W_H^k(M)$ and $S_k^2(M)$ agree as spaces and carry the same topology.*

PROOF. First, if $I \in \cup_{j=1}^k \mathbb{N}_m^j$ then the differential operator X_I is bounded from $W_H^k(M)$ to $L^2(M)$, so we get the estimate,
$$(5.5.11) \quad \|u\|_{S^k}^2 = \|u\|_{L^2}^2 + \sum_{1 \leq j \leq k} \sum_{I \in \mathbb{N}_m^j} \|X_I u\|_{L^2}^2 \leq C_k^2 \|u\|_{W_H^k} \quad u \in C^\infty(M).$$

On the other hand, the vector fields X_j's and their adjoints X_j^*'s give rise to bounded linear maps from $S_{l+1}^2(M)$ to $S_l^2(M)$ for any $l \in \mathbb{N}$. Therefore, for any integer p, the operator $(1 + \Delta_X)^p$ is bounded from $S_{l+2p}^2(M)$ to $S_l^2(M)$. It follows that, when k is even, we have
$$(5.5.12) \quad \|u\|_{W_H^k} = \|(1 + \Delta_X)^{\frac{k}{2}} u\|_{L^2} \leq C_k \|u\|_{S^k}, \quad u \in C^\infty(M).$$

Moreover, for any $u \in C^\infty(M)$ we have
$$(5.5.13) \quad \|(1 + \Delta_X)^{\frac{1}{2}} u\|_{L^2}^2 = \langle (1 + \sum_{1 \leq j \leq m} X_j^* X_j) u, u \rangle_{L^2}$$
$$= \|u\|_{L^2}^2 + \sum_{1 \leq j \leq m} \|X_j^2\|_{L^2}^2 = \|u\|_{S_1^2}^2.$$

Thus $(1 + \Delta_X)^{\frac{1}{2}}$ is bounded from $S_1^2(M)$ to $L^2(M)$. Therefore, if k is odd, say $k = 2p + 1$, then the operator $(1 + \Delta_X)^{\frac{k}{2}} = (1 + \Delta_X)^p (1 + \Delta_X)^{\frac{1}{2}}$ is bounded from $S_k^2(M)$ to $L^2(M)$. Hence (5.5.13) is valid in the odd case as well. Together with (5.5.11) this implies that $W_H^k(M)$ and $S_k^2(M)$ agree as spaces and carry the same topology. $\square$

Now, let $\mathcal{E}$ be a Hermitian vector bundle over M. We can also define weighted Sobolev spaces of sections of $\mathcal{E}$ as follows. Let $\nabla : C^\infty(M, \mathcal{E}) \to C^\infty(M, T^*M \times \mathcal{E})$ be a connection on $\mathcal{E}$ and define
$$(5.5.14) \quad \Delta_{\nabla, X} = \nabla_{X_1}^* \nabla_{X_1} + \ldots + \nabla_{X_m}^* \nabla_{X_m}.$$

As in the scalar case, the complex powers $(1+\Delta_{\nabla,X})^s$, $s \in \mathbb{C}$, form an analytic 1-parameter group of Ψ_HDOs such that $\mathrm{ord}(1+\Delta_{\nabla,X})^s = 2s$ for any $s \in \mathbb{C}$.

DEFINITION 5.5.6. $W_H^s(M,\mathcal{E})$, $s \in \mathbb{R}$, consists of all distributional sections $u \in \mathcal{D}'(M,\mathcal{E})$ such that $(1+\Delta_{\nabla,X})^{\frac{s}{2}}u \in L^2(M,\mathcal{E})$. It is endowed with the Hilbertian norm given by

$$(5.5.15) \qquad \|u\|_{W_H^s} = \|(1+\Delta_{\nabla,X})^{\frac{s}{2}}u\|_{L^2}, \qquad u \in W_H^s(M,\mathcal{E}).$$

Along similar lines as that of the proof of Proposition 5.5.2 we can prove:

PROPOSITION 5.5.7. *1) As a topological space $W_H^s(M,\mathcal{E})$, $s \in \mathbb{R}$, is independent of the choice of the connection ∇ and of the vector fields $X_1, \ldots, X_m$.*

2) We have the following continuous embeddings:

$$(5.5.16) \qquad \begin{array}{llll} L_s^2(M,\mathcal{E}) & \hookrightarrow & W_H^s(M,\mathcal{E}) & \hookrightarrow & L_{s/2}^2(M,\mathcal{E}) & \text{if } s \geq 0, \\ L_{s/2}^2(M,\mathcal{E}) & \hookrightarrow & W_H^s(M,\mathcal{E}) & \hookrightarrow & L_s^2(M,\mathcal{E}) & \text{if } s < 0. \end{array}$$

As a consequence we obtain the equalities of topological spaces,

$$(5.5.17) \quad C^\infty(M,\mathcal{E}) = \cap_{s \in \mathbb{R}} W_H^s(M,\mathcal{E}) \quad \text{and} \quad \mathcal{D}'(M,\mathcal{E}) = \cup_{s \in \mathbb{R}} W_H^s(M,\mathcal{E}).$$

Notice that we can also define Folland-Stein spaces $S_k^2(M,\mathcal{E})$, $k = 1,2,\ldots$, as in the scalar case, by using the differential operators $\nabla_{X_I} = \nabla_{X_{i_1}} \cdots \nabla_{X_{i_k}}$, $I \in \cup_{j=1}^k \mathbb{N}_m^j$. Then, by arguing as in the proof of Proposition 5.5.5, we can show that the spaces $W_H^k(M,\mathcal{E})$ and $S_k^2(M,\mathcal{E})$ agree and bear the same topology.

Now, the Sobolev spaces $W_H^s(M,\mathcal{E})$ yield sharp regularity results for Ψ_HDOs.

PROPOSITION 5.5.8. *Let $P : C^\infty(M,\mathcal{E}) \to C^\infty(M,\mathcal{E})$ be a Ψ_HDO of order m and set $k = \Re m$. Then, for any $s \in \mathbb{R}$, the operator P extends to a continuous linear mapping from $W_H^{s+k}(M,\mathcal{E})$ to $W_H^s(M,\mathcal{E})$.*

PROOF. As $P_s = (1+\Delta_{\nabla,X})^{\frac{s}{2}} P (1+\Delta_{\nabla,X})^{-\frac{(s+k)}{2}}$ is a Ψ_HDO with purely imaginary order, Proposition 3.1.8 tells us that it gives rise to a bounded operator on $L^2(M,\mathcal{E})$. Therefore, we have

$$(5.5.18) \quad \|Pu\|_{W_H^s} = \|P_s(1+\Delta_{\nabla,X})^{s+k}u\|_{L^2} \leq C_s \|u\|_{W_H^{s+k}}, \qquad u \in C^\infty(M,\mathcal{E}),$$

It then follows that P extends to a continuous linear mapping from $W_H^{s+k}(M,\mathcal{E})$ to $W_H^s(M,\mathcal{E})$. □

PROPOSITION 5.5.9. *Let $P : C^\infty(M,\mathcal{E}) \to C^\infty(M,\mathcal{E})$ be a Ψ_HDO of order m such that P satisfies the Rockland condition at every point and set $k = \Re m$. Then for any $u \in \mathcal{D}'(M,\mathcal{E})$ we have*

$$(5.5.19) \qquad Pu \in W_H^s(M,\mathcal{E}) \Longrightarrow u \in W_H^{s+k}(M,\mathcal{E}).$$

In fact, for any $s' \in \mathbb{R}$ we have the estimate,

$$(5.5.20) \quad \|u\|_{W_H^{s+k}} \leq C_{ss'}(\|Pu\|_{W_H^s} + \|u\|_{W_H^{s'}}), \qquad u \in W_H^{s+k}(M,\mathcal{E}).$$

PROOF. Since P satisfies the Rockland condition at every point and the bracket condition $H + [H,H] = TM$ holds, it follows from the proof of Proposition 5.4.2 that P admits a left parametrix in $\Psi_H^{-m}(M,\mathcal{E})$, i.e., there exist Q in $\Psi_H^{-m}(M,\mathcal{E})$ and R in $\Psi^{-\infty}(M,\mathcal{E})$ such that $QP = 1 - R$. Therefore, for any $u \in \mathcal{D}'(M,\mathcal{E})$ we have $u = QPu + Ru$. Thanks to the first part we know that Q maps $W_H^s(M,\mathcal{E})$ to

$W_H^{s+k}(m, \mathcal{E})$. Since R is smoothing, and so Ru always is smooth, it follows that if Pu is in $W_H(M, \mathcal{E})$ then u must be in $W_H(M, \mathcal{E})$.

In fact, as Q is actually bounded from W_H^s to W_H^{s+k} and (5.5.17) implies that R is bounded from any space $W_H^{s'}(M, \mathcal{E})$ to $W_H^{s+k}(M, \mathcal{E})$, on $W_H^{s+k}(M, \mathcal{E})$ we have estimates,

$$(5.5.21) \qquad \|u\|_{W_H^{s+k}} \leq \|QPu\|_{W_H^{s+k}} + \|Ru\|_{W_H^{s+k}} \leq C_{ss'}(\|Pu\|_{W_H^s} + \|u\|_{W_H^{s'}}).$$

The proof is thus complete. □

REMARK 5.5.10. By combining Proposition 5.5.9 with the embeddings (5.5.3) and (5.5.16) we recover the Sobolev regularity of Ψ_HDOs as in Proposition 3.1.8 as well as the hypoelliptic estimates (3.3.2).

REMARK 5.5.11. When $\Re m$ is an integer it follows from Proposition 5.5.5 that the estimates (5.5.20) are equivalent to maximal hypoellipticity in the sense of [**HN3**].

On the other hand, the spaces $W_H^s(M, \mathcal{E})$ can be localized as follows.

DEFINITION 5.5.12. *We say that $u \in \mathcal{D}'(M, \mathcal{E})$ is W_H^s near a point $x_0 \in M$ whenever there exists $\varphi \in C^\infty(M)$ such that $\varphi(x_0) \neq 0$ and φu is in $W_H^s(M, \mathcal{E})$.*

This definition depends only on the germ of u at x_0 because we have:

LEMMA 5.5.13. *Let $u \in \mathcal{D}'(M, \mathcal{E})$ be W_H^s near x_0. Then for any $\varphi \in C^\infty(M)$ with a small enough support about x_0 the distribution φu lies in $W_H^s(M, \mathcal{E})$.*

PROOF. Let $\varphi \in C^\infty(M)$ be such that $\varphi(x_0) \neq 0$ and φu is in $W_H^s(M, \mathcal{E})$ and let $\psi \in C^\infty(M)$ be so that $\psi(x_0) \neq 0$ and $\operatorname{supp} \psi \cap \varphi^{-1}(0) \neq \emptyset$. Then $\chi := \frac{\psi}{\varphi}$ is in $C^\infty(M)$ and $(1 + \Delta_{\nabla, X})^{\frac{s}{2}} \psi u$ is equal to

$$(5.5.22) \qquad (1 + \Delta_{\nabla, X})^{\frac{s}{2}} \chi \varphi u = (1 + \Delta_{\nabla, X})^{\frac{s}{2}} \chi (1 + \Delta_{\nabla, X})^{-\frac{s}{2}} . (1 + \Delta_{\nabla, X})^{\frac{s}{2}} \varphi u.$$

Since $(1 + \Delta_{\nabla, X})^{\frac{s}{2}} \varphi u$ is in $L^2(M, \mathcal{E})$ and $(1 + \Delta_{\nabla, X})^{\frac{s}{2}} \chi (1 + \Delta_{\nabla, X})^{-\frac{s}{2}}$ is a zero'th order Ψ_HDO, so maps $L^2(M, \mathcal{E})$ to itself, it follows that ψu lies in $W_H^s(M, \mathcal{E})$. Hence the lemma. □

We can now get a localized version of (5.5.19).

PROPOSITION 5.5.14. *Let $P \in \Psi_H^m(M, \mathcal{E})$ have an invertible principal symbol, set $k = \Re m$ and let $s \in \mathbb{R}$. Then for any $u \in \mathcal{D}'(M, \mathcal{E})$ we have*

$$(5.5.23) \qquad Pu \text{ is } W_H^s \text{ near } x_0 \implies u \text{ is } W_H^s \text{ near } x_0.$$

PROOF. Assume that Pu is W_H^s near x_0 and let $\varphi \in C^\infty(M)$ be such that $\varphi(x_0) \neq 0$ and φu is in $W_H^s(M, \mathcal{E})$. Thanks to Lemma 5.5.13 we may assume that $\varphi = 1$ near x_0. Let $\psi \in C^\infty(M)$ be such that $\psi(x_0) \neq 0$ and $\varphi = 1$ near $\operatorname{supp} \psi$. Since the principal symbol of P is invertible there exist Q in $\Psi_H^{-m}(M, \mathcal{E})$ and R in $\Psi^{-\infty}(M, \mathcal{E})$ such that $QP = 1 - R$. Thus for any $u \in \mathcal{D}'(M, \mathcal{E})$ we have

$$(5.5.24) \qquad \psi u = \psi Q Pu + \psi R u = \psi Q \varphi Pu + \psi Q(1 - \varphi) Pu + \psi Ru.$$

In the above equality Ru is a smooth function since R is a smoothing operator. Similarly, as ψ and $1 - \varphi$ have disjoint supports, the operator $\psi Q(1 - \varphi) P$ is a smoothing operator and so $\psi Q(1 - \varphi) Pu$ is a smooth function. In addition, since φPu is in $W_H^s(M, \mathcal{E})$ and ψQ is a Ψ_HDO of order $-m$, and so maps $W_H^s(M, \mathcal{E})$ to $W_H^{s+k}(M, \mathcal{E})$, it follows that ψu is in W_H^{s+k}. Hence u is W_H^{s+k} near x_0. □

Next, Proposition 5.5.8 admits a generalization to holomorphic families.

PROPOSITION 5.5.15. *Let $\Omega \subset \mathbb{C}$ be open and let $(P_z)_{z\in\Omega} \in \mathrm{Hol}(\Omega, \Psi_H^*(M, \mathcal{E}))$ be such that $m := \sup_{z\in\Omega} \mathrm{Rord} P_z < \infty$. Then for any $s \in \mathbb{R}$ the family $(P_z)_{z\in\Omega}$ defines a holomorphic family with values in $\mathcal{L}(W_H^{s+m}(M, \mathcal{E}), W_H^s(M, \mathcal{E}))$.*

PROOF. Let $V \subset \mathbb{R}^{d+1}$ be a Heisenberg chart with H-frame $Y_0, Y_1, \ldots, Y_d$ and let $(Q_z)_{z\in\Omega} \in \mathrm{Hol}(\Omega, \Psi_H^*(V))$ be such that $\mathrm{Rord} Q_z \leq 0$. Then we can write

$$(5.5.25) \qquad Q_z = p_z(x, -iY) + R_z,$$

for some families $(p_z)_{z\in\Omega} \in \mathrm{Hol}(\Omega, S^*(V \times \mathbb{R}^{d+1}))$ and $(R_z)_{z\in\Omega} \in \mathrm{Hol}(\Omega, \Psi^\infty(V))$. Since $\mathrm{Rord} p_z = \mathrm{Rord} Q_z \leq 0$ we see that $(p_z)_{z\in\Omega}$ belongs to $\mathrm{Hol}(\Omega, S_\|^0(U \times \mathbb{R}^{d+1}))$.

Next, for $j = 1, \ldots, d$ let σ_j denote the standard symbol of $-iY_j$ and set $\sigma = (\sigma_0, \ldots, \sigma_d)$. Then it follows from the proof of Proposition 4.2.5 that the family $(p_z(x, \sigma(x, \xi)))_{z\in\Omega}$ lies in $\mathrm{Hol}(\Omega, S_{\frac{1}{2},\frac{1}{2}}(V \times \mathbb{R}^{d+1}))$. Since by [**Hw**] the quantization map $q \to q(x, D_x)$ is continuous from $S_{\frac{1}{2},\frac{1}{2}}(U \times \mathbb{R}^{d+1})$ to $\mathcal{L}(L^2_{\mathrm{loc}}(U), L^2(U))$, we deduce that $(p_z(x, -iX))_{z\in\Omega}$ and $(Q_z)_{z\in\Omega}$ are holomorphic families with values in $\mathcal{L}(L^2_{\mathrm{loc}}(U), L^2(U))$.

It follows from the above result that any family $(Q_z)_{z\in\Omega} \in \mathrm{Hol}(\Omega, \Psi_H^*(M, \mathcal{E}))$ such that $\mathrm{Rord} Q_z \leq 0$ gives rise to a holomorphic family with values in $\mathcal{L}(L^2(M, \mathcal{E}))$. This applies in particular to the family,

$$(5.5.26) \qquad Q_z^{(s)} = (1 + \Delta_{\nabla, X})^{\frac{s}{2}} Q_z (1 + \Delta_{\nabla, X})^{-\frac{m+s}{2}}, \quad z \in \Omega.$$

As $Q_z = (1 + \Delta_{\nabla, X})^{-\frac{s}{2}} Q_z^{(s)} (1 + \Delta_{\nabla, X})^{\frac{m+s}{2}}$ it follows that $(Q_z)_{z\in\Omega}$ gives rise to a holomorphic family with values in $\mathcal{L}(W_H^{s+m}(M, \mathcal{E}), W_H^s(M, \mathcal{E}))$. □

Combining Proposition 5.5.15 with Theorem 5.3.4 we immediately get:

PROPOSITION 5.5.16. *Let $P : C^\infty(M, \mathcal{E}) \to C^\infty(M, \mathcal{E})$ be a positive differential operator of even (Heisenberg) order v such that P satisfies the Rockland condition at every point. Then, for any reals m and s, we have*

$$(5.5.27) \qquad (P^z)_{\Re z < m} \in \mathrm{Hol}(\Re z < m, \mathcal{L}(W_H^{s+mv}(M, \mathcal{E}), W_H^s(M, \mathcal{E}))).$$

CHAPTER 6

Spectral Asymptotics for Hypoelliptic Operators

In this chapter we apply the results of the previous chapters to derive spectral asymptotics for hypoelliptic differential operators on Heisenberg manifolds.

In Section 6.1 we get spectral asymptotics for general hypoelliptic differential operators. We then give precise geometric expressions for these asymptotics. First, in Section 6.2 we deal with the Kohn Laplacian and the horizontal sublaplacian on a κ-strictly pseudoconvex CR manifold, as well as for the Gover-Graham operators in the strictly pseudoconvex case. Second, we deal with the horizontal sublaplacian and the contact Laplacian on a contact manifold in Section 6.3.

6.1. Spectral asymptotics for hypoelliptic operators

In this section we explain how Theorem 5.4.10 enables us to derive spectral asymptotics for hypoelliptic operators on Heisenberg manifolds. Throughout this section we let (M^{d+1}, H) denote a compact Heisenberg manifold equipped with a smooth density > 0 and let $\mathcal{E}$ be a Hermitian vector bundle over M.

Let $P : C^\infty(M, \mathcal{E}) \to C^\infty(M, \mathcal{E})$ be a selfadjoint differential operator of even Heisenberg order m which is bounded from below and such that the principal symbol of $P + \partial_t$ is an invertible Volterra-Heisenberg symbol. Recall that by Proposition 5.2.9 and Theorem 5.4.10 the latter occurs when at least one of the following conditions holds:

- P is a sublaplacian and satisfies the condition (3.4.22) at every point.

- The bracket condition $H + [H, H] = TM$ holds and P satisfies the Rockland condition at every point.

As an immediate consequence of Theorem 5.1.26 we get:

PROPOSITION 6.1.1 ([**BGS**, Thm. 5.6]). *As $t \to 0^+$ we have*

$$(6.1.1) \qquad \operatorname{Tr} e^{-tP} \sim t^{-\frac{d+2}{m}} \sum t^{\frac{2j}{m}} A_j(P), \qquad A_j(P) = \int_M \operatorname{tr}_\mathcal{E} a_j(P)(x),$$

where the density $a_j(P)(x)$ is the coefficient of $t^{\frac{j-d+2}{m}}$ in the heat kernel asymptotics (5.1.20) for P.

Let $\lambda_0(P) \leq \lambda_1(P) \leq \ldots$ denote the eigenvalues of P counted with multiplicity and let $N(P; \lambda)$ be the counting function of P, that is,

$$(6.1.2) \qquad N(P; \lambda) = \#\{k \in \mathbb{N};\ \lambda_k(P) \leq \lambda\}, \qquad \lambda \in \mathbb{R}.$$

PROPOSITION 6.1.2. *1) We have $A_0(P) > 0$.*

2) As $\lambda \to \infty$ we have

$$(6.1.3) \qquad N(P; \lambda) \sim \nu_0(P) \lambda^{\frac{d+2}{m}}, \qquad \nu_0(P) = \Gamma(1 + \frac{d+2}{m})^{-1} A_0(P).$$

3) As $k \to \infty$ we have

$$(6.1.4) \qquad \lambda_k(P) \sim \left(\frac{k}{\nu_0(P)}\right)^{\frac{m}{d+2}}.$$

PROOF. First, we have $A_0(P) = \lim_{t \to 0^+} t^{\frac{d+2}{m}} \operatorname{Tr} e^{-tP} \geq 0$, so if we can show that $A_0(P) \neq 0$ then we get $A_0(P) > 0$ and the asymptotics (6.1.3) would follow from (6.1.1) by Karamata's Tauberian theorem (see, e.g., [**Ha**, Thm. 108]). This would also give (6.1.4), since the latter is equivalent to (6.1.3) (see, e.g., [**Sh**, Sect. 13]). Therefore, the bulk of the proof is to show that we have $A_0(P) \neq 0$.

Second, notice that there is at least one integer $< \frac{m}{d+2}$ such that $A_j(P) \neq 0$. Otherwise, by (6.1.1) there would exist a constant $C > 0$ such that $\operatorname{Tr} e^{-tP} \leq C$ for $0 \leq t \leq 1$. Thus,

$$(6.1.5) \qquad k e^{-t\lambda_k(P)} \leq \sum_{j<k} e^{-t\lambda_j(P)} \leq \operatorname{Tr} e^{-tP} \leq C, \qquad 0 < t < 1.$$

Therefore, letting $t \to 0^+$ would give $k \leq C$ for every $k \in \mathbb{N}$, which is not possible since P is unbounded.

Next, when $m \geq d+2$ the only integer $< \frac{m}{d+2}$ is $j = 0$, so in this case we must have $A_0(P) \neq 0$.

Suppose now that we have $m < d+2$ and $A_0(P) = 0$. Let $\mu = \frac{d+2}{m} - j_0$, where j_0 is the smallest integer j such that $A_j(P) \neq 0$. Notice that since $1 \leq j_0 < \frac{d+2}{m}$, we have $0 < \mu \leq \frac{d+2}{m} - 1$. Moreover, the asymptotics (6.1.1) becomes

$$(6.1.6) \qquad \operatorname{Tr} e^{-tP} = A_{j_0}(P) t^{-\mu} + \operatorname{O}(t^{1-\mu}) \qquad \text{as } t \to 0^+.$$

This implies that we have $A_{j_0}(P) \geq 0$. Since we have $A_{j_0}(P) \neq 0$ by definition of j_0, we get $A_{j_0}(P) > 0$. Therefore, as alluded to above, it follows from Karamata's Tauberian theorem that as $k \to \infty$ we have

$$(6.1.7) \qquad \lambda_k(P) \sim \left(\frac{k}{\beta}\right)^{\frac{1}{\mu}}, \qquad \beta = \Gamma(1+\mu)^{-1} A_{j_0}(P).$$

It follows that $\lambda_k(P^{-\frac{d+2}{2m}}) = \operatorname{O}(k^{-\frac{1}{2}-\delta})$, with $\delta = \frac{1}{2}(\frac{1}{\mu}\frac{d+2}{m} - 1) > 0$. In particular, we have $\sum_{k \geq 0} \lambda_k(P^{-\frac{d+2}{2m}})^2 < \infty$, that is, $P^{-\frac{d+2}{2m}}$ is a Hilbert-Schmidt operator on $L^2(M, \mathcal{E})$.

In addition, observe that $P^{-\frac{d+2}{2m}}$ is a $\Psi_H\text{DO}$ of order $-\frac{(d+2)}{2}$ and that any operator $Q \in \Psi_H^{-\frac{(d+2)}{2}}(M, \mathcal{E})$ can be written as

$$(6.1.8) \qquad Q = \Pi_0(P)Q + (1 - \Pi_0(P))P^{-\frac{d+2}{2m}} P^{\frac{d+2}{2m}} Q,$$

where $\Pi_0(P)$ denotes the orthogonal projection onto $\ker P$. Recall that the space of Hilbert-Schmidt operators is a two-sided ideal of $\mathcal{L}(L^2(M < \mathcal{E})$. Observe that in (6.1.8) the projection $\Pi_0(P)$ is a smoothing operator, hence is a Hilbert-Schmidt operator. Moreover, $P^{\frac{d+2}{2m}}Q$ is a bounded operator on $L^2(M, \mathcal{E})$, since this is a zero'th order $\Psi_H\text{DO}$. Therefore, we see that any $Q \in \Psi_H^{-\frac{(d+2)}{2}}(M, \mathcal{E})$ is a Hilbert-Schmidt operator on $L^2(M, \mathcal{E})$.

We now get a contradiction as follows. Let $\kappa : U \to V$ be a Heisenberg chart over which there is a trivialization $\tau : \mathcal{E}_U \to U \times \mathbb{C}^r$ of $\mathcal{E}$ and such that the open $V \subset \mathbb{R}^{d+1}$ is bounded. Let $\varphi \in C_c^\infty(\mathbb{R}^{d+1})$ have non-empty support L, let

$\chi \in C_c^\infty(V \times V)$ be such that $\chi(x,y) = 1$ near $L \times L$, and let $Q : C_c^\infty(V) \to C^\infty(V)$ be given by the kernel,

(6.1.9) $$k_Q(x,y) = |\varepsilon_x'|\varphi(x)\|\varepsilon_x(y)\|^{-\frac{d+2}{2}}\chi(x,y).$$

The kernel $k_Q(x,y)$ has a compact support contained in $V \times V$ and, as $\varphi(x)(1-\chi(x,y))$ vanishes near the diagonal of $V \times V$, we have

(6.1.10) $$k_Q(x,y) = |\varepsilon_x'|\varphi(x)\| - \varepsilon_x(y)\|^{-\frac{d+2}{2}} \mod C^\infty(V \times V).$$

Since $\varphi(x)\|y\|^{-\frac{d+2}{2}}$ belongs to $\mathcal{K}^{-\frac{d+2}{2}}(V \times \mathbb{R}^{d+1})$, it follows from Proposition 3.1.16 that Q is a Ψ_HDO of order $-\frac{d+2}{2}$.

Let $Q_0 = \tau^*\kappa^*(Q \otimes 1)$. Then Q_0 belongs to $\Psi_H^{-\frac{d+2}{2}}(M, \mathcal{E})$, hence is a Hilbert-Schmidt operator on $L^2(M, \mathcal{E})$. This implies that Q is a Hilbert-Schmidt operator on $L^2(V)$, so by [**GK**, p. 109] its kernel lies in $L^2(V \times V)$. This cannot be true, however, because we have

(6.1.11) $$\int_{V \times V} |k_Q(x,y)|^2 dxdy \geq \int_{L \times L} |\varepsilon_x'|^2|\varphi(x)|^2\|\varepsilon_x(y)\|^{-(d+2)}dxdy$$
$$= \int_L |\varepsilon_x'||\varphi(x)|^2 (\int_{\varepsilon_x(L)} \|y\|^{-(d+2)}dy)dx = \infty.$$

This gives a contradiction, so we must have $A_0(P) \neq 0$. The proof is therefore complete. $\square$

EXAMPLE 6.1.3. Proposition 6.1.2 is valid for the following operators:

(a) Real selfadjoint sublaplacian $\Delta = \nabla_{X_1}^*\nabla_{X_1} + \ldots + \nabla_{X_m}^*\nabla_{X_m}^2 + \mu(x)$, where ∇ is connection on $\mathcal{E}$, the vector fields $X_1, \ldots, X_m$ span H and $\mu(x)$ is a selfadjoint section of $\text{End}\,\mathcal{E}$, provided that the Levi form of (M, H) is non-vanishing;

(b) The Kohn Laplacian on a compact CR manifold and acting on (p,q)-forms when the condition $Y(q)$ holds everywhere;

(c) The horizontal sublaplacian on a compact Heisenberg manifold acting on horizontal forms of degree k when the condition $X(k)$ holds everywhere;

(d) The horizontal sublaplacian on a compact CR manifold acting on (p,q)-forms when the condition $X(p,q)$ holds everywhere;

(e) The Gover-Graham operators $\square_\theta^{(k)}$ on a strictly pseudoconvex CR manifold M^{2n+1} with $k \neq n+1$;

(f) The contact Laplacian on a compact orientable contact manifold.

Several authors have obtained Weyl asymptotics closely related to (1.5.3) for bicharacteristic hypoelliptic operators (see [**II**], [**Me1**], [**MS**]), including sublaplacians on Heisenberg manifolds, and for more general hypoelliptic operators (see [**Mo1**], [**Mo2**]) using different approaches involving other pseudodifferential calculi.

These authors obtained their results in settings more general than the Heisenberg setting but, as far as the Heisenberg setting is concerned, our approach using the Volterra-Heisenberg calculus presents the following advantages:

(i) The pseudodifferential analysis is somewhat simpler, since the Volterra-Heisenberg calculus yields for free a heat kernel asymptotics, once the principal symbol of the heat operator is shown to be invertible, for which it is enough to use the Rockland condition in many cases;

(ii) As the Volterra-Heisenberg calculus takes fully into account the underlying Heisenberg geometry of the manifold and is invariant by change of Heisenberg coordinates, we can explicitly compute the coefficient $\nu_0(P)$ in (6.1.3) for operators admitting a normal form. This is illustrated below by Proposition 6.1.5, which will be used in the next two sections to give geometric expressions for the Weyl asymptotics (6.1.3) for geometric operators on CR and contact manifolds.

Next, prior to dealing with operators admitting a normal form we need the following lemma.

LEMMA 6.1.4. *Let $P : C^\infty(M, \mathcal{E}) \to C^\infty(M, \mathcal{E})$ be a selfadjoint differential operator of Heisenberg order m such that P is bounded from below and the principal symbol of $P + \partial_t$ is an invertible Volterra-Heisenberg symbol. Then, for any $a \in M$, the following hold.*

1) The model operator $P^a + \partial_t$ admits a unique fundamental solution $K^a(x, t)$ which, in addition, belongs to the class $\mathcal{K}_{v, -(d+2)}(G_a M \times \mathbb{R}_{(m)})$.

2) In Heisenberg coordinates centered at a we have

(6.1.12) $$a_0(P)(0) = K^a(0, 1).$$

PROOF. First, it is enough to prove the result in a trivializing Heisenberg chart U near a. Furthermore, we may as well assume that P is scalar because the proof for systems follows along similar lines.

Let $Q \in \Psi_{H,v}^{-m}(U \times \mathbb{R}_{(m)})$ be a parametrix for $P + \partial_t$ on $U \times \mathbb{R}$. Let $\sigma_{-m}(Q) \in S_{v,-m}(\mathfrak{g}^* U \times \mathbb{R}_{(m)})$ be its global principal symbol, so that $\sigma_{-m}(Q)$ the inverse of $\sigma_m(P) + i\tau$, and let Q^a be the left-invariant Volterra-Ψ_HDO on $G_a U \times \mathbb{R}$ with symbol $\sigma_{-m}(Q)(a, ., .)$. In particular, the operator Q^a is the inverse of $P^a + \partial_t$ on $\mathcal{S}(G_a U \times \mathbb{R})$ and it agrees with the left-convolution by $K^a = [\sigma_{-m}(a, ., t)]^\vee_{(\xi, \tau) \to (y, t)}$. Therefore, the left-convolution operator with $[(P^a + \partial_t)K^a]$ agrees with $(P^a + \partial_t)Q^a = 1$. Thus,

(6.1.13) $$(P^a + \partial_t)K^a(y, t) = \delta(y, t),$$

that is, $K^a(y, t)$ is a fundamental solution for $P^a + \partial_t$.

Let $K \in \mathcal{S}'(G_a U \times \mathbb{R})$ be another fundamental solution for $P^a + \partial_t$. Then the left-convolution operator Q by K is a right-inverse for $P^a + \partial_t$, so agrees with Q^a. Hence $K = K^a$, which shows that K^a is the unique fundamental solution of $P^a + \partial_t$.

In addition, K^a is in the class $\mathcal{K}_{v,-(d+2)}(G_a M \times \mathbb{R}_{(m)})$ because this the inverse Fourier transform of a symbol in $S_{v,-m}(\mathfrak{g}^* U \times \mathbb{R}_{(m)})$.

Finally, let $q_{-m} \in S_{v,-m}(U \times \mathbb{R}^{d+1} \times \mathbb{R}_{(m)})$ be the local principal symbol of the parametrix Q. As Q has global principal symbol $\sigma_{-m}(Q)$ it follows from Proposition 5.1.19 that $q_{-m}(0, ., .) = \sigma_{-m}(Q)(a, ., .)$. Moreover, since we are in the Heisenberg coordinates centered at a, the map ε_0 to the Heisenberg coordinates centered at 0 is just the identity. Combining this with (6.1.3) then gives

(6.1.14) $$a_0(P)(0) = |\varepsilon'_0| \check{q}_m(0, 0, 1) = [\sigma_{-m}(Q)]_{(\xi, \tau) \to (y, t)}(a, 0, 1) = K^a(0, 1).$$

The proof is thus achieved. □

Now, assume that the Levi form of (M, H) has constant rank $2n$. Therefore, by Proposition 2.1.6 the tangent Lie group bundle GM is a fiber bundle with typical fiber $G = \mathbb{H}^{2n+1} \times \mathbb{R}^{d-2n}$.

Let $P : C^\infty(M, \mathcal{E}) \to C^\infty(M, \mathcal{E})$ be a selfadjoint differential operator of Heisenberg order m such that P is bounded from below and satisfies the Rockland condition at every point.

We further assume that the density $d\rho(x)$ of M and the model operator of P at a point admit normal forms. By this it is meant that there exist $\rho_0 > 0$ and a differential operator $P_0 : C^\infty(G, \mathbb{C}^r) \to C^\infty(G, \mathbb{C}^r)$ such that for any point $a \in M$ there exist trivializing Heisenberg coordinates, herewith called normal trivializing Heisenberg coordinates centered at a, with respect to which the following hold:

(i) We have $d\rho(x)|_{x=0} = [\rho_0 dx]|_{x=0}$;

(ii) At $x = 0$ the tangent group is G and the model operator of P is P_0.

In the sequel, we let $\mathrm{vol}_\rho M$ denote the volume of M with respect to ρ, that is,

$$(6.1.15) \qquad \mathrm{vol}_\rho M = \int_M d\rho(x).$$

As it turns out the assumptions (i) and (ii) allows us to relate the Weyl asymptotics (6.1.3) for P to $\mathrm{vol}_\rho M$ as follows.

PROPOSITION 6.1.5. *Under the assumptions (i) and (ii) as $\lambda \to \infty$ we have*

$$(6.1.16) \quad N(P; \lambda) \sim \frac{\nu_0(P_0)}{\rho_0}(\mathrm{vol}_\rho M)\lambda^{\frac{d+2}{m}}, \quad \nu_0(P_0) = \Gamma(1 + \frac{d+2}{m})^{-1} \mathrm{tr}_{\mathbb{C}^r} K_0(0, 1),$$

where $K_0(x, t)$ denotes the fundamental solution of $P_0 + \partial_t$.

PROOF. By Proposition 6.1.2 as $\lambda \to \infty$ we have

$$(6.1.17) \quad N(P; \lambda) \sim \nu_0(P)\lambda^{\frac{d+2}{m}}, \quad \nu_0(P) = \Gamma(1 + \frac{d+2}{m})^{-1} \int_M \mathrm{tr}_{\mathcal{E}}\, a_0(P)(x).$$

On the other hand, by Lemma 6.1.4 in normal trivializing Heisenberg coordinates centered at point $a \in M$ we have
$$(6.1.18)$$
$$[\mathrm{tr}_{\mathbb{C}^r} a_0(P)(x)dx]|_{x=0} = [\mathrm{tr}_{\mathbb{C}^r} K_0(0,1)dx]|_{x=0} = \mathrm{tr}_{\mathbb{C}^r} K_0(0,1)\rho_0^{-1}[d\rho(x)]|_{x=0}.$$

Hence $\mathrm{tr}_{\mathcal{E}}\, a_0(P)(x) = \rho_0^{-1} \mathrm{tr}_{\mathbb{C}^r} K_0(0,1)d\rho(x)$. Thus,

$$(6.1.19) \quad \nu_0(P) = \Gamma(1 + \frac{d+2}{m})^{-1} \int_M \rho_0^{-1} \mathrm{tr}_{\mathbb{C}^r} K_0(0,1)d\rho(x) = \rho_0^{-1}\nu_0(P_0)\,\mathrm{vol}_\rho M,$$

with $\nu_0(P_0) = \Gamma(1 + \frac{d+2}{m})^{-1} \mathrm{tr}_{\mathbb{C}^r} K_0(0, 1)$. Combining this (6.1.17) then proves the lemma. □

6.2. Weyl asymptotics and CR geometry

The aim of this section is to express in geometric terms the Weyl asymptotics (6.1.3) for the Kohn Laplacian, the horizontal sublaplacian and the Gover-Graham on CR manifolds with Levi metrics.

Let M^{2n+1} be a κ-strictly pseudoconvex manifold, $0 \leq \kappa \leq \frac{n}{2}$, equipped with a pseudohermitian contact form θ. Let $T_{1,0} \subset T_{\mathbb{C}}M$ be the CR tangent bundle of M, set $T_{0,1} = \overline{T_{1,0}}$ and $H = \Re(T_{1,0} \oplus T_{0,1})$ and let L_θ denote the Levi form on $T_{1,0}$ so that

$$(6.2.1) \qquad L_\theta(Z, W) = -id\theta(Z, \bar{W}) = i\theta([Z, \bar{W}]) \qquad \forall Z, W \in C^\infty(M, T_{1,0}).$$

Let X_0 be the Reeb vector field of θ, so that $\imath_{X_0}\theta = 1$ and $\imath_{X_0}d\theta = 0$. This gives rise to the splitting,

$$(6.2.2) \qquad T_{\mathbb{C}}M = T_{1,0} \oplus T_{0,1} \oplus (\mathbb{C}X_0).$$

For $p,q = 0,\ldots,n$ we let $\Lambda^{p,q}$ denote the bundle of (p,q)-forms associated to this splitting. Moreover, we have

$$(6.2.3) \quad [Z,\bar{W}] = -iL_\theta(Z,W)T \mod T_{1,0} \oplus T_{0,1} \qquad \forall Z,W \in C^\infty(M, T_{1,0}).$$

We endow M with a Levi metric as follows (see also [**FS1**]). Let $\tilde{h}$ be a (positive definite) Hermitian metric on $T_{1,0}$. Then there exists a Hermitian-valued section A of $\mathrm{End}_{\mathbb{C}}\, T_{1,0}$ such that

$$(6.2.4) \qquad L_\theta(Z,W) = \tilde{h}(AZ,W) \qquad \forall Z,W \in C^\infty(M, T_{1,0}).$$

Let $A = U|A|$ be the polar decomposition of A. Since by assumption the Hermitian form L_θ has signature $(n-\kappa, \kappa, 0)$, hence is nondegenerate, the section A is invertible and the section U is an orthogonal matrix with only the eigenvalues 1 and -1 with multiplicities $n-\kappa$ and κ respectively. Therefore, we have the splitting,

$$(6.2.5) \qquad T_{1,0} = T_{1,0}^+ \oplus T_{1,0}^-, \qquad T_{1,0}^\pm = \ker(U \mp 1),$$

where the restriction of L_θ to $T_{1,0}^+$ (resp. $T_{1,0}^-$) is positive definite (resp. negative definite). We then get a Levi metric on $T_{1,0}$ by letting

$$(6.2.6) \qquad h(Z,W) = \tilde{h}(|A|Z,W), \qquad Z,W \in C^\infty(M, T_{1,0}).$$

In particular, on the direct summands $T_{1,0}^\pm$ of the splitting (6.2.5) we have

$$(6.2.7) \qquad L_\theta(Z,W) = h(UZ,W) = \pm h(Z,W), \qquad Z,W \in C^\infty(M, T_{1,0}^\pm).$$

We now extend h into a Hermitian metric h on $T_{\mathbb{C}}M$ by making the following requirements:

$$(6.2.8) \qquad h(X_0, X_0) = 1, \qquad h(Z,W) = \overline{h(\bar{Z},\bar{W})} \quad \forall Z,W \in T_{0,1},$$

$(6.2.9)$ The splitting $T_{\mathbb{C}}M = \mathbb{C}T \oplus T_{1,0} \oplus T_{0,1}$ is orthogonal with respect to h.

This allows us to express the Levi form $\mathcal{L}_{\mathbb{C}} : (H \otimes \mathbb{C}) \times (H \otimes \mathbb{C}) \to T_{\mathbb{C}}M/(H \otimes \mathbb{C})$ as follows. Since $\theta(T) = 1$ we have

$$(6.2.10) \qquad \mathcal{L}_{\mathbb{C}}(X,Y) = \theta([X,Y])T = -d\theta(X,Y)T, \qquad X,Y \in C^\infty(M, H \otimes \mathbb{C}).$$

Therefore, if follows from (6.2.7) that we have

$$(6.2.11) \qquad \mathcal{L}(Z,\bar{W}) = -iL_\theta(Z,W) = h(Z, iUW), \qquad Z,W \in C^\infty(M, T_{1,0}).$$

Since $\mathcal{L}$ is antisymmetric and the integrability condition $[T_{1,0}, T_{1,0}] \subset T_{1,0}$ implies that $\mathcal{L}_{\mathbb{C}}$ vanishes on $T_{1,0} \times T_{1,0}$ and on $T_{0,1} \times T_{0,1}$, we get

$$(6.2.12) \qquad \mathcal{L}_{\mathbb{C}}(X,Y) = h(X,LY), \qquad X,Y \in C^\infty(M, H \otimes \mathbb{C}),$$

where L is the antilinear antisymmetric section of $\mathrm{End}_{\mathbb{R}}(H \otimes \mathbb{C})$ such that

$$(6.2.13) \qquad L(Z+\bar{W}) = iUW - i(\overline{UZ}) \qquad \forall Z,W \in C^\infty(M, T_{1,0}).$$

In particular, since $U^* = U$ and $U^2 = 1$ we have $|L| = 1$.

Let $Z_1,\ldots,Z_n$ be a local orthonormal frame for $T_{1,0}$ (with respect to h) and such that $Z_1,\ldots,Z_{n-\kappa}$ span $T_{1,0}^+$ and $Z_{n-\kappa+1},\ldots,Z_n$ span $T_{1,0}^-$. Then $\{T, Z_j, \bar{Z}_j\}$

is an orthonormal frame. In the sequel we will call such a frame an *admissible orthonormal frame of* $T_{\mathbb{C}}M$. Then from (6.2.7) we get:

(6.2.14) $$L_\theta(Z_j, \bar{Z}_k) = \epsilon_j h(Z_j, Z_k) = \epsilon_j \delta_{jk},$$

where $\epsilon_j = 1$ for $j = 1, \ldots, n - \kappa$ and $\epsilon_j = -1$ for $j = n - \kappa + 1, \ldots, n$.

Let $\{\theta, \theta^j, \theta^{\bar{j}}\}$ be the dual coframe of $T_{\mathbb{C}}^*$ associated to $\{T, Z_j, \bar{Z}_j\}$. Then the volume form of h is locally given by

(6.2.15) $$\sqrt{h(x)}dx = i^n \theta \wedge \theta^1 \wedge \theta^{\bar{1}} \wedge \cdots \wedge \theta^n \wedge \theta^{\bar{n}}.$$

Furthermore, because of (6.2.14) we have $d\theta = i \sum_{j=1}^n \epsilon_j \theta^j \wedge \theta^{\bar{j}}$ mod $\theta \wedge T^*M$, so as $\epsilon_1 \ldots \epsilon_n = (-1)^\kappa$ we have $\theta \wedge d\theta^n = n! i^n (-1)^\kappa \theta \wedge \theta^1 \wedge \theta^{\bar{1}} \wedge \cdots \wedge \theta^n \wedge \theta^{\bar{n}}$. Therefore, we get the following global formula for the volume form,

(6.2.16) $$\sqrt{h(x)}dx = \frac{(-1)^\kappa}{n!} \theta \wedge d\theta^n.$$

In particular, we see that the volume form depends only on θ and not on the choice of the Levi metric.

DEFINITION 6.2.1. *The pseudohermitian volume of* (M, θ) *is*

(6.2.17) $$\text{vol}_\theta M = \frac{(-1)^\kappa}{n!} \int_M \theta \wedge d\theta^n.$$

We shall now relate the asymptotics (6.1.3) for the Kohn Laplacian, the horizontal sublaplacian and the Gover-Graham operators to the volume $\text{vol}_\theta M$. To this end consider the Heisenberg group $\mathbb{H}^{2n+1} = \mathbb{R} \times \mathbb{R}^{2n}$ together with the left-invariant basis of $\mathfrak{h}^{2n+1}$ given by

(6.2.18) $$X_0^0 = \frac{\partial}{\partial x_0}, \qquad X_j^0 = \frac{\partial}{\partial x_j} + x_{n+j} \frac{\partial}{\partial x_0},$$

(6.2.19) $$X_{n+j}^0 = \frac{\partial}{\partial x_{n+j}} - x_j \frac{\partial}{\partial x_0}, \quad j = 1, \ldots, n.$$

For $\mu \in \mathbb{C}$ let $\mathcal{L}_\mu$ denote the Folland-Stein sublaplacian,

(6.2.20) $$\mathcal{L}_\mu = -\frac{1}{2}((X_1^0)^2 + \ldots + (X_{2n}^0)^2) - i\mu X_0^0.$$

For this sublaplacian the condition (3.4.14) reduces to $\mu \neq \pm n, \pm(n+2), \ldots$, so in this case the operators $\mathcal{L}_\mu$ and $\mathcal{L}_\mu + \partial_t$ are hypoelliptic and admit unique fundamental solutions since their symbols are invertible.

In the sequel it will be convenient to use the variable $x' = (x_1, \ldots, x_{2n})$.

LEMMA 6.2.2. *For* $|\Re\mu| < n$ *the fundamental solution* $k_\mu(x_0, x', t)$ *of* $\mathcal{L}_\mu + \partial_t$ *is equal to*

(6.2.21) $$\chi(t)(2\pi t)^{-n} \int_{-\infty}^\infty e^{ix_0\xi_0 - \mu t\xi_0} \left(\frac{t\xi_0}{\sinh t\xi_0}\right)^n \exp\left[-\frac{1}{t}\frac{t\xi_0}{\tanh t\xi_0}|x'|^2\right] d\xi_0.$$

where $\chi(t)$ *denotes the characteristic function of* $(0, \infty)$.

PROOF. The fundamental solution $k_\mu(x_0, x', t)$ is solution to the equation,

(6.2.22) $$(\mathcal{L}_\mu + \partial_t)k(x_0, x', t) = \delta(x_0) \otimes \delta(x') \otimes \delta(t).$$

Let us make a Fourier transform with respect to x_0, so that letting $\hat{k}(\xi_0, x', t) = \hat{k}_{x_0 \to \xi_0}(\xi_0, x', t)$ the equation (6.2.22) becomes

$$(6.2.23) \qquad (\hat{\mathcal{L}}_0 + \mu\xi_0 + \partial_t)\hat{k}_\mu = \delta(x') \otimes \delta(t),$$

$$(6.2.24) \qquad \hat{\mathcal{L}}_0 = -\frac{1}{2}\sum_{j=1}^{n}(\frac{\partial}{\partial x_j} - ix_{n+j}\xi_0)^2 - \frac{1}{2}\sum_{j=1}^{n}(\frac{\partial}{\partial x_{n+j}} + ix_j\xi_0)^2.$$

Notice that $\hat{\mathcal{L}}_0 = \frac{1}{2}H_{A(\xi_0)}$, where $H_{A(\xi_0)} = -\sum_{j=1}^{2n}(\partial_j - \sum_{k=1}^{2n}iA(\xi_0)_{jk}x_k)^2$ is the harmonic oscillator associated to the real antisymmetric matrix,

$$(6.2.25) \qquad A(\xi_0) = \xi_0 J, \qquad J = \begin{pmatrix} 0 & -I_n \\ I_n & 0 \end{pmatrix}.$$

Therefore, the fundamental solution of $\hat{\mathcal{L}}_0 + \mu\xi_0 + \partial_t$ is given by a version of the Mehler formula (see, e.g., [**GJ**], [**Po4**, p. 225]), that is, $\hat{k}_0(\xi_0, x', t)$ is equal to

$$(6.2.26) \quad \chi(t)(2\pi t)^{-n} \det{}^{\frac{1}{2}}(\frac{itA(\xi_0)}{\sinh(itA(\xi_0))}) \exp[-\frac{1}{2t}\langle\frac{itA(\xi_0)}{\tanh(itA(\xi_0))}x', x'\rangle],$$

$$= \chi(t)(2\pi t)^{-n}(\frac{t\xi_0}{\sinh(t\xi_0)})^n \exp[-\frac{1}{2t}\frac{t\xi_0}{\tanh(t\xi_0)}|x'|^2].$$

A solution of (6.2.23) is now given by $\hat{k}_\mu(\xi_0, x', t) = e^{-\mu\xi_0 t}\hat{k}_0(\xi_0, x', t)$. Moreover, as we have

$$(6.2.27) \qquad |h_0(\xi_0, x', t)| \leq \pi^{-n}|\xi_0|^n e^{-tn|\xi_0|},$$

we see that for $|\Re\mu| < n$ the function $e^{-t\mu\xi_0}\hat{k}_0$ is integrable with respect to ξ_0. Since $k_\mu(x_0, x', t)$ is the inverse Fourier transform with respect to ξ_0 of $\hat{k}_\mu(\xi_0, x', t)$ it follows that $k_\mu(x_0, x', t)$ is equal to

$$(6.2.28) \qquad \chi(t)(2\pi t)^{-n}\int_{-\infty}^{\infty} e^{ix_0\xi_0 - \mu t\xi_0}(\frac{t\xi_0}{\sinh t\xi_0})^n \exp[-\frac{1}{t}\frac{t\xi_0}{\tanh t\xi_0}|x'|^2]d\xi_0.$$

The lemma is thus proved. $\square$

Next, for $|\Re\mu| < n$ we let

$$(6.2.29) \qquad \nu(\mu) = \frac{1}{(n+1)!}k_\mu(0,0,1) = \frac{(2\pi)^{-(n+1)}}{(n+1)!}\int_{-\infty}^{\infty} e^{-\mu\xi_0}(\frac{\xi_0}{\sinh \xi_0})^n d\xi_0.$$

LEMMA 6.2.3. *Let* $\Delta : C^\infty(M, \mathcal{E}) \to C^\infty(M, \mathcal{E})$ *be a selfadjoint sublaplacian which is bounded from below and assume there exists* $\mu \in (-n, n)$ *such that near any point of M there is an admissible orthonormal frame* $Z_1, \ldots, Z_n$ *of* $T_{1,0}$ *with respect to which Δ takes the form,*

$$(6.2.30) \qquad \Delta = -\sum_{j=1}^{n}(\overline{Z}_j Z_j + Z_j \overline{Z}_j) - i\mu X_0 + \mathrm{O}_H(1).$$

Then as $\lambda \to \infty$ *we have*

$$(6.2.31) \qquad N(\Delta; \lambda) \sim 2^n \nu(\mu) \operatorname{rk}\mathcal{E}(\mathrm{vol}_\theta M)\lambda^{n+1}.$$

PROOF. Let $Z_1, \ldots, Z_n$ be a local admissible orthonormal frame of $T_{1,0}$. Then from (6.2.1) and (6.2.7) we obtain

$$(6.2.32) \qquad [Z_j, \overline{Z}_k] = -L_\theta(Z_j, Z_k)T = -i\epsilon_j \delta_{jk}T \mod T_{1,0} \oplus T_{0,1}.$$

In addition, we let $X_1, \ldots, X_{2n}$ be the vector fields in H such that

(6.2.33) $$Z_j = \begin{cases} \frac{1}{2}(X_j - iX_{n+j}) & \text{for } j = 1, \ldots, n - \kappa, \\ \frac{1}{2}(X_{n+j} - iX_j) & \text{for } j = n - \kappa + 1, \ldots, n. \end{cases}$$

Then $X_1, \ldots, X_{2n}$ is a local frame of $H = \Re(T_{1,0} \oplus T_{0,1})$ and from (6.2.32) we get

(6.2.34) $$[X_j, X_{n+k}] = -2\delta_{jk}X_0 \mod H,$$
(6.2.35) $$[X_0, X_j] = [X_j, X_k] = [X_{n+j}, X_{n+k}] = 0 \mod H.$$

Moreover, in terms of the vector fields $X_1, \ldots, X_{2n}$ the formula (6.2.30) becomes

(6.2.36) $$\Delta = -\frac{1}{2}(X_1^2 + \ldots + X_{2n}^2) + i\mu X_0 + \mathrm{O}_H(1).$$

Combining this with (6.2.13) shows that the condition (3.4.22) for Δ, is given in terms of the eigenvalues of $|L| = 1$ and becomes

(6.2.37) $$\mu \notin \{\pm(n+2k);\ k = 0, 1, 2, \ldots\}.$$

Since by assumption we have $\mu \in (-n, n)$, the condition (3.4.22) is fulfilled at every point, so by Proposition 5.2.9 the principal symbol of $\Delta + \partial_t$ is an invertible Volterra-Heisenberg symbol. This allows us to apply Proposition 6.1.2 to deduce that as $\lambda \to \infty$ we have

(6.2.38) $$N(\Delta; \lambda) \sim \nu_0(\Delta)\lambda^{n+1},$$

where $\nu_0(P)$ is given by (6.1.3).

Let us now work in Heisenberg coordinates centered at a point $a \in M$ related to a local H-frame $X_0, X_1, \ldots, X_{2n}$ as above. Because of (6.2.34) and (6.2.35) the model vector fields $X_0^a, X_1^a \ldots, X_{2n}^a$ coincide with the vector fields (6.2.18)–(6.2.19), so $G_a M$ agrees with the Heisenberg group $\mathbb{H}^{2n+1}$ in the Heisenberg coordinates. In addition, using (6.2.36) we see that the model operator of Δ is

(6.2.39) $$\Delta^a = -\frac{1}{2}((X_1^a)^2 + \ldots + (X_{2n}^a)^2) - i\mu X_0^a = \mathcal{L}_\mu.$$

As $\mu \in (-n, n)$ it follows that Δ satisfies the Rockland condition at every point.

On the other hand, since we are in Heisenberg coordinates we have $X_j(0) = \frac{\partial}{\partial x_j}$. In particular, we have $\theta(0) = dx_0$. Moreover, as $d\theta(X, Y) = -\theta([X, Y])$ we deduce from (6.2.34)–(6.2.35) that, for $j = 0, 1, \ldots, n$ and $k = 1, \ldots, n$, we have

(6.2.40) $$d\theta(X_j, X_{n+k}) = 2\delta_{jk}, \qquad d\theta(X_j, X_k) = 0.$$

It then follows that $d\theta(0) = 2\sum_{j=1}^n dx_j \wedge dx_{n+j}$. Therefore, the volume form is given by

(6.2.41) $$\frac{(-1)^\kappa}{n!}\theta \wedge d\theta^n(0) = (-1)^\kappa 2^n dx_0 \wedge dx_1 \wedge dx_{n+1} \wedge \ldots \wedge dx_n \wedge dx_{2n}.$$

Note also that because of (6.2.33) the orientation of M is that given by $i^n T \wedge Z_1 \wedge \bar{Z}_1 \wedge \ldots \wedge Z_n \wedge \bar{Z}_n = (-1)^\kappa T \wedge X_1 \wedge X_{n+1} \wedge \ldots \wedge X_n \wedge X_{2n}$. Therefore, at $x = 0$ the volume form of M is in the same orientation class as $(-1)^\kappa dx_0 \wedge dx_1 \wedge dx_{n+1} \wedge \ldots \wedge dx_n \wedge dx_{2n}$. Thus,

(6.2.42) $$\frac{(-1)^\kappa}{n!}\theta \wedge d\theta^n(0) = 2^n dx_{|x=0}.$$

All this shows that in the above Heisenberg coordinates the volume form and the model operator of Δ have normal forms in the sense of Proposition 6.1.5 with $\rho_0 = 2^n$ and $P_0 = \mathcal{L}_\mu \otimes 1_{\mathbb{C}^r}$. Therefore, we may apply (6.1.16) to get

$$(6.2.43) \qquad \nu_0(\Delta) = 2^n \nu_0(\mathcal{L}_\mu \otimes 1_{\mathbb{C}^r}) \int_M \frac{(-1)^\kappa}{n!} \theta \wedge d\theta^n = 2^n \nu_0(\mathcal{L}_\mu) \operatorname{rk} \mathcal{E} \operatorname{vol}_\theta M,$$

where $\nu_0(\mathcal{L}_\mu) = \frac{1}{(n+1)!} k_\mu(0,0,1) = \nu(\mu)$. The lemma is thus proved. $\square$

We are now ready to relate the asymptotics (6.1.3) for the Kohn Laplacian to the pseudohermitian volume of M.

THEOREM 6.2.4. *Assume that M is endowed with a Levi metric and consider the Kohn Laplacian $\Box_{b;p,q} : C^\infty(M, \Lambda^{p,q}) \to C^\infty(M, \Lambda^{p,q})$ acting on (p,q)-forms with $q \neq \kappa$ and $q \neq n - \kappa$. Then as $\lambda \to \infty$ we have*

$$(6.2.44) \qquad N(\Box_{b;p,q}; \lambda) \sim \alpha_{n\kappa pq} (\operatorname{vol}_\theta M) \lambda^{n+1},$$

where $\alpha_{n\kappa pq}$ is equal to

$$(6.2.45) \qquad \sum_{\max(0, q-\kappa) \leq k \leq \min(q, n-\kappa)} \frac{1}{2} \binom{n}{p} \binom{n-\kappa}{k} \binom{\kappa}{q-k} \nu(n - 2(\kappa - q + 2k)).$$

In particular $\alpha_{n\kappa pq}$ is a universal constant depending only on n, κ, p and q.

PROOF. As M is κ-strictly pseudoconvex the $Y(q)$-condition reduces to $q \neq \kappa$ and $q \neq n - \kappa$, so in this case $\Box_{b;p,q}$ has an invertible principal symbol. Since $\Box_{b;p,q}$ is a positive it follows from Proposition 6.1.2 that as $\lambda \to \infty$ we have

$$(6.2.46) \qquad N(\Box_{b;p,q}; \lambda) \sim \nu_0(\Box_{b;p,q}) \lambda^{n+1}.$$

It then remains to show that $\nu_0(\Box_{b;p,q}) = \alpha_{n\kappa pq} \operatorname{vol}_\theta M$ with $\alpha_{n\kappa pq}$ given by (6.2.45).

Let $\{T, Z_j, \bar{Z}_j\}$ be a local admissible orthonormal frame of $T_\mathbb{C} M$ and let $\{\theta, \theta^j, \theta^{\bar{j}}\}$ be the dual coframe of $T_\mathbb{C}^* M$. For ordered subsets $J = \{j_1, \ldots, j_p\}$ and $K = \{k_1, \ldots, k_q\}$ of $\{1, \ldots, n\}$ with $j_1 < \ldots < j_p$ and $k_1 < \ldots < k_q$ we let

$$(6.2.47) \qquad \theta^{J,\bar{K}} = \theta^{j_1} \wedge \ldots \wedge \theta^{j_p} \wedge \theta^{\bar{k}_1} \wedge \ldots \wedge \theta^{\bar{k}_q}.$$

Then $\{\theta^{J,\bar{K}}\}$ form an orthonormal frame of $\Lambda^{*,*}$ and, as shown in [**BG**, Sect. 20], with respect to this frame $\Box_b$ takes the form,

$$(6.2.48) \qquad \Box_b = \operatorname{diag}\{\Box_{J\bar{K}}\} + \mathrm{O}_H(1),$$

$$(6.2.49) \qquad \Box_{J\bar{K}} = -\frac{1}{2} \sum_{1 \leq j \leq n} (Z_j \bar{Z}_j + \bar{Z}_j Z_j) + \frac{1}{2} \sum_{j \in K} [Z_j, \bar{Z}_j] - \frac{1}{2} \sum_{j \notin K} [Z_j, \bar{Z}_j].$$

Moreover, using (6.2.32) we see that the leading part of $\Box_{J\bar{K}}$ is equal to

$$(6.2.50) \qquad -\frac{1}{2} \sum_{1 \leq j \leq n} (Z_j \bar{Z}_j + \bar{Z}_j Z_j) - \frac{i}{2} \mu_K X_0, \qquad \mu_K = \sum_{j \in K} \epsilon_j - \sum_{j \notin K} \epsilon_j.$$

Notice that since $\epsilon_j = 1$ for $j = 1, \ldots, n - \kappa$ and $\epsilon_j = -1$ for $j = n - \kappa + 1, \ldots, n$ we have $\mu_K \in [-n, n]$ and $\mu_K = \pm n$ if, and only if, we have $K = \{1, \ldots, n - \kappa\}$ or $K = \{n - \kappa + 1, \ldots, n\}$. Thus if $|K| = q$ with $q \notin \{\kappa, n - \kappa\}$ then $\Box_{J\bar{K}}$ is two times a sublaplacian of the form (6.2.30) with $\mu = \mu_K$ in $(-n, n)$.

On the other hand, complex conjugation is an isometry, so from the orthogonal splitting (6.2.5) we get the orthogonal splitting $T_{0,1} = T_{0,1}^+ \oplus T_{0,1}^-$ with $T_{0,1}^\pm = \overline{T_{1,0}^\pm}$. By duality these splittings give rise to the orthogonal decompositions,

$$(6.2.51) \qquad \Lambda^{1,0} = \Lambda_+^{1,0} \oplus \Lambda_-^{1,0}, \qquad \Lambda^{0,1} = \Lambda_+^{0,1} \oplus \Lambda_-^{0,1}.$$

Therefore, letting $\Lambda^{p;q,k} = \Lambda^{p,0} \wedge (\Lambda_+^{0,1})^k \wedge (\Lambda_-^{0,1})^{q-k}$ we have

$$(6.2.52) \qquad \Lambda^{p,q} = \bigoplus_{\max(0,q-\kappa) \leq k \leq \min(q,n-\kappa)} \Lambda^{p;q,k}.$$

Let $\Pi_{p;qk}$ be the orthogonal projection onto $\Lambda^{p;q,k}$ and set $\square_{pqk} = \Pi_{pqk} \square_b \Pi_{pqk}$. As (6.2.48) and (6.2.49) show that $\square_{b;p,q}$ is scalar up to first order terms we have

$$(6.2.53) \qquad \square_{b;p,q} = \sum_{\max(0,q-\kappa) \leq k \leq \min(q,n-\kappa)} \square_{p;qk} + \mathrm{O}_H(1).$$

In particular, as $\nu_0(\square_{b;p,q})$ depends only on the principal symbol of $\square_{b;p,q}$ we get

$$(6.2.54) \qquad \nu_0(\square_{b;p,q}) = \sum_{\max(0,q-\kappa) \leq k \leq \min(q,n-\kappa)} \nu_0(\square_{p;qk}).$$

We are thus reduced to express each coefficient $\nu_0(\square_{p;qk})$ in terms of $\mathrm{vol}_\theta M$.

Next, if $\theta^{J\bar{K}}$ is a section of $\Lambda^{p,q,k}$ then we have

$$(6.2.55) \qquad \#K \cap \{1,\ldots,n-\kappa\} = k, \qquad \#K \cap \{n-\kappa+1,\ldots,n\} = q-k,$$
$$(6.2.56) \qquad \#K^c \cap \{1,\ldots,n-\kappa\} = n-\kappa-k,$$
$$(6.2.57) \qquad \#K^c \cap \{n-\kappa+1,\ldots,n\} = \kappa-q+k,$$

from which we get $\mu_K = k - (q-k) - [(n-\kappa-k) - (\kappa-q+k)] = n + 2q - 2\kappa - 4k$. Combining this with (6.2.48)–(6.2.50) then gives

$$(6.2.58) \qquad \square_{p;qk} = -\frac{1}{2} \sum_{1 \leq j \leq n} (Z_j \bar{Z}_j + \bar{Z}_j Z_j) - \frac{i}{2}(n + 2q - 2\kappa - 4k)X_0 + \mathrm{O}_H(1).$$

Now, we can apply Lemma 6.2.3 to $2\square_{p;qk}$ to get that $\nu_0(\square_{p;qk})$ is equal to

$$(6.2.59) \quad 2^{-(n+1)}\nu_0(2\square_{p;qk}) = 2^{-(n+1)} 2^n (\mathrm{rk}\, \Lambda^{p;q,k})\nu(n + 2q - 2\kappa - 4k).\mathrm{vol}_\theta M$$
$$= \frac{1}{2}\binom{n}{p}\binom{n-\kappa}{k}\binom{\kappa}{q-k}\nu(n + 2q - 2\kappa - 4k).\mathrm{vol}_\theta M.$$

Combining this with (6.2.46) and (6.2.54) then shows that as $\lambda \to \infty$ we have

$$(6.2.60) \qquad N(\square_{b;p,q}; \lambda) \sim \alpha_{n\kappa pq}(\mathrm{vol}_\theta M)\lambda^{n+1},$$

with $\alpha_{n\kappa pq}$ given by (6.2.45). The proof is thus achieved. $\square$

Next, we turn to the horizontal sublaplacian on (p,q)-forms.

THEOREM 6.2.5. *Assume that M is endowed with a Levi metric and consider the horizontal sublaplacian $\Delta_b : C^\infty(M, \Lambda^{p,q}) \to C^\infty(M, \Lambda^{p,q})$ acting on (p,q)-forms with $(p,q) \neq (\kappa, n-\kappa)$ and $(p,q) \neq (n-\kappa, \kappa)$. Then as $\lambda \to \infty$ we have*

$$(6.2.61) \qquad N(\Delta_{b;p,q}; \lambda) \sim \beta_{n\kappa pq}(\mathrm{vol}_\theta M)\lambda^{n+1},$$

where $\beta_{n\kappa pq}$ is equal to
(6.2.62)
$$\sum_{\substack{\max(0,q-\kappa)\leq k\leq \min(q,n-\kappa) \\ \max(0,p-\kappa)\leq l\leq \min(p,n-\kappa)}} 2^n \binom{n-\kappa}{l}\binom{\kappa}{p-l}\binom{n-\kappa}{k}\binom{\kappa}{q-k}\nu(2(q-p)+4(l-k)).$$

In particular $\beta_{n\kappa pq}$ is a universal constant depending only on n, κ, p and q.

PROOF. Thanks to (2.2.13) we know that we have

(6.2.63) $$\Delta_b = \Box_{b;p,q} + \overline{\Box}_{b;p,q},$$

where $\overline{\Box}_b$ denotes the conjugate operator of $\Box_b$, that is, $\overline{\Box}_b \eta = \overline{\Box_b \bar{\eta}}$, (or, equivalently, the Laplacian of the $\bar{\partial}_b$-complex).

As in (6.2.52) we have orthogonal splittings,

(6.2.64) $$\Lambda^{p,q} = \bigoplus_{\max(0,p-\kappa)\leq l\leq \min(p,n-\kappa)} \Lambda^{p,l;q} = \bigoplus_{\substack{\max(0,q-\kappa)\leq k\leq \min(q,n-\kappa)\\ \max(0,p-\kappa)\leq l\leq \min(p,n-\kappa)}} \Lambda^{p,l;q,k},$$

(6.2.65) $$\Lambda^{p,l;q} = (\Lambda^{1,0}_+)^l \wedge (\Lambda^{1,0}_-)^{p-l} \wedge \Lambda^{0,q}, \qquad \Lambda^{p,l;q,k} = \Lambda^{p,l;0} \wedge \Lambda^{p,l;0}.$$

Since $\Box_{b;p,q}$ is a scalar operator modulo lower order terms, the same is true for $\overline{\Box}_b$ and Δ_b. Therefore, on $\Lambda^{p,q}$ we can write

(6.2.66) $$\overline{\Box}_b = \sum_{\max(0,p-\kappa)\leq l\leq \min(p,n-\kappa)} \overline{\Box}_{p,l;q} + \mathrm{O}_H(1), \qquad \overline{\Box}_{p,l;q} = \Pi_{p,l;q}\overline{\Box}_b \Pi_{p,l;q},$$

where $\Pi_{p,l;q}$ denote the orthogonal projection onto $\Lambda^{p,l;q}$. Therefore, if we let $\Pi_{p,l;q,k}$ denote the orthogonal projection onto $\Lambda^{p,l;q,k}$ and set $\Delta_{p,l;q,k} = \Pi_{p,l;q,k}\Delta_b \Pi_{p,l;q,k}$, then we have

(6.2.67) $$\Delta_b = \sum_{\substack{\max(0,q-\kappa)\leq k\leq \min(q,n-\kappa)\\ \max(0,p-\kappa)\leq l\leq \min(p,n-\kappa)}} \Delta_{p,l;q,k} + \mathrm{O}_H(1).$$

In particular, as in (6.2.68) we have

(6.2.68) $$\nu_0(\Delta_{b;p,q}) = \sum_{\substack{\max(0,q-\kappa)\leq k\leq \min(q,n-\kappa)\\ \max(0,p-\kappa)\leq l\leq \min(p,n-\kappa)}} \nu_0(\Delta_{p,l;q,k}).$$

Next, let $\{X_0, Z_j, \bar{Z}_j\}$ be a local admissible orthonormal frame for $T_\mathbb{C} M$. Since $\overline{\Box}_{p,l;q} = \overline{\Box_{q;p,l}}$, using (6.2.58) we see that on $\Lambda^{p,l;q}$ we have

(6.2.69) $$\overline{\Box}_{p,l;q} = -\frac{1}{2}\sum_{1\leq j\leq n}(Z_j\bar{Z}_j + \bar{Z}_j Z_j) + \frac{i}{2}(n+2p-2\kappa-4l)X_0 + \mathrm{O}_H(1).$$

Therefore, on $\Lambda^{p,l;q,k}$ we can write

(6.2.70) $$\Delta_{p,l;q,k} = \Box_{p;q,k} + \overline{\Box}_{p,l;q}$$
$$= -\sum_{1\leq j\leq n}(Z_j\bar{Z}_j + \bar{Z}_j Z_j) - i(2(q-p)+4(l-k))X_0 + \mathrm{O}_H(1).$$

Thanks to (6.2.70) we can apply Lemma 6.2.3 to get

(6.2.71) $\nu_0(\Delta_{p,l;q,k}) = 2^n \operatorname{rk} \Lambda^{p,l;q,k} \nu(2(q-p) + 4(l-k)) \operatorname{vol}_\theta M$

$$= 2^n \binom{n-\kappa}{l}\binom{\kappa}{p-l}\binom{n-\kappa}{k}\binom{\kappa}{q-k} \nu(2(q-p)+4(l-k))\operatorname{vol}_\theta M.$$

Combining this with (6.2.68) then shows that as $\lambda \to \infty$ we have

(6.2.72) $$N(\Delta_b; \lambda) \sim \beta_{n\kappa pq}(\operatorname{vol}_\theta M)\lambda^{n+1},$$

with $\beta_{n\kappa pq}$ given by (6.2.62). The theorem is thus proved. $\square$

Finally, suppose that M is strictly pseudoconvex, i.e., we have $\kappa = 0$, and for $k = 1, \ldots, n+1, n+2, n+4, \ldots$ let $\square_\theta^{(k)} : C^\infty(M) \to C^\infty(M)$ be the Gover-Graham operator of order k. Then we have:

THEOREM 6.2.6. *Assume $k \neq n+1$. Then there exists a universal constant $\nu_n^{(k)} > 0$ depending only on n and k such that as $\lambda \to \infty$ we have*

(6.2.73) $$N(\square_\theta^{(k)}; \lambda) \sim \nu_n^{(k)}(\operatorname{vol}_\theta M)\lambda^{\frac{n+1}{k}}.$$

PROOF. Let $Z_1, \ldots, Z_n$ be an orthonormal frame of $T_{1,0}$ over an open $U \subset M$. Since $\kappa = 0$ this is an admissible orthonormal frame. Therefore, as shown in the proof of Lemma 6.2.3, if for $j = 1, \ldots, n$ we let $Z_j = \frac{1}{2}(X_j - iX_{n+j})$ then $X_0, X_1, \ldots, X_{2n}$ is an H-frame such that in associated Heisenberg coordinates centered at a point $a \in U$ we have $G_0M = \mathbb{H}^{2n+1}$ and $\theta \wedge d\theta^n(0) = 2^n dx$ and the model operator at $x = 0$ of $\Delta_\mu = -\sum(\overline{Z}_j Z_j + Z_j \overline{Z}_j) - i\mu T$, $\mu \in \mathbb{C}$, is the Folland-Stein sublaplacian $\mathcal{L}_\mu$ in (6.2.20).

By Proposition 3.5.7 the operator $\square_\theta^{(k)}$ is of the form,

(6.2.74) $(\Delta_{b;0} + i(k-1)X_0)(\Delta_{b;0} + i(k-3)X_0)\cdots(\Delta_{b;0} - i(k-1)X_0) + O_H(2k-1).$

Moreover, by (3.5.24) we have $\Delta_{b;0} = -\sum(\overline{Z}_j Z_j + Z_j \overline{Z}_j) + O_H(1)$, so in Heisenberg coordinates as above the model operator at $x = 0$ of $\Delta_{b;0} - i\mu X_0$, $\mu \in \mathbb{C}$, is just $\mathcal{L}_\mu$. Incidentally, the model operator of $\square_\theta^{(k)}$ is $\mathcal{L}_{-k+1}\mathcal{L}_{-k+3}\cdots\mathcal{L}_{k-1}$.

All this shows that $\square_\theta^{(k)}$ admits a normal form in the sense of Proposition 6.1.4 and that we can make use of this proposition as in the proof of Lemma 6.2.3 to deduce that as $\lambda \to \infty$ we have

(6.2.75) $N(\square_\theta^{(k)}; \lambda) \sim \nu_n^{(k)}(\operatorname{vol}_\theta M)\lambda^{\frac{n+1}{k}}, \qquad \nu_n^{(k)} = 2^n\Gamma(1 + \frac{2n+2}{k})^{-1}K^{(k)}(0,1),$

where $K^{(k)}(x,t)$ denotes the fundamental solution of $\mathcal{L}_{-k+1}\mathcal{L}_{-k+3}\cdots\mathcal{L}_{k-1} + \partial_t$. In particular, we see that $\nu_n^{(k)}$ is universal constant depending only on k and n. Furthermore, it follows from Proposition 6.1.2 that $\nu_n^{(k)}$ is > 0. $\square$

6.3. Weyl asymptotics and contact geometry

In this section we express in more geometric terms the Weyl asymptotics for the horizontal sublaplacian and the contact Laplacian on a compact orientable contact manifold (M^{2n+1}, H) equipped with a contact form θ.

Let X_0 be the Reeb vector field associated to θ, so that $\imath_{X_0} d\theta = 0$ and $\imath_{X_0}\theta = 1$. Since M is orientable H admits a calibrated almost complex structure $J \in C^\infty(M, \operatorname{End} H)$, $J^2 = -1$, so that for any nonzero section X of H we have $d\theta(X, JX) = -d\theta(JX, X) > 0$. We then endow M with the orientation defined by

θ and the almost complex structure J, so that $\theta \wedge d\theta^n > 0$, and we equip it with the Riemannian metric,

$$(6.3.1) \qquad g_{\theta,J} = d\theta(., J.) + \theta^2.$$

A local orthonormal frame $X_1, \ldots, X_{2n}$ of H will be called admissible when we have $X_{n+j} = JX_j$ for $j = 1, \ldots, n$. If $\theta^1, \ldots, \theta^{2n}$ denotes the dual frame then we have $d\theta = \sum_{j=1}^n \theta^j \wedge \theta^{n+j}$, so the volume form of g_θ is equal to

$$(6.3.2) \qquad \theta^1 \wedge \theta^{n+1} \wedge \ldots \wedge \theta^n \wedge \theta^{2n} \wedge \theta = \frac{1}{n!} d\theta^n \wedge \theta.$$

In particular, the volume form is independent of the choice of the almost complex structure and depends only on the contact form.

DEFINITION 6.3.1. *The contact volume of (M^{2n+1}, θ) is given by*

$$(6.3.3) \qquad \text{vol}_\theta M = \frac{1}{n!} \int_M d\theta^n \wedge \theta.$$

LEMMA 6.3.2. *Let $\Delta : C^\infty(M, \mathcal{E}) \to C^\infty(M, \mathcal{E})$ be a selfadjoint sublaplacian such that Δ is bounded from below. We further assume that there exists $\mu \in (-n, n)$ so that near any point of M there is an admissible orthonormal frame $X_1, \ldots, X_{2n}$ of H with respect to which Δ takes the form,*

$$(6.3.4) \qquad \Delta = -(X_1^2 + \ldots + X_{2n}^2) - i\mu X_0 + \mathrm{O}_H(1).$$

Then as $\lambda \to \infty$ we have

$$(6.3.5) \qquad N(\Delta; \lambda) \sim 2^n \nu(\mu) \, \mathrm{rk}\, \mathcal{E}(\text{vol}_\theta M) \lambda^{n+1}.$$

PROOF. Let $X_1, \ldots, X_{2n}$ be an admissible local orthonormal frame of TM and for $j = 1, \ldots, 2n$ let $\tilde{X}_j = \sqrt{2} X_j$. Then $X_0, \tilde{X}_1, \ldots, \tilde{X}_{2n}$ is a local H-frame of TM with respect to which Δ takes the form,

$$(6.3.6) \qquad \Delta = -\frac{1}{2}(\tilde{X}_1^2 + \ldots + \tilde{X}_{2n}^2) - i\mu(x) T + \mathrm{O}_H(1).$$

Moreover, for $j, k = 1, \ldots, 2n$ we have

$$(6.3.7) \qquad \theta([\tilde{X}_j, J\tilde{X}_k]) = -2d\theta(X_j, JX_k) = -2g_\theta(X_j, X_k) = -2\delta_{jk}.$$

Therefore, for $j, k = 1, \ldots, n$ we get:

$$(6.3.8) \qquad [X_j, X_{n+k}] = -2\delta_{jk} X_0 \mod H,$$

$$(6.3.9) \qquad [T, X_j] = [X_j, X_k] = [X_{n+j}, X_{n+k}] = 0 \mod H.$$

The equalities (6.3.6)–(6.3.9) are the same as (6.2.34)–(6.2.36) in the case $\kappa = 0$. Therefore, along the same lines as that of the proof Lemma 6.2.3 we get

$$(6.3.10) \qquad \nu_0(\Delta) = 2^n \frac{\nu(\mu)}{n!} \int_M d\theta^n \wedge \theta = 2^n \nu(\mu) \text{vol}_\theta M.$$

Combining this with Proposition 6.1.2 then proves the asymptotics (6.3.5). $\square$

THEOREM 6.3.3. *Let $\Delta_{b;k} : C^\infty(M, \Lambda_\mathbb{C}^k H^*) \to C^\infty(M, \Lambda_\mathbb{C}^{k+1} H^*)$ be the horizontal sublaplacian on M in degree k with $k \neq n$. Then as $\lambda \to \infty$ we have*

$$(6.3.11) \quad N(\Delta_{b;k}; \lambda) \sim \gamma_{nk}(\text{vol}_\theta M) \lambda^{n+1}, \qquad \gamma_{nk} = \sum_{p+q=k} 2^n \binom{n}{p}\binom{n}{q} \nu(p-q).$$

In particular γ_{nk} is universal constant depending on n and k only.

PROOF. As explained in Section 2.2 the almost complex structure of H gives rise to an orthogonal decomposition $\Lambda_{\mathbb{C}}^k H^* = \bigoplus_{p+q=k} \Lambda^{p,q}$. If $X_1,\ldots,X_{2n}$ is a local admissible orthonormal frame of H then, as shown by Rumin [**Ru**, Prop. 2], on $\Lambda^{p,q}$ the operator Δ_b takes the form,

$$(6.3.12) \qquad \Delta_b = -(X_1^2 + \ldots + X_{2n}^2) + i(p-q)X_0 + \mathrm{O}_H(1).$$

where the lower order part is not scalar. Therefore, modulo lower order terms, Δ_b preserves the bidegree. We thus may write

$$(6.3.13) \qquad \Delta_{b;k} = \sum_{p+q=k} \Delta_{p,q} + \mathrm{O}_H(1), \qquad \Delta_{p,q} = \Pi_{p,q}\Delta_b\Pi_{p,q},$$

where $\Pi_{p,q}$ denotes the orthogonal projection of $\Lambda_{\mathbb{C}}^* H^*$ onto $\Lambda^{p,q}$. In particular,

$$(6.3.14) \qquad \nu_0(\Delta_{b;k}) = \sum_{p+q=k} \nu_0(\Delta_{p,q}).$$

Moreover, since $\Delta_{p,q}$ takes the form (6.3.12) with respect to any admissible orthonormal frame of H, we may apply Lemma 6.3.2 to get

$$(6.3.15) \qquad \nu_0(\Delta_{p,q}) = 2^n \sum_{p+q=k} \binom{n}{p}\binom{n}{q}\nu(p-q).$$

Combining this with (6.3.14) then gives the asymptotics (6.3.11). □

Finally, in the case of the contact Laplacian we can prove:

THEOREM 6.3.4. *1) Let $\Delta_{R;k} : C^\infty(M,\Lambda^k) \to C^\infty(M,\Lambda^k)$ be the contact Laplacian in degree $k \neq n$. Then there exists a universal constant $\nu_{nk} > 0$ depending only on n and k such that as $\lambda \to \infty$ we have*

$$(6.3.16) \qquad N(\Delta_{R;k}) \sim \nu_{nk}(\mathrm{vol}_\theta M)\lambda^{n+1}.$$

2) For $j = 1,2$ consider the contact Laplacian $\Delta_{R;nj} : C^\infty(M,\Lambda_j^n) \to C^\infty(M,\Lambda_j^n)$. Then there exists a universal constant $\nu_{n,j} > 0$ depending only on n and j such that as $\lambda \to \infty$ we have

$$(6.3.17) \qquad N(\Delta_{R;nj}) \sim \nu_{n,j}(\mathrm{vol}_\theta M)\lambda^{\frac{n+1}{2}}.$$

PROOF. Let $a \in M$ and consider a chart around a together with an admissible orthonormal frame $X_1,\ldots,X_{2n}$ of H. Since $T,X_1,\ldots,X_{2n}$ form a H-frame this chart is a Heisenberg chart. Moreover, as shown in the proofs of Lemma 6.2.3 and Lemma 6.3.2 the following hold:

(i) We have $X_j^a = X_j^0$, where $X_0^0,\ldots,X_{2n}^0$ denote the left-invariant vector fields (6.2.18) and (6.2.19) on $\mathbb{H}^{2n+1}$. In particular, we have $G_a M = \mathbb{H}^{2n+1}$ and $H_a = H_0^0$, where H_0^0 is the left-invariant Heisenberg structure of $\mathbb{H}^{2n+1}$.

(ii) We have $\theta(0) = dx_0 = \theta^0(0)$ and $d\theta(0) = 2\sum_{j=1}^n dx_j \wedge dx_{n+j} = d\theta^0(0)$, where $\theta^0 = dx_0 + \sum_{j=1}^n (x_j dx_{n+j} - x_{n+j} dx_j)$ is the standard left-invariant contact form of $\mathbb{H}^{2n+1}$.

(iii) The density on M given by the contact volume form $\frac{1}{n!}\theta \wedge d\theta^n$ agrees at $x = 0$ with the density $2^n dx$ on $\mathbb{R}^{d+1}$.

On the other hand, for $k = 0,1,\ldots,2n$ the fiber at a of the bundle Λ_*^k depends only on H_a and on the values of θ and $d\theta$ at a. Therefore, it follows from the statements (i) and (ii) that in the Heisenberg coordinates centered at a the fibers at $x=0$ of the bundles $\Lambda^* \oplus \Lambda^n_*$ of M and $\mathbb{H}^{2n+1}$ agree.

Next, let Δ_R^0 denote the contact Laplacian on $\mathcal{H}^{2n+1}$. Then we have:

LEMMA 6.3.5. *In the Heisenberg coordinates centered at a the model operators of $\Delta_{R;k}^a$ and $\Delta_{R;nj}$ agree with $\Delta_{R;k}^0$ and $\Delta_{R;nj}^0$ respectively.*

PROOF OF THE LEMMA. First, note that in view of the formulas (3.5.38)–(3.5.42) for Δ_R and of Proposition 3.2.9 and Proposition 3.2.11, we only have to show that in the Heisenberg coordinates centered at a the model operators $d_{R;k}^a$ and $D_{R;n}^a$ agree with the operators $d_{R;k}^0$ and $D_{R;n}^0$ on $\mathbb{H}^{2n+1}$.

Let $\theta^1, \ldots, \theta^{2n}$ be the coframe of H^* dual to $X_1, \ldots, X_{2n}$. This coframe gives rise to a trivialization of $\Lambda_{\mathbb{C}}^* H^*$ over the chart, in which we have $d_b = \sum_{j=1}^{2n} \varepsilon(\theta^j) X_j$. Furthermore, since $X_j(0) = \frac{d}{dx_j}$ we have $\theta_j(0) = dx_j$, so the model operator of d_b is $d_b^a = \sum_{j=1}^{2n} \varepsilon(dx_j) X_j^0 = d_b^0$, where d_b^0 is the d_b-operator on $\mathbb{H}^{2n+1}$. In particular, as for $k = n+1, \ldots, 2n$ we have $d_{R;k} = d_{b;k}$ on Λ^k, we get $d_{R;k}^a = d_{b;k}^a = d_{b;k}^0 = d_{R;k}^0$.

On the other hand, by definition for $k = 0, \ldots, n-1$ we have $d_{R;k}^* = d_{b;k}^*$ on Λ^k. Thus $(d_{R;k}^a)^* = (d_{R;k}^*)^a = (d_{b;k}^*)^a = (d_{b;k}^a)^* = (d_{b;k}^0)^* = (d_{R;k}^0)^*$, which by taking adjoints gives $d_{R;k}^a = d_{R;k}^0$.

Finally, as $D_{R;n} = \mathcal{L}_{X_0} + d_{b;n-1}\varepsilon(d\theta)^{-1} d_{b;n}$ the model operator $D_{R;n}^a$ is equal to $\mathcal{L}_{T^a} + d_{b;n-1}^a (\varepsilon(d\theta)^a)^{-1} d_{b;n}^a = \mathcal{L}_{T^0} + d_{b;n-1}^0 \varepsilon(d\theta^0)^* d_{b;n}^0 = D_{R;n}^0$. The proof is therefore achieved. □

Thanks to the statements (i) and (iii) and the claim above we may apply Proposition 6.1.5. Letting $K_{0;k}(x,t)$ be the fundamental solution of $\Delta_{R;k}^0 + \partial_t$ we then deduce that as $\lambda \to \infty$ we have

(6.3.18) $\qquad N(\Delta_{R;k}) \sim \nu_{nk}(\text{vol}_\theta M)\lambda^{n+1}, \qquad \nu_{nk} = \frac{2^n}{(n+1)!} \operatorname{tr}_{\Lambda^{k,0}} K_{0;k}(0,1).$

In particular, the constant ν_{nk} depends only on n and k, hence is a universal constant, and it follows from Proposition 6.1.2 that ν_{nk} is a positive number.

Similarly, let $K_{0;nj}(x,t)$ be the fundamental solution of $\Delta_{R;nj}^0 + \partial_t$. Then as $\lambda \to \infty$, we have

(6.3.19) $\qquad\qquad N(\Delta_{R;nj}) \sim \nu_{n,j}(\text{vol}_\theta M)\lambda^{\frac{n+1}{2}},$

where $\nu_{n,j} = 2^n \Gamma(1 + \frac{n+1}{2})^{-1} \operatorname{tr}_{\Lambda_j^{n,0}} K_{0|_{\Lambda_j^{n,0}}}(0,1)$. In particular $\nu_{n,j}$ is a universal constant depending only on n and j. The proof is thus complete. □

APPENDIX A

Proof of Proposition 3.1.18

First, we need the lemma below.

LEMMA A.1. *For $\Re m > 0$ we have $\mathcal{K}^m(U \times \mathbb{R}^{d+1}) \subset C^\infty(U) \hat{\otimes} C^{[\frac{\Re m}{2}]}(\mathbb{R}^{d+1})$.*

PROOF. Let $N = [\frac{\Re m}{2}]$ and let α be a multi-order such that $|\alpha| \leq N$. As $\langle\alpha\rangle \leq 2|\alpha| \leq -(k+d+2)$ the multiplication by ξ^α maps $S^{\hat{m}}(U \times \mathbb{R}^{d+1})$ to $C^\infty(U) \hat{\otimes} L^1(\mathbb{R}^{d+1})$. Composing it with the inverse Fourier transform with respect to ξ then shows that the map $p \to \partial_y^\alpha \check{p}_{\xi \to y}$ maps $S^{\hat{m}}(U \times \mathbb{R}^{d+1})$ to $C^\infty(U) \hat{\otimes} C^0(\mathbb{R}^{d+1})$. It then follows that for any $p \in S^{\hat{m}}(U \times \mathbb{R}^{d+1})$ the transform $\check{p}_{\xi \to y}(x,y)$ belongs to $C^\infty(U) \hat{\otimes} C^N(\mathbb{R}^{d+1})$.

Now, if $K \in \mathcal{K}^m(U \times \mathbb{R}^{d+1})$ then by Lemma 3.1.14 there exists a symbol $p \in S^{\hat{m}}(U \times \mathbb{R}^{d+1})$, $\hat{m} = -(m+d+2)$, such that $K(x,y)$ is equal to $\check{p}_{\xi \to y}(x,y)$ modulo a smooth function. Hence $K(x,y)$ belongs to $C^\infty(U) \hat{\otimes} C^N(\mathbb{R}^{d+1})$. The lemma is thus proved. □

We are now ready to prove Proposition 3.1.18. Let $\tilde{U}$ be an open subset of $\mathbb{R}^{d+1}$ together with a hyperplane bundle $\tilde{H} \subset T\tilde{U}$ and a $\tilde{H}$-frame of $T\tilde{U}$ and let $\phi : (U, H) \to (\tilde{U}, \tilde{H})$ be a Heisenberg diffeormorphism. Let $\tilde{P} \in \Psi_{\tilde{H}}^m(V)$ and set $P = \phi^* \tilde{P}$. We need to show that P is a $\Psi_H \mathrm{DO}$ of order m on U.

First, by Proposition 3.1.16 the distribution kernel of $\tilde{P}$ takes the form,

$$(A.1) \qquad k_{\tilde{P}}(\tilde{x}, \tilde{y}) = |\tilde{\varepsilon}'_{\tilde{x}}| K_{\tilde{P}}(\tilde{x}, -\tilde{\varepsilon}_{\tilde{x}}(\tilde{y})) + \tilde{R}(\tilde{x}, \tilde{y}),$$

with $K_{\tilde{P}}(\tilde{x}, \tilde{y})$ in $\mathcal{K}^{\hat{m}}(\tilde{U} \times \mathbb{R}^{d+1})$ and $\tilde{R}(\tilde{x}, \tilde{y})$ in $C^\infty(\tilde{U} \times \tilde{U})$. Therefore, the distribution kernel of $P = \phi^* P$ is given by

$$(A.2) \qquad k_P(x,y) = |\phi'(y)| k_{\tilde{P}}(\phi(x), \phi(y)) = |\varepsilon'_x| K(x, -\varepsilon_x(y)) + \tilde{R}(\phi(x), \phi(y)),$$

where K is the distribution on $\mathcal{U} = \{(x,y) \in U \times \mathbb{R}^{d+1};\ \varepsilon_x^{-1}(-y) \in U\} \subset U \times \mathbb{R}^{d+1}$ given by

$$(A.3) \quad K(x,y) = |\partial_y \Phi(x,y)| K_{\tilde{P}}(\phi(x), \Phi(x,y)), \quad \Phi(x,y) = -\tilde{\varepsilon}_{\phi(x)} \circ \phi \circ \varepsilon_x^{-1}(-y).$$

Next, it follows from [**Po6**, Props. 3.16, 3.18] that we have

$$(A.4) \qquad \Phi(x,y) = -\phi'_H(x)(-y) + \Theta(x,y) = \phi'_H(x)(y) + \Theta(x,y),$$

where $\Theta(x,y)$ is a smooth function on $\mathcal{U}$ with a behavior near $y = 0$ of the form

$$(A.5) \qquad \Theta(x,y) = (\mathrm{O}(\|y\|^3), \mathrm{O}(\|y\|^2), \ldots, \mathrm{O}(\|y\|^2)).$$

Then a Taylor expansion around $\tilde{y} = \phi'_H(x)y$ gives

$$(A.6) \quad K(x,y) = \sum_{\langle\alpha\rangle < N} |\partial_y \Phi(x,y)| \frac{\Theta(x,y)^\alpha}{\alpha!} (\partial_{\tilde{y}}^\alpha K_{\tilde{P}})(\phi(x), \phi'_H(x)y) + R_N(x,y),$$

where $R_N(x,y)$ is equal to

$$(A.7) \quad R_N(x,y) = \sum_{\langle\alpha\rangle=N} |\partial_y \Phi(x,y)| \frac{\Theta(x,y)^\alpha}{\alpha!} \int_0^1 (t-1)^{N-1} \partial_{\tilde{y}}^\alpha K_{\tilde{P}}(\phi(x), \Phi_t(x,y)) dt.$$

and we have let $\Phi_t(x,y) = \phi'_H(x)y + t\Theta(x,y)$.

Set $f_\alpha(x,y) = |\partial_y\Phi(x,y)|\Theta(x,y)^\alpha$. Then (A.5) implies that near $y=0$ we have

$$(A.8) \qquad f_\alpha(x,y) = O(\|y\|^{3\alpha_0 + 2(\alpha_1+\cdots+\alpha_d)}) = O(\|y\|^{\frac{3}{2}\langle\alpha\rangle}).$$

Thus all the homogeneous components of degree $< \frac{3}{2}\langle\alpha\rangle$ in the Taylor expansion for $f_\alpha(x,y)$ at $y=0$ must be zero. Therefore, we can write

$$(A.9) \qquad f_\alpha(x,y) = \sum_{\frac{3}{2}\langle\alpha\rangle \leq \langle\beta\rangle < \frac{3}{2}N} f_{\alpha\beta}(x) \frac{y^\beta}{\beta!} + \sum_{\langle\beta\rangle \doteq \frac{3}{2}N} r_{N\alpha\beta}(x,y) y^\beta,$$

where we have let $f_{\alpha\beta}(x) = \partial_y^\beta f_\alpha(x,0)$, the functions $r_{M\alpha\beta}(x,y)$ are in $C^\infty(\mathcal{U})$ and the notation $\langle\beta\rangle \doteq \frac{3}{2}N$ means that $\langle\beta\rangle$ is equal to $\frac{3}{2}N$ if $\frac{3}{2}N$ is an integer and to $\frac{3}{2}N + \frac{1}{2}$ otherwise. Thus,

$$(A.10) \qquad K(x,y) = \sum_{\langle\alpha\rangle<N} \sum_{\frac{3}{2}\langle\alpha\rangle \leq \langle\beta\rangle < \frac{3}{2}N} K_{\alpha\beta}(x,y) + \sum_{\langle\alpha\rangle<N} R_{N\alpha}(x,y) + R_N(x,y),$$

where we have let

$$(A.11) \qquad K_{\alpha\beta}(x,y) = f_{\alpha\beta}(x) y^\beta (\partial_{\tilde{y}}^\alpha K_{\tilde{P}})(\phi(x), \phi'_H(x)y),$$

$$(A.12) \qquad R_{N\alpha}(x,y) = \sum_{\langle\beta\rangle \doteq \frac{3}{2}N} r_{M\alpha\beta}(x,y) y^\beta (\partial_{\tilde{y}}^\alpha K_{\tilde{P}})(\phi(x), \phi'_H(x)y).$$

As in the proof of Proposition 3.1.16 the smoothness of $\phi'_H(x)y$ and the fact that $\phi'_H(x)(\lambda.y) = \lambda.\phi_{H'}(x)y$ for any $\lambda \in \mathbb{R}$ imply that $K_{\alpha\beta}(x,y)$ belongs to $\mathcal{K}^{\hat{m}-\langle\alpha\rangle+\langle\beta\rangle}(U \times \mathbb{R}^{d+1})$. Notice that if $\frac{3}{2}\langle\alpha\rangle \leq \langle\beta\rangle \doteq \frac{3}{2}N$ then we have $\Re\hat{m} - \langle\alpha\rangle + \langle\beta\rangle \geq \Re\hat{m} + \frac{1}{3}\langle\beta\rangle \geq \Re\hat{m} + \frac{N}{2}$. It then follows from Lemma A.1 that, for any integer J, the remainder term $R_{N\alpha}$ is in $C^J(U \times \mathbb{R}^{d+1})$ as soon as N is large enough.

Let $\pi_x: U \times \mathbb{R}^{d+1} \to U$ denote the projection on the first coordinate. In the sequel we will say that a distribution $K(x,y) \in \mathcal{D}'(U \times \mathbb{R}^{d+1})$ is properly supported with respect to x when $\pi_x|_{\text{supp } K}$ is a proper map, i.e., for any compact $L \subset U$ the set $\text{supp } K \cap (L \times \mathbb{R}^{d+1})$ is compact.

In order to deal with the regularity of $R_N(x,y)$ in (A.7) we need the lemma below.

LEMMA A.2. *There exists a function $\chi_n \in C_c^\infty(\mathcal{U})$ properly supported with respect to x such that $\chi(x,y) = 1$ near $y=0$ and, for any multi-order α, we can write*

$$(A.13) \qquad \chi(x,y)\Theta(x,y)^\alpha = \sum_{\langle\beta\rangle \doteq \frac{3}{2}\langle\alpha\rangle} \theta_{\alpha\beta}(t,x,y) \Phi_t(x,y)^\beta$$

where the functions $\theta_{\alpha\beta}(t,x,y)$ are in $C^\infty([0,1] \times U \times \mathbb{R}^{d+1})$.

PROOF OF THE LEMMA. Let U' be a relatively compact open subset of U and let $(t_0, x_0) \in [0,1] \times U'$. Since $\Phi_{t_0}(x_0, 0) = 0$ and $\partial_y \Phi_{t_0}(x_0, 0) = \phi'_H(x_0)$ is invertible the implicit function theorem implies that there exist an open interval I_{x_0} containing t_{x_0}, an open subset U_{x_0} of U containing x_0, open subsets V_{x_0} and $\tilde{V}_{x_0}$ of $\mathbb{R}^{d+1}$

containing 0 and a smooth map $\Psi_{x_0}(t,x,\tilde{y})$ from $I_{x_0} \times U_{x_0} \times \tilde{V}_{x_0}$ to V_{x_0} such that $U_{x_0} \times V_{x_0}$ is contained in $\mathcal{U}$ and for any (t,x,y) in $I_{x_0} \times U_{x_0} \times V_{x_0}$ and any $\tilde{y}$ in $\tilde{V}_{x_0}$ we have

$$\text{(A.14)} \qquad \tilde{y} = \Phi_t(x,y) \iff y = \Psi_{x_0}(t,x,\tilde{y}).$$

Since $[0,1] \times \overline{U'}$ is compact we can cover it by finitely many products $I_{x_k} \times U_{x_k}$, $k = 1,..,p$, with $(t_k, x_k) \in [0,1] \times \overline{U'}$. In particular, the sets $I = \cup_k I_k$ and $U'' = \cup U_k$ are open neighborhoods of I and $\overline{U'}$ respectively. Thanks to (A.14) we have $\Psi_{x_k} = \Psi_{x_l}$ on $(I_{x_k} \times U_{x_k} \times V_{x_k}) \cap (I_{x_l} \times U_{x_l} \times V_{x_l})$. Therefore, setting $V = \cap_k V_k$ and $\tilde{V} = \cap_k \tilde{V}_k$ we have $U'' \times V \subset \mathcal{U}$ and there exists a smooth map Ψ from $I \times U'' \times \tilde{V}$ such that for any (t,x,y) in $I \times U'' \times V$ and any $\tilde{y}$ in $\tilde{V}$ we have

$$\text{(A.15)} \qquad \tilde{y} = \Phi_t(x,y) \iff y = \Psi(t,x,\tilde{y}).$$

Furthermore, as $\partial_{\tilde{y}} \Psi(t,x,0) = [\partial_y \Phi_t(x,0)]^{-1} = \phi'_H(x)^{-1}$ and for any $\lambda \in \mathbb{R}$ we have $\phi'_H(x)^{-1}(\lambda.y) = \lambda \phi'_H(x)^{-1}(y)$, the function $\Theta(x, \Psi(t,x,\tilde{y}))$ behaves near $\tilde{y} = 0$ as in (A.5), so as in (A.8) and (A.9) for any multi-order α we can write

$$\text{(A.16)} \qquad \Theta(x, \Psi(t,x,\tilde{y}))^\alpha = \sum_{\langle \beta \rangle \doteq \frac{3}{2} \langle \alpha \rangle} \tilde{\theta}_{\alpha\beta}(t,x,\tilde{y}) \tilde{y}^\beta,$$

for some functions $\tilde{\theta}_{\alpha\beta}(t,x,\tilde{y}) \in C^\infty(I \times U'' \times \tilde{V})$. Setting $\tilde{y} = \Phi_t(x,y)$ then gives

$$\text{(A.17)} \qquad \Theta(x,y)^\alpha = \sum_{\langle \beta \rangle \doteq \frac{3}{2} \langle \alpha \rangle} \theta_{\alpha\beta}(t,x,y) \Phi_t(x,y)^\beta,$$

for some functions $\theta_{\alpha\beta}(t,x,y) \in C^\infty(I \times U'' \times V)$.

All this allows us to construct locally finite coverings $(U'_n)_{n \geq 0}$ and $(U''_n)_{n \geq 0}$ of U by relatively compact open subsets in such way that, for each integer n, the open U''_n contains $\overline{U'_n}$ and there exists an open $V_n \subset \mathbb{R}^{d+1}$ containing 0 so that, for any multiorder α, on $[0,1] \times U''_n \times V_n$ we have

$$\text{(A.18)} \qquad \Theta(x,y)^\alpha = \sum_{\langle \beta \rangle \doteq \frac{3}{2} \langle \alpha \rangle} \theta^{(n)}_{\alpha\beta}(t,x,y) \Phi_t(x,y)^\beta,$$

for some functions $\theta^{(n)}_{\alpha\beta}(t,x,y) \in C^\infty([0,1] \times U''_n \times V_n)$.

For each n let $\varphi_n \in C^\infty_c(U''_n)$ be such that $\varphi_n = 1$ on U'_n and let $\psi_n \in C^\infty_c(V_n)$ be such that $\psi_n = 1$ on a neighborhood V'_n of 0. Then we construct a locally finite family $(\chi_n)_{n \geq 0} \subset C^\infty_c(\mathcal{U})$ as follows: for $n = 0$ we set $\chi_0(x,y) = \varphi_0(x)\psi_0(y)$ and for $n \geq 1$ we let

$$\text{(A.19)} \qquad \chi_n(x,y) = (1 - \varphi_0(x)\psi_0(y)) \ldots (1 - \varphi_{n-1}(x)\psi_{n-1}(y))\varphi_n(x)\psi_n(y).$$

Then $\chi = \sum_{n \geq 0} \chi_n$ is a well defined smooth function on $U \times \mathbb{R}^{d+1}$ supported on $\cup_{n \geq 0}(U''_n \times V_n) \subset \mathcal{U}$, hence properly supported with respect to x. Also, as $\chi(x,y)$ is equal to 1 on each product $U'_n \times V'_n$ we see that $\chi(x,y) = 1$ on a neighborhood of $U \times \{0\}$. In addition, thanks to (A.18) on $[0,1] \times \mathcal{U}$ we have

$$\text{(A.20)} \qquad \chi(x,y)\Theta(x,y)^\alpha = \sum_{n \geq 0} \chi_n(x,y)\Theta(x,y)^\alpha = \sum_{\langle \beta \rangle \doteq \frac{3}{2} \langle \alpha \rangle} \theta_{\alpha\beta}(t,x,y) \Phi_t(x,y)^\beta,$$

where $\theta_{\alpha\beta}(t,x,y) := \sum_n \chi_n(x,y)\theta^{(n)}_{\alpha\beta}(t,x,y)$ belongs to $C^\infty([0,1] \times U \times \mathbb{R}^{d+1})$. The lemma is thus proved. $\square$

Let us go back to the proof of Proposition 3.1.18. Thanks to (A.7) and (A.13) we see that $\chi(x,y)R_N(x,y)$ is equal to

$$(A.21) \qquad \sum_{\langle\alpha\rangle=N}\sum_{\langle\beta\rangle\doteq\frac{3}{2}N}\int_0^1 r_{\alpha\beta}(t,x,y)(\tilde{y}^\beta\partial_{\tilde{y}}^\alpha K_{\tilde{P}})(\phi(x),\Phi_t(x,y))dt,$$

with $r_{\alpha\beta}(t,x,y)$ in $C^\infty([0,1\times U\times\mathbb{R}^{d+1})$. Since $\tilde{y}^\beta\partial_{\tilde{y}}^\alpha K_{\tilde{P}}$ is in $\mathcal{K}^{\hat{m}+N/2}(\tilde{U}\times\mathbb{R}^{d+1})$ it follows from Lemma A.1 that, for any integer $J\geq 0$, as soon as N is taken large enough $\tilde{y}^\beta\partial_{\tilde{y}}^\alpha K_{\tilde{P}}$ is in $C^J(\tilde{U}\times\mathbb{R}^{d+1})$ and so $\chi(x,y)R_N(x,y)$ is in $C^J(U\times\mathbb{R}^{d+1})$.

On the other hand, set $K_P(x,y)=\chi(x,y)K(x,y)=\sum\chi_n(x,y)K(x,y)$. Since $\chi(x,y)$ is supported in $\mathcal{U}$ and is properly supported with respect to x this defines a distribution on $U\times\mathbb{R}^{d+1}$. Moreover, using (A.10) we get

$$(A.22) \qquad K_P(x,y)=\sum_{\langle\alpha\rangle<N}\sum_{\frac{3}{2}\langle\alpha\rangle\leq\langle\beta\rangle<\frac{3}{2}N}K_{\alpha\beta}(x,y)+\sum_{j=1}^3 R_N^{(j)},$$

where the remainder terms $R_{N,z}^{(j)}$, $j=1,2,3$ are given by

$$(A.23) \qquad R_N^{(1)}=\chi(x,y)R_N(x,y),\quad R_N^{(2)}=\sum_{\langle\alpha\rangle<N}\chi(x,y)R_{N\alpha}(x,y),$$

$$(A.24) \qquad R_N^{(2)}(x,y)=\sum_{\langle\alpha\rangle<N}\sum_{\frac{3}{2}\langle\alpha\rangle\leq\langle\beta\rangle<\frac{3}{2}N}(1-\chi(x,y))K_{\alpha\beta}(x,y).$$

Each term $K_{\alpha\beta}(x,y)$ belongs to $\mathcal{K}^{\hat{m}-\langle\alpha\rangle+\langle\beta\rangle}(U\times\mathbb{R}^{d+1})$ and, as $\hat{m}+\langle\beta\rangle-\langle\alpha\rangle=\hat{m}+j$ and $\frac{3}{2}\langle\alpha\rangle\leq\langle\beta\rangle$ imply $\langle\alpha\rangle\leq 2j$ and $\langle\beta\rangle\leq\frac{4}{3}j$, in the r.h.s. (A.22) there are only finitely many such distributions in a given space $\mathcal{K}^{\hat{m}+j}(U\times\mathbb{R}^{d+1})$ as α and β range over all multi-orders such that $\frac{3}{2}\langle\alpha\rangle\leq\langle\beta\rangle$.

Furthermore, the reminder term $R_N^{(3)}$ is smooth and the other remainder terms $R_N^{(j)}$, $j=1,2$, are in $C^J(U\times\mathbb{R}^{d+1})$ as soon as N is large enough. Thus,

$$(A.25) \qquad K_P(x,y)\sim\sum_{\frac{3}{2}\langle\alpha\rangle\leq\langle\beta\rangle}K_{\alpha\beta}(x,y),$$

which implies that K_P belongs to $\mathcal{K}^{\hat{m}}(U\times\mathbb{R}^{d+1})$ and satisfies (A.25).

Finally, from (A.1) and the very definition of $\Phi(x,y)$ on $U\times U$, we deduce that the distribution kernel of P differs from $|\varepsilon_x'|K_P(x,-\varepsilon_x(y))$ by the smooth function

$$(A.26) \qquad [1-\chi(x,\varepsilon_x(y))]|\tilde{\varepsilon}_{\phi(x)}'|K_{\tilde{P}}(\phi(x),-\tilde{\varepsilon}_{\phi(x)}\circ\phi(y))+\tilde{R}(\phi(x),\phi(y)).$$

Combining this with Proposition 3.1.16 and the fact that $K_P(x,y)$ satisfies (A.25) proves Proposition 3.1.18.

APPENDIX B

Proof of Proposition 3.1.21

Let $P : C_c^\infty(U) \to C^\infty(U)$ be a Ψ_HDO of order m and let us show that its transpose operator $P^t : C_c^\infty(U) \to C^\infty(U)$ is a Ψ_HDO of order m. By Proposition 3.1.16 the distribution kernel of P is of the form,

(B.1) $$k_P(x,y) = |\varepsilon_x'| K_P(x, -\varepsilon_x(y)) + R(x,y),$$

with $K_P(x,y)$ in $\mathcal{K}^m(U \times \mathbb{R}^{d+1})$ and $R(x,y)$ in $C^\infty(U \times U)$. Thus the distribution kernel $k_{P^t}(x,y) = k_P(y,x)$ of P^t is equal to

(B.2) $$|\varepsilon_y'| K_P(y, -\varepsilon_y(x)) + R(y,x) = |\varepsilon_x'| K(x, -\varepsilon_x(y)) + R(y,x),$$

where K is the distribution on the open $\mathcal{U} = \{(x,y);\ \varepsilon_x^{-1}(-y) \in U\}$ given by

(B.3) $$K(x,y) = |\varepsilon_x'|^{-1} |\varepsilon_y'| K_P(\varepsilon_x^{-1}(-y), -\varepsilon_{\varepsilon_x^{-1}(-y)}(x)).$$

LEMMA B.1. *On $\mathcal{U}$ we have*

(B.4) $$\varepsilon_{\varepsilon_x^{-1}(-y)}(x) = y - \Theta(x,y),$$

where $\Theta : \mathcal{U} \to \mathbb{R}^{d+1}$ is a smooth map with a behavior near $y=0$ of the form (A.5).

PROOF. Let $(x,y) \in \mathcal{U}$ and $Y \in G_x U$ let $\lambda_y(Y) = y.Y$, that is, λ_y is the left multiplication by y on $G_x U$. Then by [**Po6**, Eq. (3.32)] for Y small enough we have

(B.5) $$\lim_{t \to 0} \varepsilon_x \circ \varepsilon_{\varepsilon_x^{-1}(t.-y)}^{-1}(t.Y) = \lambda_{-y}(Y) = \lambda_y^{-1}(Y).$$

Since $\varepsilon_x \circ \varepsilon_{\varepsilon_x^{-1}(t.-y)}^{-1}(t.Y)$ is a smooth function of (t,Y) near $(0,0)$, it follows from the implicit function theorem that for Y small enough we have

(B.6) $$\lim_{t \to 0} t^{-1}.\varepsilon_{\varepsilon_x^{-1}(t.-y)} \circ \varepsilon_x^{-1}(Y) = \lambda_y(Y).$$

In particular, for $Y = 0$ we get

(B.7) $$\lim_{t \to 0} t^{-1}.\varepsilon_{\varepsilon_x^{-1}(t.-y)}(x) = y.$$

Now, the function $\varepsilon_{\varepsilon_x^{-1}(-y)}(x)$ depends smoothly on $(x,y) \in \mathcal{U}$, so (B.7) allows us to put it into the form,

(B.8) $$\varepsilon_{\varepsilon_x^{-1}(-y)}(x) = y - \Theta(x,y),$$

where $\Theta = (\Theta_0, \ldots, \Theta_d)$ is smooth map from $\mathcal{U}$ to $\mathbb{R}^{d+1}$ with a behavior near $y = 0$ of the form

(B.9) $\Theta_0(x,y) = O(|y_0|^2 + |y_0||y| + |y|^3), \qquad \Theta_j(x,y) = O(|y|^2), \quad j = 1, \ldots, d.$

In particular, near $y = 0$ the map Θ has a behavior of the form (A.5). $\square$

Next, a Taylor expansion around $(\varepsilon_x^{-1}(-y), -y)$ gives

(B.10) $\quad K(x,y) = \sum_{\langle\alpha\rangle < N} |\varepsilon_x'|^{-1}|\varepsilon_y'|\frac{\theta(x,y)^\alpha}{\alpha!}(\partial_y^\alpha K_P)(\varepsilon_x^{-1}(-y), -y) + R_N(x,y),$

where $R_N(x,y)$ is equal to

(B.11) $\quad \sum_{\langle\alpha\rangle = N} |\varepsilon_x'|^{-1}|\varepsilon_y'|\frac{\theta(x,y)^\alpha}{\alpha!}\int_0^1 (1-t)^{N-1} \partial_y^\alpha K_P)(\varepsilon_x^{-1}(-y), \Phi_t(x,y)),$

and we have let $\Phi_t(x,y) = -y + t\Theta(x,y)$.

Let $a_\alpha(x,y) = |\varepsilon_x'|^{-1}|\varepsilon_y'|\frac{\theta(x,y)^\alpha}{\alpha!}$. Thanks to (A.5) the same arguments used to prove (A.9) show that there exist functions $r_{N\alpha}(x,y) \in C^\infty(\mathcal{U})$, $\langle\beta\rangle \doteq \frac{3}{2}N$, so that

(B.12) $\quad a_\alpha(x,y) = \sum_{\frac{3}{2}\langle\alpha\rangle \leq \langle\beta\rangle < \frac{3}{2}N} a_{\alpha\beta}(x)y^\beta + \sum_{\langle\beta\rangle \doteq \frac{3}{2}N} r_{N\alpha}(x,y)y^\beta,$

where we have let $a_{\alpha\beta}(x) = \frac{1}{\beta!}\partial^\beta a_\alpha(x,0)$. Therefore, we get

(B.13) $\quad K(x,y) = \sum_{\langle\alpha\rangle < N} \sum_{\frac{3}{2}\langle\alpha\rangle \leq \langle\beta\rangle < \frac{3}{2}N} a_{\alpha\beta}(x)y^\beta(\partial_y^\alpha K_P)(\varepsilon_x^{-1}(-y), -y)$
$$+ \sum_{\langle\alpha\rangle < N} R_{N\alpha}(x,y) + R_N(x,y),$$

where we have let

(B.14) $\quad R_{N\alpha}(x,y) = \sum_{\langle\beta\rangle \doteq \frac{3}{2}N} r_{N\alpha}(x,y)y^\beta(\partial_y^\alpha K_P)(\varepsilon_x^{-1}(-y), -y).$

Next, a further Taylor expansion gives

(B.15) $\quad (\partial_y^\alpha K_P)(\varepsilon_x^{-1}(-y), -y) = \sum_{|\gamma| < N} \frac{1}{\gamma!}(\varepsilon_x^{-1}(-y) - x)^\gamma (\partial_x^\gamma \partial_y^\alpha K_P)(x, -y)$
$$+ \sum_{|\gamma|=N} \int_0^1 (1-t)^{N-1}(\partial_x^\gamma \partial_y^\alpha K_P)(\varepsilon_t(x,y), -y),$$

where we have let $\varepsilon_t(x,y) = x + t(\varepsilon_x^{-1}(-y) - x)$. Since $\varepsilon_x^{-1}(-y) - x$ is polynomial in y of degree 2 and vanishes for $y = 0$, we can write

(B.16) $\quad \frac{1}{\gamma!}(\varepsilon_x^{-1}(-y) - x)^\gamma = \sum_{|\gamma| \leq |\delta| \leq 2|\gamma|} b_{\gamma\delta}(x)y^\delta,$

where we have let $b_{\gamma\delta}(x) = \frac{1}{\gamma!\delta!}[\partial_y(\varepsilon_x^{-1}(-y) - x)^\gamma]_{y=0}$. Thus,

(B.17) $\quad K(x,y) = \sum_{\alpha,\beta,\gamma,\delta}^{(N)} K_{\alpha\beta\gamma\delta}(x,y) + \sum_{\langle\alpha\rangle < N} \sum_{\frac{3}{2}\langle\alpha\rangle \leq \langle\beta\rangle < \frac{3}{2}N} R_{N\alpha\beta}(x,y)$
$$+ \sum_{\langle\alpha\rangle < N} R_{N\alpha}(x,y) + R_N(x,y),$$

where the first summation goes over all the multi-orders α, β, γ and δ such that $\langle\alpha\rangle < N$, $\frac{3}{2}\langle\alpha\rangle \leq \langle\beta\rangle < \frac{3}{2}N$ and $|\gamma| \leq |\delta| \leq 2|\gamma| < 2N$, and

(B.18) $\quad K_{\alpha\beta\gamma\delta}(x,y) = f_{\alpha\beta\gamma\delta}(x)y^{\beta+\delta}(\partial_x^\gamma \partial_y^\alpha K_P)(x, -y),$

with $f_{\alpha\beta\gamma\delta}(x) = a_{\alpha\beta}(x)b_{\gamma\delta}(x)$ and $R_{N\alpha\beta}(x,y)$ is equal to

$$\text{(B.19)} \qquad \sum_{|\gamma|=N} \sum_{N \leq |\delta| \leq 2N} a_{\alpha\beta\gamma\delta}(x) y^{\beta+\delta} \int_0^1 (1-t)^{N-1} (\partial_x^\gamma \partial_y^\alpha K_P)(\varepsilon_t(x,y), -y).$$

Now, the distribution $y^\beta K_P(x, -y)$ belongs to $\mathcal{K}^{\hat{m}-\langle\alpha\rangle+\langle\beta\rangle}(U \times \mathbb{R}^{d+1})$. In particular, if $\frac{3}{2}\langle\alpha\rangle \leq \langle\beta\rangle \doteq \frac{3}{2}N$ then $\Re\hat{m} - \langle\alpha\rangle + \langle\beta\rangle \geq \Re\hat{m} + \frac{1}{3}\langle\beta\rangle \geq \Re\hat{m} + \frac{1}{2}N$. Therefore, for any given integer J Lemma A.1 tells us that $y^\beta K_P(x, -y)$ is in $C^J(U \times \mathbb{R}^{d+1})$ as soon as N is large enough. It follows that all the remainder terms $R_{N\alpha}(x,y)$, $\langle\alpha\rangle < N$, belong to $C^J(\mathcal{U})$ for N large enough.

Similarly, if $\frac{3}{2}\langle\alpha\rangle \leq \langle\beta\rangle$ and $|\gamma| = N \leq |\delta| \leq 2N$ then $\Re\hat{m} - \langle\alpha\rangle + \langle\beta\rangle + \langle\delta\rangle \geq \Re\hat{m} + \langle\delta\rangle \geq \Re\hat{m} + \frac{1}{2}N$, so using Lemma A.1 we see that $y^{\beta+\delta}(\partial_x^\gamma \partial_y^\alpha K_P)(x, -y)$ is in $C^J(U \times \mathbb{R}^{d+1})$ for N large enough. It then follows that for N large enough the remainder terms $R_{N\alpha\beta}(x,y)$ with $\langle\alpha\rangle < N$ and $\frac{3}{2}\langle\alpha\rangle \leq \langle\beta\rangle \doteq \frac{3}{2}N$ are all in $C^J(\mathcal{U})$ as soon as N is chosen large enough.

In order to deal with the last remainder term $R_N(x,y)$ notice that, along the same lines as that of the proof of Lemma A.2, one can show that there exists a $\chi \in C^\infty(\mathcal{U})$ properly supported with respect to x such that $\chi(x,y) = 1$ near $y = 0$ and, for any multi-order α, we can write

$$\text{(B.20)} \qquad \chi(x,y)\Theta(x,y)^\alpha = \sum_{\langle\beta\rangle \doteq \frac{3}{2}\langle\alpha\rangle} \theta_{\alpha\beta}(t,x,y) \Phi_t(x,y)^\beta,$$

where the functions $\theta_{\alpha\beta}(t,x,y)$ are in $C^\infty([0,1] \times U \times \mathbb{R}^{d+1})$. Therefore $\chi(x,y) R_N(x,y)$ is equal to

$$\text{(B.21)} \qquad \sum_{\langle\alpha\rangle=N} \sum_{\langle\beta\rangle \doteq \frac{3}{2}\langle\alpha\rangle} |\varepsilon_x'|^{-1} |\varepsilon_y'| \int_0^1 r_{N\alpha\beta}(t,x,y)(y^\beta \partial_y^\alpha K_P)(\varepsilon_x^{-1}(-y), \Phi_t(x,y)),$$

for some functions $r_{N\alpha\beta}(t,x,y)$ in $C^\infty([0,1] \times U \times \mathbb{R}^{d+1})$. Since $(y^\beta \partial_y^\alpha K_P)$ is in $\mathcal{K}^{\hat{m}-\langle\alpha\rangle+\langle\beta\rangle}(U \times \mathbb{R}^{d+1})$ and we have $\Re\hat{m} - \langle\alpha\rangle + \langle\beta\rangle \geq \Re\hat{m} + \frac{1}{2}\langle\alpha\rangle = \Re\hat{m} + \frac{1}{2}N$, using Lemma A.1 we see that for N large enough $\chi(x,y) R_N(x,y)$ is in $C^J(\mathcal{U})$. As $\chi(x,y) R_N(x,y)$ is a properly supported with respect to x this shows that it belongs to $C^J(U \times \mathbb{R}^{d+1})$.

Let $K_{P^t}(x,y) = \chi(x,y) K(x,y)$. This defines a distribution on $U \times \mathbb{R}^{d+1}$ since χ is properly supported with respect to x. Moreover, we have

$$\text{(B.22)} \qquad K_{P^t}(x,y) = \sum_{\alpha,\beta,\gamma,\delta}^{(N)} K_{\alpha\beta\gamma\delta}(x,y) + \sum_{j=1}^4 R_N^{(j)}(x,y),$$

where the remainder terms $R_N^{(j)}$, $j=1,\ldots,4$, are given by the formulas,

$$\text{(B.23)} \qquad R_N^{(1)} = \chi(x,y) R_N(x,y), \qquad R_N^{(2)} = \sum_{\langle\alpha\rangle < N} \chi(x,y) R_{N\alpha}(x,y),$$

$$\text{(B.24)} \qquad R_N^{(3)} = \sum_{\langle\alpha\rangle < N} \sum_{\frac{3}{2}\langle\alpha\rangle \leq \langle\beta\rangle < \frac{3}{2}N} \chi(x,y) R_{N\alpha\beta}(x,y),$$

$$\text{(B.25)} \qquad R_N^{(4)} = \sum_{\alpha,\beta,\gamma,\delta}^{(N)} (1 - \chi(x,y)) K_{\alpha\beta\gamma\delta}(x,y).$$

Note that $K_{\alpha\beta\gamma\delta}(x,y)$ belongs to $\mathcal{K}^{\hat{m}-\langle\alpha\rangle+\langle\gamma\rangle+\langle\delta\rangle}(U\times\mathbb{R}^{d+1})$ and there are finitely many terms of a given order as α, β, γ and δ ranges over all the multi-orders such that $\frac{3}{2}\langle\alpha\rangle\leq\langle\beta\rangle$ and $|\gamma|\leq|\delta|\leq 2|\gamma|$.

On the other hand, the remainder term $R_N^{(4)}$ is smooth and it follows from the above observations that the other remainder terms are in $C^J(U\times\mathbb{R}^{d+1})$ as soon as N is large enough. Thus,

$$(B.26) \qquad K_{P^t}(x,y) \sim \sum_{\frac{3}{2}\langle\alpha\rangle\leq\langle\beta\rangle} \sum_{|\gamma|\leq|\delta|\leq 2|\gamma|} K_{\alpha\beta\gamma\delta}(x,y),$$

which incidentally shows that $K_{P^t}(x,y)$ belongs to $\mathcal{K}^{\hat{m}}(U\times\mathbb{R}^{d+1})$.

Finally, thanks to (B.2) we can put the kernel of P^t into the form,

$$(B.27) \quad k_{P^t}(x,y) = |\varepsilon'_x|K_P(x,-\varepsilon_x(y)) + |\varepsilon'_x|[(1-\chi)K](x,-\varepsilon_x(y)) + R(y,x)$$
$$= |\varepsilon'_x|K_P(x,-\varepsilon_x(y)) \mod C^\infty(U\times U).$$

It then follows from Proposition 3.1.16 that P^t is a $\Psi_H\mathrm{DO}$ of order m. Moreover, working out the expression for $K_{\alpha\beta\gamma\delta}$ shows that the asymptotics expansion (B.26) reduces to (3.1.38). The proof of Proposition 3.1.21 is thus achieved.

Bibliography

[AS1]   Atiyah, M.F.; Singer, I.M.: *The index of elliptic operators. I.* Ann. of Math. (2) **87** (1968), 484–530.
[AS2]   Atiyah, M.F, Singer, I.M: *The index of elliptic operators. III.* Ann. of Math. (2) **87**, 546–604 (1968).
[APS]   Atiyah, M., Patodi, V., Singer, I.: *Spectral asymmetry and Riemannian geometry. III.* Math. Proc. Camb. Philos. Soc. **79**, 71–99 (1976).
[BEG]   Bailey, T.N.; Eastwood, M.G.; Gover, A.R.: *Thomas's structure bundle for conformal, projective and related structures.* Rocky Mountain J. Math. **24** (1994), no. 4, 1191–1217.
[Bea]   Beals, R.: *Opérateurs invariants hypoelliptiques sur un groupe de Lie nilpotent.* Séminaire Goulaouic-Schwartz (1976-77), Exposé No. 19. Ecole Polytech., Palaiseau, 1977.
[BG]   Beals, R.; Greiner, P.C.: *Calculus on Heisenberg manifolds.* Annals of Mathematics Studies, vol. 119. Princeton University Press, Princeton, NJ, 1988.
[BGS]   Beals, R.; Greiner, P.C.; Stanton, N.K.: *The heat equation on a CR manifold.* J. Differential Geom. **20** (1984), no. 2, 343–387.
[BS1]   Beals, R.; Stanton, N.K.: *The heat equation for the $\overline{\partial}$-Neumann problem. I.* Comm. Partial Differential Equations **12** (1987), no. 4, 351–413.
[BS2]   Beals, R.; Stanton, N.K.: *The heat equation for the $\overline{\partial}$-Neumann problem. II.* Canad. J. Math. **40** (1988), no. 2, 502–512.
[Be]   Bellaïche, A.: *The tangent space in sub-Riemannian geometry. Sub-Riemannian geometry*, 1–78, Progr. Math., 144, Birkhäuser, Basel, 1996.
[Bo1]   Boutet de Monvel, L.: *Hypoelliptic operators with double characteristics and related pseudo-differential operators.* Comm. Pure Appl. Math. **27** (1974), 585–639.
[Bo2]   Boutet de Monvel, L.: *Logarithmic trace of Toeplitz projectors.* Math. Res. Lett. **12** (2005), 401–412.
[BGH]   Boutet de Monvel, L.; Grigis, A.; Helffer, B.: *Parametrixes d'opérateurs pseudo-différentiels à caractéristiques multiples.* Journées EDP (Rennes, 1975) pp. 93–121. Astérisque, No. 34-35, Soc. Math. France, Paris, 1976.
[Ch1]   Christ, M.: *On the regularity of inverses of singular integral operators.* Duke Math. J. **57** (1988), no. 2, 459–484.
[Ch2]   Christ, M.: *Inversion in some algebras of singular integral operators.* Rev. Mat. Iberoamericana **4** (1988), no. 2, 219–225.
[CGGP] Christ, M.; Geller, D.; Głowacki, P.; Polin, L.: *Pseudodifferential operators on groups with dilations.* Duke Math. J. **68** (1992) 31–65.
[Co]   Connes, A.: *Noncommutative geometry.* Academic Press, Inc., San Diego, CA, 1994.
[CM]   Connes, A.; Moscovici, H.: *The local index formula in noncommutative geometry.* Geom. Funct. Anal. **5** (1995), no. 2, 174–243.
[Dy1]   Dynin, A.: *Pseudodifferential operators on the Heisenberg group.* Dokl. Akad. Nauk SSSR **225** (1975) 1245–1248.
[Dy2]   Dynin, A.: *An algebra of pseudodifferential operators on the Heisenberg groups. Symbolic calculus.* Dokl. Akad. Nauk SSSR **227** (1976), 792–795.
[EM]   Epstein, C.L.; Melrose, R.B.: *The Heisenberg algebra, index theory and homology.* Preprint, 1998. Available online at http://www-math.mit.edu/~rbm.
[EMM]   Epstein, C.L.; Melrose, R.B.; Mendoza, G.: *Resolvent of the Laplacian on strictly pseudoconvex domains.* Acta Math. **167** (1991), no. 1-2, 1–106.
[ET]   Eliashberg, Y.; Thurston, W.: *Confoliations.* University Lecture Series, 13, AMS, Providence, RI, 1998.

[FH] Fefferman, C.; Hirachi, K.: *Ambient metric construction of Q-curvature in conformal and CR geometries*. Math. Res. Lett. **10** (2003), no. 5-6, 819–831.

[Fo] Folland, G. *Subelliptic estimates and function spaces on nilpotent Lie groups*. Ark. Mat. **13** (1975) 161–207.

[FS1] Folland, G.; Stein, E.: *Estimates for the $\bar{\partial}_b$ complex and analysis on the Heisenberg group*. Comm. Pure Appl. Math. **27** (1974) 429–522.

[FS2] Folland, G.; Stein, E.: *Hardy spaces on homogeneous groups*. Mathematical Notes, 28. Princeton University Press, Princeton, 1982.

[Ge1] Geller, D.: *Liouville's theorem for homogeneous groups*. Comm. Partial Differential Equations **8** (1983) 1665–1677.

[Ge2] Geller, D.: *Complex powers of convolution operators on the Heisenberg group*. Analysis, geometry, number theory: the mathematics of Leon Ehrenpreis (Philadelphia, PA, 1998), 223–242, Contemp. Math., 251, Amer. Math. Soc., Providence, RI, 2000.

[Gi] Gilkey, P.B.: *Invariance theory, the heat equation, and the Atiyah-Singer index theorem*. Mathematics Lecture Series, 11. Publish or Perish, Inc., Wilmington, Del., 1984.

[GJ] Glimm, J.; Jaffe, A.: *Quantum physics. A functional integral point of view*. Springer-Verlag, 1987.

[Gł1] Głowacki, P.: *Stable semigroups of measures as commutative approximate identities on nongraded homogeneous groups*. Invent. Math. **83** (1986), no. 3, 557–582.

[Gł2] Głowacki, P.: *The Rockland condition for nondifferential convolution operators. II*. Studia Math. **98** (1991), no. 2, 99–114.

[GK] Gohberg, I.C.; Kreĭn, M.G.: *Introduction to the theory of linear nonselfadjoint operators*. Trans. of Math. Monographs 18, AMS, Providence, 1969.

[GG] Gover, A.R.; Graham, C.R.: *CR invariant powers of the sub-Laplacian*. J. Reine Angew. Math. **583** (2005), 1–27.

[GJMS] Graham, C.R.; Jenne, R.; Mason, L.J.; Sparling, G.A.: *Conformally invariant powers of the Laplacian. I. Existence*. J. London Math. Soc. (2) **46** (1992), no. 3, 557–565.

[Gr] Greenleaf, A.: *The first eigenvalue of a sub-Laplacian on a pseudo-Hermitian manifold*. Comm. Partial Differential Equations **10** (1985), no. 2, 191–217.

[Gre] Greiner, P. *An asymptotic expansion for the heat equation*. Arch. Rational Mech. Anal. **41** (1971) 163–218.

[Gri] Grigis, A.: *Hypoellipticité et paramétrix pour des opérateurs pseudodifférentiels à caractéristiques doubles*. Journées EDP (Rennes, 1975) pp. 183–205. Astérisque, No. 34-35, Soc. Math. France, Paris, 1976.

[Gro] Gromov, M.: *Carnot-Carathéodory spaces seen from within*. Sub-Riemannian geometry, 79–323, Progr. Math., 144, Birkhäuser, Basel, 1996.

[Gu1] Guillemin, V.W.: *A new proof of Weyl's formula on the asymptotic distribution of eigenvalues*. Adv. Math. **55** (1985), no. 2, 131–160.

[Gu2] Guillemin, V.W.: *Gauged Lagrangian distributions*. Adv. Math. **102** (1993), no. 2, 184–201.

[Ha] Hardy, G.H.: *Divergent Series*. Oxford, at the Clarendon Press, 1949.

[HN1] Helffer, B.; Nourrigat, F.: *Hypoellipticité pour des groupes nilpotents de rang de nilpotence 3*. Comm. Partial Differential Equations **3** (1978), no. 8, 643–743.

[HN2] Helffer, B.; Nourrigat, J.: *Caracterisation des opérateurs hypoelliptiques homogènes invariants à gauche sur un groupe de Lie nilpotent gradué*. Comm. Partial Differential Equations **4** (1979), no. 8, 899–958.

[HN3] Helffer, B.; Nourrigat, J.: *Hypoellipticité maximale pour des opérateurs polynômes de champs de vecteurs*. Prog. Math., No. 58, Birkhäuser, Boston, 1986.

[HS] Hilsum, M.; Skandalis, G.: *Morphismes K-orientés d'espaces de feuilles et fonctorialité en théorie de Kasparov (d'après une conjecture d'A. Connes)*. Ann. Sci. École Norm. Sup. (4) **20** (1987), no. 3, 325–390.

[Hi] Hirachi, K.: *Logarithmic singularity of the Szegö kernel and a global invariant of strictly pseudoconvex domains*. E-print, arXiv, Sep. 03, 17 pages. To appear in Ann. of Math..

[Hö1] Hörmander, L.: *Pseudo-differential operators and hypoelliptic equations. Singular integrals* (Proc. Sympos. Pure Math., Vol. X, Chicago, Ill., 1966), pp. 138–183. Amer. Math. Soc., Providence, R.I., 1967.

[Hö2] Hörmander, L.: *Hypoelliptic second order differential equations*. Acta Math. **119** (1967), 147–171.

[Hö3] Hörmander, L.: *A class of hypoelliptic pseudodifferential operators with double characteristics.* Math. Ann. **217** (1975), no. 2, 165–188.

[Hw] I.L. Hwang *The L^2-boundedness of pseudodifferential operators.* Trans. Amer. Math. Soc. **302** (1987), no. 1, 55–76.

[II] Iwasaki, C.; Iwasaki, N.: *Parametrix for a degenerate parabolic equation and its application to the asymptotic behavior of spectral functions for stationary problems.* Publ. Res. Inst. Math. Sci. **17** (1981), no. 2, 577–655.

[JL1] Jerison, D.; Lee, J.M.: *The Yamabe problem on CR manifolds.* J. Differential Geom. **25** (1987), no. 2, 167–197.

[JL2] Jerison, D.; Lee, J.M.: *Extremals for the Sobolev inequality on the Heisenberg group and the CR Yamabe problem.* J. Amer. Math. Soc. **1** (1988), no. 1, 1–13.

[JL3] Jerison, D.; Lee, J.M.: *Intrinsic CR normal coordinates and the CR Yamabe problem.* J. Differential Geom. **29** (1989), no. 2, 303–343.

[JK] Julg, P.; Kasparov, G.: *Operator K-theory for the group $SU(n, 1)$.* J. Reine Angew. Math. **463** (1995), 99–152.

[KS] Knapp, A.W.; Stein, E.M.: *Intertwining operators for semisimple groups.* Ann. of Math. **93** (1971), 489–578.

[Ko] Koenig, K.: *On maximal Sobolev and Hlder estimates for the tangential Cauchy-Riemann operator and boundary Laplacian.* Amer. J. Math. **124** (2002), no. 1, 129–197.

[Koh1] Kohn, J.J.: *Boundaries of complex manifolds.* 1965 Proc. Conf. Complex Analysis (Minneapolis, 1964) pp. 81–94. Springer, Berlin.

[Koh2] Kohn, J.J.: *Superlogarithmic estimates on pseudoconvex domains and CR manifolds.* Ann. of Math. (2) **156** (2002), no. 1, 213–248.

[Koh3] Kohn, J.J.: *Hypoellicitity and loss of derivatives.* Preprint, 2003, 43 pages. To appear in Ann. of Math..

[KR] Kohn, J.J.; Rossi, H.: *On the extension of holomorphic functions from the boundary of a complex manifold.* Ann. of Math. **81** (1965) 451–472.

[KV] Kontsevich, M.; Vishik, S.: *Geometry of determinants of elliptic operators.* Functional analysis on the eve of the 21st century, Vol. 1 (New Brunswick, NJ, 1993), 173–197, Progr. Math., 131, Birkhäuser Boston, Boston, MA, 1995.

[Kr1] Krainer, T.: *Parabolic pseudodifferential operators and long-time asymptotics of solutions.* PhD dissertation, University of Potsdam, 2000.

[Kr2] Krainer, T.: *Volterra families of pseudodifferential operators* and *The calculus of Volterra Mellin pseudodifferential operators with operator-valued symbols. Parabolicity, Volterra calculus, and conical singularities*, pp. 1–45 and 47–91. Oper. Theory Adv. Appl., 138, Birkhäuser, Basel, 2002.

[KSc] Krainer, T.; Schulze, B.W.: *On the inverse of parabolic systems of partial differential equations of general form in an infinite space-time cylinder.* in *Parabolicity, Volterra calculus, and conical singularities*, pp. 93–278. Oper. Theory Adv. Appl., 138, Birkhäuser, Basel, 2002.

[Le] Lee, J.M.: *The Fefferman metric and pseudo-Hermitian invariants.* Trans. Amer. Math. Soc. **296** (1986), no. 1, 411–429.

[MMS] Mathai, V.; Melrose, R.B.; Singer, I.M: *Fractional analytic index.* E-print, ArXiv, Feb. 04.

[MS] Menikoff, A.; Sjöstrand, J.: *On the eigenvalues of a class of hypoelliptic operators. II.* Global analysis (Proc. Biennial Sem. Canad. Math. Congr., Univ. Calgary, Calgary, Alta., 1978), pp. 201–247, Lecture Notes in Math., 755, Springer, Berlin, 1979.

[Me1] Melrose, R.B.: *Hypoelliptic operators with characteristic variety of codimension two and the wave equation.* Séminaire Goulaouic-Schwartz, 1979–1980 (French), Exp. No. 11, 13 pp., Ecole Polytech., Palaiseau, 1980.

[Me2] Melrose, R.B.: *The Atiyah-Patodi-Singer index theorem.* A.K. Peters, Boston, 1993.

[Me3] Melrose, R.B.: *Star products and local line bundles.* Preprint, 2004.

[Mo1] Mohamed, A.: *Etude spectrale d'opérateurs hypoelliptiques à caractéristiques multiples. I.* Ann. Inst. Fourier (Grenoble) **32** (1982), no. 3, ix, 39–90.

[Mo2] Mohamed, A.: *Etude spectrale d'opérateurs hypoelliptiques à caractéristiques multiples. II.* Comm. Partial Differential Equations **8** (1983), no. 3, 247–316.

[Ni] Nicoara, A.: *Global regularity for $\overline{\partial}_b$ on weakly pseudoconvex CR manifolds.* Adv. in Math. **199** (2006), no. 2, 356–447.

[Pi] Piriou, A.: *Une classe d'opérateurs pseudo-différentiels du type de Volterra.* Ann. Inst. Fourier **20** (1970) 77–94.

[Po1] Ponge, R.: *Calcul hypoelliptique sur les variétés de Heisenberg, résidu non commutatif et géométrie pseudo-hermitienne.* PhD Thesis, Univ. Paris-Sud (Orsay), Dec. 2000.

[Po2] Ponge, R.: *Calcul fonctionnel sous-elliptique et résidu non commutatif sur les variétés de Heisenberg.* C. R. Acad. Sci. Paris, Série I, **332** (2001) 611–614.

[Po3] Ponge, R.: *Géométrie spectrale et formules d'indices locales pour les variétés CR et contact.* C. R. Acad. Sci. Paris, Série I, **332** (2001) 735–738.

[Po4] Ponge, R.: *A new short proof of the local index formula and some of its applications.* Comm. Math. Phys. **241** (2003) 215–234. *Erratum* Comm. Math. Phys. **248** (2004) 639.

[Po5] Ponge, R.: *On the asymptotic completeness of the Volterra calculus.* Appendix by H. Mikaleyan and R. Ponge. J. Anal. Math. **94** (2004) 249–263.

[Po6] Ponge, R.: *The tangent groupoid of a Heisenberg manifold.* E-print, arXiv, Apr. 04, 17 pages. To appear in Pacific Math. J..

[Po7] Ponge, R.: *A new proof of the regularity of the eta invariant for Dirac operators.* E-print, arXiv, Nov. 04, 11 pages.

[Po8] Ponge, R.: *Functional calculus and spectral asymptotics for hypoelliptic operators on Heisenberg manifolds. I..* E-print, arXiv, Feb. 05.

[Po9] Ponge, R.: *New invariants for CR and contact manifolds.* E-print, arXiv, Oct. 05, 23 pages.

[Po10] Ponge, R.: *Remarks on Julg-Kasparov's proof of the Baum-Connes conjecture for the group* $SU(n,1)$. E-print, arXiv, Dec. 05.

[Po11] Ponge, R.: *Noncommutative geometry, Heisenberg calculus and CR geometry.* E-print, arXiv, Jan. 06.

[Po12] Ponge, R.: *Functional calculus and spectral asymptotics for hypoelliptic operators on Heisenberg manifolds* In preparation.

[Ro1] Rockland, C.: *Hypoellipticity on the Heisenberg group-representation-theoretic criteria.* Trans. Amer. Math. Soc. **240** (1978) 1–52.

[Ro2] Rockland, C.: *Intrinsic nilpotent approximation.* Acta Appl. Math. **8** (1987), no. 3, 213–270.

[RS] Rothschild, L.; Stein, E.: *Hypoelliptic differential operators and nilpotent groups.* Acta Math. **137** (1976) 247–320.

[Ru] Rumin, M.: *Formes différentielles sur les variétés de contact.* J. Differential Geom. **39** (1994), no.2, 281–330.

[Se] Seeley, R.T.: *Complex powers of an elliptic operator. Singular integrals* (Proc. Sympos. Pure Math., Vol. X, Chicago, Ill., 1966), pp. 288–307. Amer. Math. Soc., Providence, R.I., 1967.

[Sh] Shubin, M.: *Pseudodifferential operators and spectral theory.* Springer Series in Soviet Mathematics. Springer-Verlag, Berlin-New York, 1987.

[Ta] Tanaka, N.: *A differential geometric study on strongly pseudo-convex manifolds.* Lectures in Mathematics, Department of Mathematics, Kyoto University, No. 9. Kinokuniya Book-Store Co., Ltd., Tokyo, 1975.

[Tay] Taylor, M.E.: *Noncommutative microlocal analysis. I.* Mem. Amer. Math. Soc. 52 (1984), no. 313,

[Va] Van Erp, E.: PhD dissertation, Penn State University, June 2004.

[We] Webster, S.: *Pseudo-Hermitian structures on a real hypersurface.* J. Differential Geom. **13** (1978), no. 1, 25–41.

[Wo1] Wodzicki, M.: *Local invariants of spectral asymmetry.* Invent. Math. **75** (1984) 143–177.

[Wo2] Wodzicki, M.: *Noncommutative residue. I. Fundamentals. K-theory, arithmetic and geometry* (Moscow, 1984–1986), 320–399, Lecture Notes in Math., 1289, Springer, 1987.

Editorial Information

To be published in the *Memoirs*, a paper must be correct, new, nontrivial, and significant. Further, it must be well written and of interest to a substantial number of mathematicians. Piecemeal results, such as an inconclusive step toward an unproved major theorem or a minor variation on a known result, are in general not acceptable for publication.

Papers appearing in *Memoirs* are generally at least 80 and not more than 200 published pages in length. Papers less than 80 or more than 200 published pages require the approval of the Managing Editor of the Transactions/Memoirs Editorial Board.

As of March 31, 2008, the backlog for this journal was approximately 17 volumes. This estimate is the result of dividing the number of manuscripts for this journal in the Providence office that have not yet gone to the printer on the above date by the average number of monographs per volume over the previous twelve months, reduced by the number of volumes published in four months (the time necessary for preparing a volume for the printer). (There are 6 volumes per year, each usually containing at least 4 numbers.)

A Consent to Publish and Copyright Agreement is required before a paper will be published in the *Memoirs*. After a paper is accepted for publication, the Providence office will send a Consent to Publish and Copyright Agreement to all authors of the paper. By submitting a paper to the *Memoirs*, authors certify that the results have not been submitted to nor are they under consideration for publication by another journal, conference proceedings, or similar publication.

Information for Authors

Memoirs are printed from camera copy fully prepared by the author. This means that the finished book will look exactly like the copy submitted.

Initial submission. The AMS uses Centralized Manuscript Processing for initial submissions. Authors should submit a PDF file using the Initial Manuscript Submission form found at www.ams.org/cgi-bin/peertrack/submission.pl, or send one copy of the manuscript to the following address: Centralized Manuscript Processing, MEMOIRS OF THE AMS, 201 Charles Street, Providence, RI 02904-2294 USA. If a paper copy is being forwarded to the AMS, indicate that it is for it Memoirs and include the name of the corresponding author, contact information such as email address or mailing address, and the name of an appropriate Editor to review the paper (see the list of Editors below).

The paper must contain a *descriptive title* and an *abstract* that summarizes the article in language suitable for workers in the general field (algebra, analysis, etc.). The *descriptive title* should be short, but informative; useless or vague phrases such as "some remarks about" or "concerning" should be avoided. The *abstract* should be at least one complete sentence, and at most 300 words. Included with the footnotes to the paper should be the 2000 *Mathematics Subject Classification* representing the primary and secondary subjects of the article. The classifications are accessible from www.ams.org/msc/. The list of classifications is also available in print starting with the 1999 annual index of *Mathematical Reviews*. The Mathematics Subject Classification footnote may be followed by a list of *key words and phrases* describing the subject matter of the article and taken from it. Journal abbreviations used in bibliographies are listed in the latest *Mathematical Reviews* annual index. The series abbreviations are also accessible from www.ams.org/publications/. To help in preparing and verifying references, the AMS offers MR Lookup, a Reference Tool for Linking, at www.ams.org/mrlookup/.

Electronically prepared manuscripts. The AMS encourages electronically prepared manuscripts, with a strong preference for $\mathcal{AMS}$-LaTeX. To this end, the Society has prepared $\mathcal{AMS}$-LaTeX author packages for each AMS publication. Author packages include instructions for preparing electronic manuscripts, samples, and a style file that generates the particular design specifications of that publication series. Though $\mathcal{AMS}$-LaTeX is the highly preferred format of TeX, author packages are also available in $\mathcal{AMS}$-TeX.

Authors may retrieve an author package from the AMS website starting from www.ams.org/tex/ or via FTP to ftp.ams.org (login as anonymous, enter username as password, and type cd pub/author-info). The *AMS Author Handbook* and the *Instruction Manual* are available in PDF format following the author packages link from www.ams.org/tex/. The author package can also be obtained free of charge by sending email to tech-support@ams.org (Internet) or from the Publication Division, American Mathematical Society, 201 Charles St., Providence, RI 02904-2294, USA. When requesting an author package, please specify $\mathcal{AMS}$-LaTeX or $\mathcal{AMS}$-TeX and the publication in which your paper will appear. Please be sure to include your complete mailing address.

After acceptance. The final version of the electronic file should be sent to the Providence office (this includes any TeX source file, any graphics files, and the DVI or PostScript file) immediately after the paper has been accepted for publication.

Before sending the source file, be sure you have proofread your paper carefully. The files you send must be the EXACT files used to generate the proof copy that was accepted for publication. For all publications, authors are required to send a printed copy of their paper, which exactly matches the copy approved for publication, along with any graphics that will appear in the paper.

Accepted electronically prepared files can be submitted via the web at www.ams.org/submit-book-journal/, sent via FTP, or sent on CD-Rom or diskette to the Electronic Prepress Department, American Mathematical Society, 201 Charles Street, Providence, RI 02904-2294 USA. TeX source files, DVI files, and PostScript files can be transferred over the Internet by FTP to the Internet node ftp.ams.org (130.44.1.100). When sending a manuscript electronically via CD-Rom or diskette, please be sure to include a message identifying the paper as a Memoir.

Electronically prepared manuscripts can also be sent via email to pub-submit@ams.org (Internet). In order to send files via email, they must be encoded properly. (DVI files are binary and PostScript files tend to be very large.)

Electronic graphics. Comprehensive instructions on preparing graphics are available at www.ams.org/jourhtml/. A few of the major requirements are given here.

Submit files for graphics as EPS (Encapsulated PostScript) files. This includes graphics originated via a graphics application as well as scanned photographs or other computer-generated images. If this is not possible, TIFF files are acceptable as long as they can be opened in Adobe Photoshop or Illustrator. No matter what method was used to produce the graphic, it is necessary to provide a paper copy to the AMS.

Authors using graphics packages for the creation of electronic art should also avoid the use of any lines thinner than 0.5 points in width. Many graphics packages allow the user to specify a "hairline" for a very thin line. Hairlines often look acceptable when proofed on a typical laser printer. However, when produced on a high-resolution laser imagesetter, hairlines become nearly invisible and will be lost entirely in the final printing process.

Screens should be set to values between 15% and 85%. Screens which fall outside of this range are too light or too dark to print correctly. Variations of screens within a graphic should be no less than 10%.

Inquiries. Any inquiries concerning a paper that has been accepted for publication should be sent to memo-query@ams.org or directly to the Electronic Prepress Department, American Mathematical Society, 201 Charles St., Providence, RI 02904-2294 USA.

Editors

This journal is designed particularly for long research papers, normally at least 80 pages in length, and groups of cognate papers in pure and applied mathematics. Papers intended for publication in the *Memoirs* should be addressed to one of the following editors. The AMS uses Centralized Manuscript Processing for initial submissions to AMS journals. Authors should follow instructions listed on the Initial Submission page found at www.ams.org/memo/memosubmit.html.

Algebra to ALEXANDER KLESHCHEV, Department of Mathematics, University of Oregon, Eugene, OR 97403-1222; email: ams@noether.uoregon.edu

Algebraic geometry and its application to MINA TEICHER, Emmy Noether Research Institute for Mathematics, Bar-Ilan University, Ramat-Gan 52900, Israel; email: teicher@macs.biu.ac.il

Algebraic geometry to DAN ABRAMOVICH, Department of Mathematics, Brown University, Box 1917, Providence, RI 02912; email: amsedit@math.brown.edu

Algebraic number theory to V. KUMAR MURTY, Department of Mathematics, University of Toronto, 100 St. George Street, Toronto, ON M5S 1A1, Canada; email: murty@math.toronto.edu

Algebraic topology to ALEJANDRO ADEM, Department of Mathematics, University of British Columbia, Room 121, 1984 Mathematics Road, Vancouver, British Columbia, Canada V6T 1Z2; email: adem@math.ubc.ca

Combinatorics to JOHN R. STEMBRIDGE, Department of Mathematics, University of Michigan, Ann Arbor, Michigan 48109-1109; email: FRS@umich.edu

Complex analysis and harmonic analysis to ALEXANDER NAGEL, Department of Mathematics, University of Wisconsin, 480 Lincoln Drive, Madison, WI 53706-1313; email: nagel@math.wisc.edu

Differential geometry and global analysis to LISA C. JEFFREY, Department of Mathematics, University of Toronto, 100 St. George St., Toronto, ON Canada M5S 3G3; email: jeffrey@math.toronto.edu

Dynamical systems and ergodic theory and complex anaysis to YUNPING JIANG, Department of Mathematics, CUNY Queens College and Graduate Center, 65-30 Kissena Blvd., Flushing, NY 11367; email: Yunping.Jiang@qc.cuny.edu

Functional analysis and operator algebras to DIMITRI SHLYAKHTENKO, Department of Mathematics, University of California, Los Angeles, CA 90095; email: shlyakht@math.ucla.edu

Geometric analysis to WILLIAM P. MINICOZZI II, Department of Mathematics, Johns Hopkins University, 3400 N. Charles St., Baltimore, MD 21218; email: trans@math.jhu.edu

Geometric analysis to MARK FEIGHN, Math Department, Rutgers University, Newark, NJ 07102; email: feighn@andromeda.rutgers.edu

Harmonic analysis, representation theory, and Lie theory to ROBERT J. STANTON, Department of Mathematics, The Ohio State University, 231 West 18th Avenue, Columbus, OH 43210-1174; email: stanton@math.ohio-state.edu

Logic to STEFFEN LEMPP, Department of Mathematics, University of Wisconsin, 480 Lincoln Drive, Madison, Wisconsin 53706-1388; email: lempp@math.wisc.edu

Number theory to JONATHAN ROGAWSKI, Department of Mathematics, University of California, Los Angeles, CA 90095; email: jonr@math.ucla.edu

Partial differential equations to GUSTAVO PONCE, Department of Mathematics, South Hall, Room 6607, University of California, Santa Barbara, CA 93106; email: ponce@math.ucsb.edu

Partial differential equations and dynamical systems to PETER POLACIK, School of Mathematics, University of Minnesota, Minneapolis, MN 55455; email: polacik@math.umn.edu

Probability and statistics to RICHARD BASS, Department of Mathematics, University of Connecticut, Storrs, CT 06269-3009; email: bass@math.uconn.edu

Real analysis and partial differential equations to DANIEL TATARU, Department of Mathematics, University of California, Berkeley, Berkeley, CA 94720; email: tataru@math.berkeley.edu

All other communications to the editors should be addressed to the Managing Editor, ROBERT GURALNICK, Department of Mathematics, University of Southern California, Los Angeles, CA 90089-1113; email: guralnic@math.usc.edu.

Titles in This Series

909 **Cameron McA. Gordon and Ying-Qing Wu,** Toroidal Dehn fillings on hyperbolic 3-manifolds, 2008

908 **J.-L. Waldspurger,** L'endoscopie tordue n'est pas si tordue, 2008

907 **Yuanhua Wang and Fei Xu,** Spinor genera in characteristic 2, 2008

906 **Raphaël S. Ponge,** Heisenberg calculus and spectral theory of hypoelliptic operators on Heisenberg manifolds, 2008

905 **Dominic Verity,** Complicial sets characterising the simplicial nerves of strict ω-categories, 2008

904 **William M. Goldman and Eugene Z. Xia,** Rank one Higgs bundles and representations of fundamental groups of Riemann surfaces, 2008

903 **Gail Letzter,** Invariant differential operators for quantum symmetric spaces, 2008

902 **Bertrand Toën and Gabriele Vezzosi,** Homotopical algebraic geometry II: Geometric stacks and applications, 2008

901 **Ron Donagi and Tony Pantev (with an appendix by Dmitry Arinkin),** Torus fibrations, gerbes, and duality, 2008

900 **Wolfgang Bertram,** Differential geometry, Lie groups and symmetric spaces over general base fields and rings, 2008

899 **Piotr Hajłasz, Tadeusz Iwaniec, Jan Malý, and Jani Onninen,** Weakly differentiable mappings between manifolds, 2008

898 **John Rognes,** Galois extensions of structured ring spectra/Stably dualizable groups, 2008

897 **Michael I. Ganzburg,** Limit theorems of polynomial approximation with exponential weights, 2008

896 **Michael Kapovich, Bernhard Leeb, and John J. Millson,** The generalized triangle inequalities in symmetric spaces and buildings with applications to algebra, 2008

895 **Steffen Roch,** Finite sections of band-dominated operators, 2008

894 **Martin Dindoš,** Hardy spaces and potential theory on C^1 domains in Riemannian manifolds, 2008

893 **Tadeusz Iwaniec and Gaven Martin,** The Beltrami Equation, 2008

892 **Jim Agler, John Harland, and Benjamin J. Raphael,** Classical function theory, operator dilation theory, and machine computation on multiply-connected domains, 2008

891 **John H. Hubbard and Peter Papadopol,** Newton's method applied to two quadratic equations in $\mathbb{C}^2$ viewed as a global dynamical system, 2008

890 **Steven Dale Cutkosky,** Toroidalization of dominant morphisms of 3-folds, 2007

889 **Michael Sever,** Distribution solutions of nonlinear systems of conservation laws, 2007

888 **Roger Chalkley,** Basic global relative invariants for nonlinear differential equations, 2007

887 **Charlotte Wahl,** Noncommutative Maslov index and eta-forms, 2007

886 **Robert M. Guralnick and John Shareshian,** Symmetric and alternating groups as monodromy groups of Riemann surfaces I: Generic covers and covers with many branch points, 2007

885 **Jae Choon Cha,** The structure of the rational concordance group of knots, 2007

884 **Dan Haran, Moshe Jarden, and Florian Pop,** Projective group structures as absolute Galois structures with block approximation, 2007

883 **Apostolos Beligiannis and Idun Reiten,** Homological and homotopical aspects of torsion theories, 2007

882 **Lars Inge Hedberg and Yuri Netrusov,** An axiomatic approach to function spaces, spec tral synthesis and Luzin approximation, 2007

881 **Tao Mei,** Operator valued Hardy spaces, 2007

TITLES IN THIS SERIES

- 880 **Bruce C. Berndt, Geumlan Choi, Youn-Seo Choi, Heekyoung Hahn, Boon Pin Yeap, Ae Ja Yee, Hamza Yesilyurt, and Jinhee Yi,** Ramanujan's forty identities for Rogers-Ramanujan functions, 2007
- 879 **O. García-Prada, P. B. Gothen, and V. Muñoz,** Betti numbers of the moduli space of rank 3 parabolic Higgs bundles, 2007
- 878 **Alessandra Celletti and Luigi Chierchia,** KAM stability and celestial mechanics, 2007
- 877 **María J. Carro, José A. Raposo, and Javier Soria,** Recent developments in the theory of Lorentz spaces and weighted inequalities, 2007
- 876 **Gabriel Debs and Jean Saint Raymond,** Borel liftings of Borel sets: Some decidable and undecidable statements, 2007
- 875 **C. Krattenthaler and T. Rivoal,** Hypergéométrie et fonction zêta de Riemann, 2007
- 874 **Sonia Natale,** Semisolvability of semisimple Hopf algebras of low dimension, 2007
- 873 **A. J. Duncan,** Exponential genus problems in one-relator products of groups, 2007
- 872 **Anthony V. Geramita, Tadahito Harima, Juan C. Migliore, and Yong Su Shin,** The Hilbert function of a level algebra, 2007
- 871 **Pascal Auscher,** On necessary and sufficient conditions for L^p-estimates of Riesz transforms associated to elliptic operators on $\mathbb{R}^n$ and related estimates, 2007
- 870 **Takuro Mochizuki,** Asymptotic behaviour of tame harmonic bundles and an application to pure twistor D-modules, Part 2, 2007
- 869 **Takuro Mochizuki,** Asymptotic behaviour of tame harmonic bundles and an application to pure twistor D-modules, Part 1, 2007
- 868 **Gelu Popescu,** Entropy and multivariable interpolation, 2006
- 867 **Vilmos Totik,** Metric properties of harmonic measures, 2006
- 866 **William Craig,** Semigroups underlying first-order logic, 2006
- 865 **Nathanial P. Brown,** Invariant means and finite representation theory of $C*$-algebras, 2006
- 864 **John M. Lee,** Fredholm operators and Einstein metrics on conformally compact manifolds, 2006
- 863 **M. Lübke and A. Teleman,** The Universal Kobayashi-Hitchin correspondence on Hermitian manifolds, 2006
- 862 **Alberto Canonaco,** The Beilinson complex and canonical rings of irregular surfaces, 2006
- 861 **Leon A. Takhtajan and Lee-Peng Teo,** Weil-Petersson metric on the universal Teichmüller space, 2006
- 860 **Thomas M. Fiore,** Pseudo limits, biadjoints and pseudo algebras: Categorical foundations of conformal field theory, 2006
- 859 **N. Arcozzi, R. Rochberg, and E. Sawyer,** Carleson measures and interpolating sequences for Besov spaces on complex balls, 2006
- 858 **Enrico Valdinoci, Berardino Sciunzi, and Vasile Ovidiu Savin,** Flat level set regularity of p-Laplace phase transitions, 2006
- 857 **Donatella Danielli, Nocola Garofalo, and Duy-Minh Nhieu,** Non-doubling Ahlfors measures, perimeter measures, and the characterization of the trace spaces of Sobolev functions in Carnot-Carathéodory spaces, 2006
- 856 **Vladimir Bolotnikov and Harry Dym,** On boundary interpolation for matrix valued Schur functions, 2006
- 855 **Yevgenia Kashina, Yorck Sommerhäuser, and Yongchang Zhu,** On higher Frobenius-Schur indicators, 2006

For a complete list of titles in this series, visit the
AMS Bookstore at **www.ams.org/bookstore/**.